MW01381449

Krill

Biology, Ecology and Fisheries

Edited by

Inigo Everson

Leader, Marine Living Resources Section
British Antarctic Survey
Cambridge

© 2000 Blackwell Science Ltd
Editorial Offices:
Osney Mead, Oxford OX2 0EL
25 John Street, London WC1N 2BS
23 Ainslie Place, Edinburgh EH3 6AJ
350 Main Street, Malden
MA 02148 5018, USA
54 University Street, Carlton
Victoria 3053, Australia
10, rue Casimir Delavigne
75006 Paris, France

Other Editorial Offices:

Blackwell Wissenschafts-Verlag GmbH
Kurfürstendamm 57
10707 Berlin, Germany

Blackwell Science KK
MG Kodenmacho Building
7–10 Kodenmacho Nihombashi
Chuo-ku, Tokyo 104, Japan

The right of the Author to be identified as the Author of this Work has been asserted in accordance with the Copyright, Designs and Patents Act 1988.

All rights reserved. No part of this publication may be reproduced, stored in a retrieval system, or transmitted, in any form or by any means, electronic, mechanical, photocopying, recording or otherwise, except as permitted by the UK Copyright, Designs and Patents Act 1988, without the prior permission of the publisher.

First published 2000

Set in 10/13pt Times
by DP Photosetting, Aylesbury, Bucks
Printed and bound in Great Britain by
MPG Books Ltd, Bodmin, Cornwall

The Blackwell Science logo is a trade mark of Blackwell Science Ltd, registered at the United Kingdom Trade Marks Registry

DISTRIBUTORS

Marston Book Services Ltd
PO Box 269
Abingdon
Oxon OX14 4YN
(*Orders:* Tel: 01235 465500
Fax: 01235 465555)

USA
Blackwell Science, Inc.
Commerce Place
350 Main Street
Malden, MA 02148 5018
(*Orders:* Tel: 800 759 6102
781 388 8250
Fax: 781 388 8255)

Canada
Login Brothers Book Company
324 Saulteaux Crescent
Winnipeg, Manitoba R3J 3T2
(*Orders:* Tel: 204 837-2987
Fax: 204 837-3116)

Australia
Blackwell Science Pty Ltd
54 University Street
Carlton, Victoria 3053
(*Orders:* Tel: 03 9347 0300
Fax: 02 9347 5001)

A catalogue record for this title is available from the British Library

ISBN 0-632-05565-0

Library of Congress
Cataloging-in-Publication Data
Krill: biology, ecology, and fisheries/edited by
Inigo Everson.
p. cm.—(Fish and aquatic resources series; 6)
Derived from a workshop held at the Fisheries Center, University of British Columbia, November 1994.
Includes bibliographical references (p.).
ISBN 0-632-05565-0) (alk. paper)
1. Krill fisheries—Congresses. 2. Krill—Congresses. I. Everson, Inigo. II. Series.
SH380.7.K75 2000
333.95′55913—dc21 00-045491

For further information on
Blackwell Science, visit our website:
www.blackwell-science.com

Contents

Please note: the plate section falls between pp. 182 and 183.

Series Foreword

Fish researchers (a.k.a. fish freaks) like to explain, to the bemused bystander, how fish have evolved an astonishing array of adaptations, so much so that it can be difficult for them to comprehend why anyone would study anything else. Yet, at the same time, fish are among the last wild creatures on our planet that are hunted by humans for food. As a consequence, few today would fail to recognize that the reconciliation of exploitation with the conservation of biodiversity provides a major challenge to our current knowledge and expertise. Even evaluating the trade-offs that are needed is a difficult task. Moreover, solving this pivotal issue calls for a multidisciplinary conflation of fish physiology, biology and ecology with social sciences such as economics and anthropology in order to probe new frontiers of applied science. The Blackwell Science Series on *Fish and Aquatic Resources* is an initiative aimed at providing key, peer-reviewed texts in this fast-moving field.

While bony fish stem from a great radiation that followed the invention of the swimbladder in the Cretaceous period 100 million years ago, some fish groups, such as the sharks, lungfish and sturgeons are more ancient beasts. Survivors from earlier eras may be more widespread than we think: the deep sea coelacanths, formerly known only from the Indian Ocean, have recently turned up in Indonesia. Also, these fishes may be more effectively adapted to specialized niches than their ancient body plans would suggest. For example, rays and angel sharks have perfected the art of the ambush predator, while most cartilaginous fishes can detect electric discharges in the nerves of their prey.

Bony fish themselves have evolved into an amazing array of habitats and niches. As well as the open sea, there are fish in lakes, ponds, rivers and rock pools; in deserts, forests, mountains, the great deeps of the sea, and the extreme cold of the Antarctic; in warm waters of high alkalinity or of low oxygen; and in habitats like estuaries or mudflats, where their physiology is challenged by continuous change. Air-breathing climbing perch (regularly found up trees), walking catfish and mangrove mudskippers are currently repeating the land invasion of their Carboniferous ancestors. We can marvel at high-speed swimming adaptations in the fins, tails, gills and muscles of marlins, sailfish and warm-blooded tunas; gliding and flapping flight in several groups of fish; swinging, protrusible jaws providing suction-assisted feeding that have evolved in parallel in groupers, carps and cods; parental care in mouth-brooding cichlids; the birth of live young in many sharks, tooth carps, rockfish and blennies; immense migrations in salmon, shads and tunas; and even the so-called four-eyed fish, with eyes divided into upper air and lower water-adapted sections.

In addition to food, recreation (and inspiration for us fish freaks), it has, moreover, recently been realized that fish are essential components of aquatic ecosystems that provide vital services to human communities. But, sadly, virtually all sectors of the stunning biodiversity of fishes are at risk from human activities. In fresh water, for example, the largest mass extinction event since the end of the dinosaurs has occurred as the introduced Nile perch in Lake Victoria eliminated over 100 species of endemic haplochromine fish. But, at the same time, precious food and income from the Nile perch fishery was created in a miserably poor region. In the oceans, we have barely begun to understand the profound changes that have accompanied a vast expansion of human fishing over the past 100 years.

Krill: biology, ecology and fisheries edited by Inigo Everson from the British Antarctic Survey team in Cambridge, UK. comprises the 6th book in the new *Blackwell Science Fish and Aquatic Resources Series*. It includes 14 chapters authored by permutations of 13 krill experts from 7 countries world-wide. The book, like the workshop held at the Fisheries Centre, University of British Columbia, in November 1995 from which it is derived (Pitcher & Chuenpagdee, 1995), aims to address the problems and issues underlying the sustainable harvesting of krill. Five inter-linked themes make up the focus of the book: krill ecology; krill sustainable harvest; the ecological implications of krill harvesting; krill resource assessment methods; and products and markets for krill.

The word krill comes from the Norwegian *kril*, meaning very small fish fry, but has been used in English since 1907 for the shrimp-like food of baleen whales (OED). Krill are defined here as the euphausids, a relatively small (about 85 species world-wide) and uniform taxon of large pelagic shrimp-like filter-feeding crustaceans (adults are from 10–80 mm long). Most krill species inhabit the upper 400 m of the ocean and shed their eggs into the sea, but a few (e.g. *Nyctiphanes*) have evolved parental care of the eggs until hatching into nauplii. Many species exhibit diurnal migration, approaching the surface to feed at night. Krill biomass world-wide is thought to exceed 300 million tonnes and they are particularly abundant in North Atlantic, Antarctic and North Pacific oceans. Krill, whose biomass often rivals that of copepods in the plankton community, have a critical ecological role. They comprise a critical link in oceanic food webs between their phytoplankton food and their fish predators, many of which are commercially important fishes. It follows that direct harvesting of such a pivotal component in the food web has an ecological impact that must be evaluated if large-scale fisheries for krill are contemplated.

Most krill species are herbivorous, and in high latitudes krill consume many diatoms, but many krill species can utilise zooplankton as well, and some (e.g. *Megayctiphanes*) are carnivorous. So for most krill species, through the 'cascade' effect, phytoplankton might be expected to increase when krill are harvested by humans. Organisms that feed on krill might decrease in abundance if krill are harvested. These include many species of fish that are themselves the subjects of substantial human fisheries. In the North Pacific, krill form an important component of the diet of herring, salmon, pollack, sardine, mackerel and capelin, all of which support important commercial fisheries. The role of krill as the food of squid is thought to have been overestimated, but there is a fear that excessive harvest of krill might lead to: algal blooms of under-harvested phytoplankton; reductions in krill-dependent predators such as baleen whales and many fish; subsequent lower abundance of commercial fish stocks through trophic cascade effects.

Abundance is the key input parameter for the evaluation of all harvest impacts and for fishery regulation. The assessment and measurement of krill abundance presents challenges to both existing sonar technology and to mathematical modelling. Estimating the biomass of krill in the plankton cannot be done with standard fisheries acoustics technology, most of which has been designed for fish, which are larger and have relatively high target strengths. Selection of equipment, selection of frequency, target identification, calibration of gear and measurement error are important topics in acoustic methodology for krill assessment.

Although krill growth appears to be well understood, modelling their population dynamics is subject to considerable uncertainty, especially in recruitment and the effect of social swarming behaviour. The design and analysis of surveys that can provide robust estimates of krill abundance is therefore critical to success. Estimation of the potential yield of krill stocks leads to problems in demographics and in the estimation of recruitment. Assessment and potential yield evaluations may subsequently be used in developing suitable management measures for krill fisheries. Moreover, krill consumption by fish may be estimated and modeled, and the impact of various levels of krill harvest forecast, estimates that must be incorporated in any ecosystem-based assessment of krill fisheries.

The processing of organisms high in oils and pigments presents both technical difficulties and economic opportunities to serve new and emerging markets. Unconventional processing of krill products is developing unexpected new markets. For example, krill pigments that may be used in the aquaculture feed industry fetch much higher prices and profits than using krill for fishmeal or using krill tails for human consumption, as has been done in Russia and Japan. Some krill enzymes have a range of biomedical applications.

This book begins the vital process of evaluating the future and potential of krill fisheries in the light of fisheries that are moving down the trophic levels of marine food webs (Pauly *et al.* 1998) as traditional table fish resources become depleted (Pitcher 2000). It will, I am sure, provide a definitive reference work on krill for some time to come, and I am delighted to welcome it to the growing list of titles in the *Blackwell Science Fish and Aquatic Resources Series.*

Professor Tony J. Pitcher
Editor, Blackwell Science Fish and Aquatic Resources Series
Director, Fisheries Centre, University of British Columbia, Vancouver, Canada

References

Pauly, D., Christensen, V., Dalsgaaard, J., Froese, R., and Torres, F. Jnr (1998) Fishing down marine food webs. *Science* 279: 860–863.
Pitcher, T.J. (2000) Rebuilding as a New Goal for Fisheries Management: Reconstructing the Past to Salvage The Future. *Ecological Applications* (in press).
Pitcher, T.J. and Chuenpagdee, R. (Eds) (1995) Harvesting Krill: Ecological impact, assessment, products and markets. *Fisheries Centre Research Reports* 3(3): 82 pp.

Acknowledgement

The Series Editor, Editor and Authors are most grateful to the industrial companies comprising the former Sustainable Ocean Resources Society of British Columbia (now incorporated in the Institute for Pacific Ocean Science and Technology), the British Columbia Ministry of Fisheries, and the Science Council of British Columbia for sponsorship of the 1995 workshop from which this book on krill and its fisheries grew.

Acknowledgements

Gestation for this book has been long and geographically widespread. It began with a Workshop Meeting, organised by Tony Pitcher in November 1995 in Vancouver when a number of the participants agreed to participate in the production of a book on krill. Our timetable, agreed as being sensible at that time, went from being realistic to elastic. The list of contributors changed although the majority stayed with the project. To them all I extend my thanks for putting so much effort into the venture and for providing me with encouragement to complete the project. I also thank all those who provided review comments on the chapters. Various librarians gave quick and courteous support and this is gratefully acknowledged. In particular I thank Christine Phillips for her sisyphean attempts at teaching me the efficient use of library systems and I am extremely grateful to her for her unerring skill at locating documents from minimal or at times wrong information. Many other people have helped with the project by freely discussing the subject matter at a wide variety of meetings particularly within the CCAMLR arena. During my career two people, Dick Laws and the late John Gulland, have provided considerable influence and guidance over the years in matters both Antarctic and fisheries, this is gratefully acknowledged. I also thank my wife, Diana, for allowing this book to permeate the household for so long.

In addition I thank the following publishers for permission to reproduce figures from their publications: Academic Press for permission to reproduce Fig. 7.3.3; Alfred-Wegener for permission to reproduce Figs 8.1 and 8.2; Cambridge University Press for permission to reproduce Fig. 3.3.1; CCAMLR for permission to reproduce Plate 2 and Figs 9.2.2 and 12.2.1; FAO for permission to reproduce Fig. 9.4.1; Fisheries Centre, UBC, Vancouver for permission to reproduce Fig. 3.2.3; Inter-Research for permission to reproduce Figs 8.7b and 8.8; The Royal Society for permission to reproduce Fig. 7.3.2; Scott Polar Research Institute for permission to reproduce Figs 2.3.3. and 2.3.4 which originally appeared in a BIOMASS Handbook; Springer-Verlag for permission to reproduce Figs 2.3.2 which originally appeared in *Polar Biology*. In addition I thank Hidehiro Kato and Steve Nicol for permission to publisher their photographs as Plates 4 and 7 respectively.

Inigo Everson

List of Contributors

David J. Agnew
Head, Falklands Group
Renewable Resources Assessment Group
T.H. Huxley School of Environment, Earth Sciences and Engineering
Imperial College
Room 406
Royal School of Mines
Prince Consort Road
London SW7 2BP

Yoshi Endo
Laboratory of Biological Oceanography
Faculty of Agriculture
Tohoku University, Sendai
Japan

Inigo Everson
Leader, Marine Living Resources Section
British Antarctic Survey
Cambridge CB3 0ET

Ian Forster
The Oceanic Institute
41–202 Kalanianole Hwy
Waimanalo, H1 96795
USA

Taro Ichii
National Research Institute of Far Seas Fisheries
Orido 5-7-1
Shimizu
Shizuoka 424
Japan

Michael C. Macaulay
Applied Physics Laboratory
College of Ocean and Fisheries Sciences
University of Washington
Seattle WA 98195
USA

Denzil G.M. Miller
Marine and Coastal Management
P/bag X2
Rogge Bay 8012
South Africa

Stephen Nicol
Australian Antarctic Division
Channel Highway
Kingston, Tasmania 7050
Australia

Langdon B. Quetin
Marine Science Institute
University of California
Santa Barbara Ca 93106
USA

Robin M. Ross
Marine Science Institute
University of California
Santa Barbara Ca 93106
USA

Volker Siegel
Institut fur Seefishcherei
Palmaille 9
22767 Hamburg
Germany

John Spence
Biozyme Systems, Inc
West Vancouver
BC V7V 1N6
Canada

R.W. Tanasichuk
Fisheries and Oceans Canada
Pacific Biological Station
Nanaimo
BC V9R 5K6
Canada

Jon L. Watkins
British Antarctic Survey
Cambridge CB3 0ET

Chapter 1
Introducing Krill

Inigo Everson

Krill, a term originally applied to 'fish fry', is now taken to refer to euphausiids, a group comprising over 80 species most of which are planktonic. They are widespread with examples to be found in all the oceans of the world. Their size and, in places high numerical density, makes them of particular importance in some marine ecosystems. That importance is enhanced in the case of the few species that are commercially harvested.

Commercial harvesting and economic importance are the two criteria which, when applied to euphausiids, provide the focus for this volume. This immediately reduces the area of interest from all the oceans and seas of the world to the consideration of a few geographical regions. In the Pacific Ocean there is the area in the north-west around Japan and in the north-east the coast of British Columbia. In the North Atlantic, interest is concentrated on the Nova Scotian shelf and in the Southern Ocean interest has been greatest in the Atlantic sector with lesser interest in the Indian and Pacific Ocean sectors. Commercial interest in harvesting krill has been reported from other regions such as the Mediterranean Sea and coastal Tasmania, although currently this appears to be of very limited extent. Several species of krill aggregate into swarms and it is this swarming behaviour which makes them attractive to commercial harvesting.

Focussing attention towards those species likely to be present in areas where fishing is taking place invites a further restriction to the surface waters down to approximately 500 m. Thus although there are different species present in deep water they have little impact on or interaction with those that are found closer to the surface. Applying these restrictions we are left with a small number of key species and others whose ranges overlap with them geographically and vertically.

1.1 Euphausiid Identification

Accurate identification is central to many ecological investigations and is a topic worthy of careful study. Convergent evolution in pelagic crustacea has resulted in a variety of species that have similar superficial appearances but different phylogenies. Thus the first step must be to ensure that the specimen under consideration is a Euphausiid and not, for example, a Decapod or Mysid. All Euphausiids have gills clearly visible below the carapace on thoracic segments seven and eight. The morphology of a typical euphausiid is shown in Fig. 1.1. There are photophores or luminous organs at the base of the abdominal pleopods, at the genital segment of the cephalothorax and also near to the mouth parts. The photophores produce a blue light which can often be seen in fresh caught live specimens while still in the net.

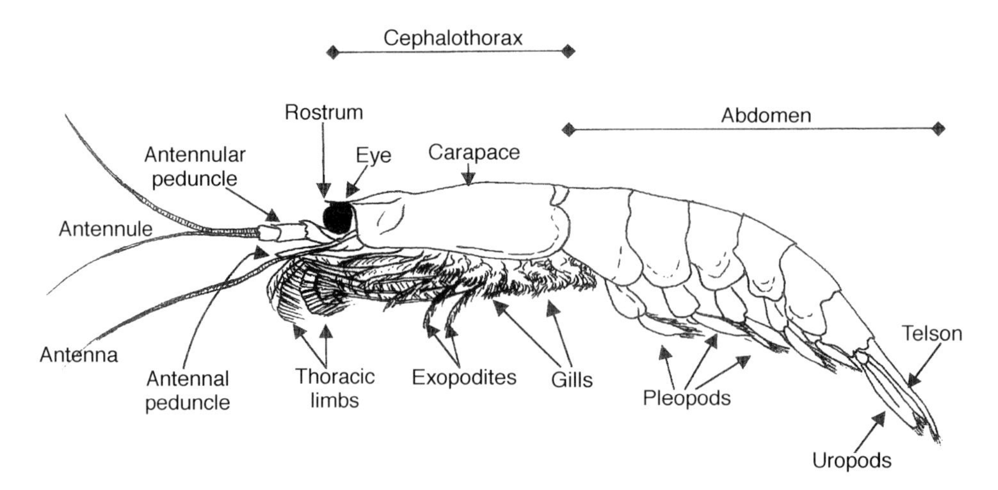

Fig. 1.1 Generalised view of a euphausiid to show the main morphological features mentioned in the text and also that are relevant to identification.

There are good dichotomous keys to be found in Baker *et al.* (1990) and Mauchline & Fisher (1969). Both of these are comprehensive and include euphausiids world-wide, the former having a very good set of clear drawings. In addition, for the north Atlantic and Southern Ocean Mauchline (1984, 1980a respectively) provide good illustrated keys both of which are a truncated version of the Mauchline & Fisher (1969) key. Although there is much commonality between these keys it is worth consulting more than one, and also source literature, when there is any doubt over the identity of a particular specimen. This is because there are differences in the way individual features are illustrated and as an added bonus many of the engraved plates of the early expedition reports, as well as possessing considerable scientific merit, are also an art form frequently to be admired.

1.2 Broad-scale distribution

The following is a brief summary by way of general introduction to describe the broad-scale distribution of euphausiids in the main areas of interest. More detailed information on the key species of commercial interest and species that are likely to be found associated with them will be found in Chapter 3. Further general information on distribution can be found in Mauchline & Fisher (1969) and Mauchline

(1980b); most of the following information has been summarised from these two publications.

North Pacific Ocean

Several species are found on both eastern and western seaboards of the North Pacific Ocean. The species of greatest commercial interest in the North Pacific Ocean is undoubtedly *Euphausia pacifica* which is found across the Bering Sea, through the Southern part of the Sea of Okhotsk and the Sea of Japan extending southwards to about 30° N. The densest concentrations are generally found in the North Pacific Drift and Aleutian Current extending southwards off the Californian Coast and within the Californian Current. The southern limit for *E. pacifica* is at the 9.5°C isotherm at a depth of 200 m. The only similar species with which it might be confused is the much smaller *E. nana* whose distribution is restricted to Southern Japan and the East China Sea.

Although rarely present in great numbers, *Nematobrachion flexipes* is found in deep oceanic water of the North Pacific Ocean south of latitude 40° N. In addition, between latitudes 35° and 55° N, *Thysanopoda acutifrons* is likely to be present although its distribution appears to be limited by the 4°C and 10°C isotherm at 100 m depth. With a latitudinal span slightly further towards the equator, *E. gibboides* is found between 30° and 45° N across the Pacific Ocean and in the eastern end of this range it extends down the Californian Current. Closer to the equator the closely related *E. hemigibba* is probably restricted to deep oceanic waters between latitudes 18° and 42° N and in the equatorial zone from 160° E to 110° W *E. paragibba* is likely to occur.

Tessarabrachion oculatum is a mesopelagic species found between latitudes 35° and 53° N. It is absent from the oceanic region south-east of Kamchatka and also from the Bering Sea although it is found around the Aleutian Islands. The genus *Thysanoessa* is characterised by species found in high latitudes. *T. longipes* is present in the North Pacific, Gulf of Alaska, Bering Sea, eastern region of the Beaufort Sea and north of the Bering Strait. *T. raschi*, a neritic species, is found as far north as 80° N in the Arctic Ocean. In the Gulf of Alaska is *T. spinifera* another neritic species whose distribution extends down to the Californian coast although it is not found in the western Pacific. *T. inermis* has a latitudinal range extending from 63° N in the Bering Sea south to latitude 43° N and although it is present in the Arctic it is not known whether there is continuity with Atlantic populations.

In the western Pacific region *Pseudeuphausia sinica* is neritic and confined to the coastal regions of the East China and Southern Yellow Seas while *Euphausia similis* extends northwards through the South and East China Seas and extends eastwards of Japan in the Kuroshio Extension.

The final genus, *Nematoscelis*, is represented by three species. *Nematobrachion gracilis* is found in the western Pacific in the Kuroshio Current and its extension east of Japan. *N. difficilis* occurs across the Pacific Ocean between latitudes 35°

and 45° North and extends southwards to 20° N in the Californian Current. *N. microps* is found north as far as 40° N in the western Pacific but is absent from the eastern Pacific.

North Atlantic

The most important species is undoubtedly *Meganyctiphanes norvegica* which in the west Atlantic is found northwards from Cape Hatteras at 35° North along the edge of the continental slope to 70° North where it is found along the coasts of Labrador, Baffin Island and West Greenland. It is also present around south and east Greenland as far as 80° North. On the eastern side of the north Atlantic it extends from the western end of the Barents Sea, and is found throughout the Norwegian Sea, North Sea and Skagerrak. It is not however to be found in the Kattegat or eastern English Channel. The main breeding areas are thought to be off the Gulf of Maine, Gulf of St Lawrence, southwestern and southern Iceland and in the Norwegian Sea northwards to about 70° North.

Several other species are found on both sides of the Atlantic Ocean. On the western Atlantic seaboard *Thysanopoda acutifrons* is found from the Gulf of Maine to 70° North in the region of south-west Iceland and the Davis Strait while on the eastern seaboard it is found from Gibraltar northwards to 70° North. In the western Atlantic *Thysanoessa longicaudata* occurs from around the Gulf of Maine at 35° North and extends as far as 70° North off West Greenland. On the eastern side, although most common between latitudes 55° and 70° North, its range overall extends from the Bay of Biscay to 83° North off north-east Greenland.

Nyctiphanes couchii is a neritic species which occurs in the eastern seaboard above the slope and shelf of the north Atlantic between Gibraltar and 60° North. Although present in the Irish Sea it does not appear to breed there or north of the Skagerrak. It occurs in the Kattegat and is the only euphausiid to penetrate the western Baltic although it does not breed there.

Thysanoessa raschi, another neritic species, is the commonest euphausiid off West Greenland and extends northwards to about 70° and in extreme cases to 78° North. It is present in the Gulf of St Lawrence and Gulf of Maine but is not found further south than 40° North. On the eastern side of the Atlantic it is found around Iceland and is common around Scotland as far as 55° North. It is also found along the Norwegian coastline north to the Barents, White and Kara Seas. *Thysanoessa inermis* occurs from the Gulf of Maine north as far as west Greenland at around 70° North. Off East Greenland it occurs as far north as 75° to 80° North although, except in the Norwegian and Barents Seas, it does not breed north of 65° to 70° North. On the eastern Atlantic seaboard it is not found south of about 50° North.

The only bottom dwelling euphausiid is *Bentheuphausia amblyops* which appears to be widespread but is confined to water deeper than 1000 m from the equator to latitude 46° 15′ North on both eastern and western sides of the Atlantic.

Southern Ocean

All the Southern Ocean euphausiids have a circumpolar distribution and are broadly separated by their latitudinal ranges. The species commonly referred to as the Antarctic Krill is *Euphausia superba*, a widespread species which frequently swarms and is the subject of significant commercial fishing (Plate 1, facing p. 182). *Euphausia vallentini* and *E. longirostris* are restricted to the subantarctic zone, the former with a southern limit of the Antarctic Polar Front while the latter is found only between latitudes 40° and 55° South. The southern limit of *E. vallentini* approximates to the northern limit of *E. triacantha* a species having a circumpolar distribution between latitudes 50° and 60° South. The southern limit of *E. triacantha* overlaps the northern limit of *E. superba* and *E. frigida* both of which are only found south of the Antarctic Polar Frontal Zone (APFZ). The most southerly species is *E. crystallorophias* which is generally neritic and is restricted to the Antarctic continental shelf although some swarms have been found in the Scotia Sea (Brierley & Brandon, 1999) and they are also present in the caldera of Deception Island.

Thysanoessa vicina and *T. macrura* are two species with a circumpolar distribution in the Southern Ocean south of about 50° to 55° South and extending into the pack-ice zone. Although they are difficult to distinguish reliably without a binocular microscope, these two species can be recognised from all others in the Southern Ocean because their eyes are clearly in two parts with a constriction between. All the other Southern Ocean euphausiids have circular eyes.

Other areas

There have been a few proposals to fish for krill in other areas. In a recent review Nicol & Endo (1997, 1999) noted that there had been interest in fishing for krill in the Mediterranean Sea, although it is not clear what the target species might be, and also a proposal to fish for *Nyctiphanes australis* off Tasmania. In preparing this chapter I was made aware of anecdotal information indicating that there was some local fishing for *Meganyctiphanes norvegica* in Norwegian waters to provide feed for the aquaculture industry; however, I have found nothing to substantiate these suggestions.

1.3 Synopsis of the book

This book has been written primarily for postgraduate students, professional scientists and administrators concerned with krill ecology and in particular fishery management. Krill was initially seen as a ready source of protein to satisfy an expanding world population although the expansion anticipated in the 1970s has only partly materialised. Arising from this the main areas of commercial fishery interest are relatively few as has already been mentioned.

In Chapter 2 the methods of sampling krill are considered. Nets provide a method of direct sampling from which information on the type and quality of the krill can be determined. That method does have the disadvantage that it is time consuming. Major improvement in the development of quantitative echo-sounders over the past 30 years has meant that this technique is the preferred method for covering large areas and also for obtaining information on the local distribution of krill. Allied to these sampling methods is the need to make biological observations on the krill and these are discussed at the end of the chapter.

Having considered sampling methods the distribution and abundance of krill in the main areas of commercial interest are discussed in Chapter 3. In the coastal waters of Japan and Western Canada there are local fisheries while in the Southern Ocean the fishery has a very much larger area over which it can be distributed. Turning from the larger-scale distribution of krill considerations of swarming and vertical migration are discussed in Chapter 4. It is these aggregation patterns which make the krill suitable for commercial fishing and at the same time affect the foraging behaviour of dependent species.

Keeping to a discussion of krill centred features in Chapter 5 we have a full consideration of population parameters that can be used in developing population models. The information in this chapter is wide ranging in order to provide the reader with a view of the range of values likely to be encountered for each parameter. This broader view is retained as the reproduction of krill is considered in Chapter 6. Here aspects of the development of germ cells are considered leading to a discussion of the reproductive biology of various members of the group.

Having set the scene with regard to krill on their own, consideration is then given in Chapter 7 to the role of krill in the ecosystems within which it is harvested. This highlights the position of krill in the food chain. Against that background Chapter 8 considers ecosystem dynamics involving krill and considers the different time and space scales that are of relevance in developing management advice. These range from the broad scale of precautionary total allowable catches down to the interactions with the dependent species alluded to earlier.

The scene having been set, the emphasis changes wholly towards the krill fisheries and their management. In Chapter 9 the krill fisheries are described in terms of fishing methods, catch rates and history of the fisheries. A natural progression from considerations of the capture of krill is discussion on the biochemical qualities that can be utilised in the development of marketable products; these are discussed in Chapter 10.

Moving from the fishery and its operation commercially Chapter 11 is the first to consider fishery management, in this case in Japanese waters. The management regime there takes account of market forces and environmental variation. Management of the krill fishery in the Southern Ocean is discussed in Chapter 12. The Convention for the Conservation of Antarctic Marine Living Resources, being the first international fishery agreement to require the management regime to take an ecosystem approach, has required the introduction of novel approaches. These have been considered by the authorities in determining how to manage the fisheries in Canadian waters described in Chapter 13. The final chapter looks to the future by considering what has been happening in the krill fisheries world-wide and considering how this might be brought forward as we move into the twenty-first century.

References

Baker, A. de C., Boden, B.P. & Brinton, E. (1990) *A Practical Guide to the Euphausiids of the World.* British Museum (Natural History), London.

Brierley, A.S. & Brandon, M.A. (1999) Potential for long-distance dispersal of *Euphausia crystallorophias* in fast current jets. *Marine Biology* **135**, 77–82.

Mauchline, J.R. (1980a) Key for the identification of Antarctic euphausiids. *BIOMASS Handbook* No 5, 4 pp.

Mauchline, J.R. (1980b) The biology of mysids and euphausiids. *Advances in Marine Biology* **18**, 1–681.

Mauchline, J.R. (1984) *Euphausiid, Stomatopod and Leptostracan Crustaceans. Keys and Notes for the Identification of the Species.* Synopses of the British Fauna. Edited by Doris M. Kermack & R.S.K. Barnes, No 30. The Linnean Society of London and The Estuarine and Brackish-Water Sciences Association.

Mauchline, J.R. & Fisher, L.R. (1969) The biology of euphausiids. *Advances in Marine Biology* **7**, 1–454.

Nicol, S. & Endo, Y. (1997) Krill fisheries of the world. *FAO Fisheries Technical Paper* **367**, 100 pp.

Nicol, S. & Endo, Y. (1999) Krill fisheries: Development, management and ecosystem implications. *Aquatic Living Resources* **12**, 105–120.

Chapter 2
Sampling Krill

2.1 Direct sampling

Jon Watkins

There has been a continuing requirement to obtain samples of krill since the earliest investigations into krill biology; in the case of Antarctic krill significant sampling began in 1925 with the Discovery investigations (Kemp & Hardy, 1929). At that time net sampling was the main way of obtaining information on distribution, abundance, population demography and behaviour of euphausiids. Indeed, with the exception of visual sightings of surface swarms of krill, it was not possible to know where krill were without undertaking some form of net sampling. Since then the techniques and equipment available to marine scientists have increased dramatically. Nowadays much of the information on distribution and abundance of krill can be derived from remote sampling techniques, such as acoustics, as well as a variety of direct sampling techniques.

The last 20 years have seen great advances with *in situ* observation and sampling techniques which include diving (Hamner *et al.*, 1983; O'Brien, 1987), remotely operated vehicles (ROVs) (Marschall, 1988), underwater photography and video (Guzman & Marin, 1982; O'Brien, 1987; Sameoto *et al.*, 1993), and towed electronic sensors such as the Optical Plankton Counter (OPC) (Herman *et al.*, 1993). However, while these modern techniques provide data that could never be collected by traditional net sampling, they have not eliminated the need for net sampling, rather the emphasis for net sampling has changed.

Modern day net sampling programmes are now undertaken for the following reasons:

- To validate or ground-truth remote sensing techniques.
- To collect animals for population demography and biological classification. For example, to assess the length, maturity stage and age of the krill.
- To get live animals for laboratory studies of behaviour and physiology.
- To get estimates of biomass and distribution. This may be the only way for some life stages such as the eggs and larvae to be sampled.

Such a wide range of objectives for net sampling means that many different types of net have been developed and a range of strategies have been devised for using these nets in the most efficient manner. In an ideal world a net would sample whatever was in front of it and the resultant sample would be totally representative of the local population. However, in reality all nets provide a biased sample. The amount and type of bias will depend on the actual sampling gear and the chosen sampling strategy in relation to the distribution, behaviour and life cycle stage of the krill being studied.

No one net or sampling strategy is likely to be suitable for all purposes and so arranging a modern-day net sampling programme can be an intimidating experience. This section discusses the major factors that need to be considered when selecting an appropriate net and sampling strategy for a given project.

Problems with direct sampling

The ultimate aim of a sampling programme is to obtain a set of results that are considered representative of the population under study. How we define representative and population are important. These will depend on each study and will have a major effect on the degree of sampling bias that occurs or is deemed acceptable (Watkins *et al.*, 1990). Every type of sampler used will have a variety of biases which ultimately affect the representativeness of the sampling. The level of importance of these various biases will depend on the objectives of the sampling programme. Sampling biases may be divided into three broad categories: availability of krill to be sampled, sampler avoidance and representativeness of sampling.

(a) Availability of krill to the sampler

Many species of krill undertake a diurnal vertical migration as discussed by Watkins (Chapter 4). It is therefore quite possible that the proportion of a population that could be found in front of the net at any one time will vary dramatically unless we are able to sample the entire distribution range. The use of independent methods, particularly acoustics, to determine the actual distribution of the krill has done much to reduce the degree of error that may be caused by this general problem. Although note that in the case of krill occurring near the surface many acoustic systems regularly used do not sample this stratum (Everson & Bone, 1986a; Macaulay Chapter 2).

(b) Net avoidance

It has long been known that not all the krill that are in front of a sampler will enter the net (Tattersall, 1924; Mackintosh, 1934; Hardy, 1936). Initially avoidance was surmised by comparing differences between day and night net catches. This was thought to be a response primarily to visual cues and so the assessment of avoidance was restricted to a relative comparison of catch rates (Brinton, 1967; Wiebe *et al.*,

1982; Hovekamp, 1989). Most investigations of net avoidance have been field studies; however, Fleminger & Clutter (1965) conducted experiments on a captive population of mysids and copepods which provided substantially more control over the results.

More recently evidence of significant net avoidance has come from comparisons with other types of sampling, such as visual, photographic or acoustic estimates of density. Thus Nicol (1986) compared the densities of *Meganyctiphanes norvegica* surface swarms estimated from samples by three different methods:

- a bag-sampler which gave numbers up to 41 000 krill m^{-3};
- photographic methods, which gave numbers up to 770 000 krill m^{-3}; and
- a plankton net which gave numbers that never exceeded 6 krill m^{-3}.

Although precisely the same swarms were not sampled by each method it is clear that the plankton net with estimated density orders of magnitude lower than the other two methods must be providing a significant underestimate.

Acoustic density estimates have also been compared with net sample estimates and the conclusion is that net samples are usually lower because of significant net avoidance (Everson, 1987; Sameoto *et al.*, 1993; Pauly *et al.*, 1997; Watkins & Murray, 1998). Of course with all such comparisons there is the possibility that the independent sampling technique used may also be biased in some way. The relative problems associated with visual and acoustic sampling have been discussed by Watkins & Murray (1998). However, in the case of acoustic techniques there is an additional advantage – it is possible to see the krill actually avoiding the net. Thus Everson & Bone (1986b) used an echo-sounder transducer mounted just in front of the net mouth to observe how Antarctic krill moved out of the way to avoid a rectangular midwater trawl (RMT8). Studies carried out by Sameoto *et al.* (1993) on *Meganyctiphanes norvegica* used a variety of independent sampling techniques together (ROV-mounted video, OPC, multi-frequency acoustics and net) to show that net density estimates were 2–3 orders of magnitude less than those obtained from the other sampling methods.

Net avoidance will affect the absolute number of krill entering the net. However, it may also affect the type of krill entering the net. Thus large krill, which can swim faster than small krill, may be able to avoid the net more successfully than small krill (Hovekamp, 1989). The type of net selectivity will depend on the animals and the net used. For any net, selectivity is likely to vary for different nets and this presents a major problem in comparing studies. It is interesting to note that Bone (1986) found no obvious size difference in the krill caught by two scientific nets, a rectangular midwater trawl (RMT8) and a large Longhurst-Hardy plankton recorder (LLHPR). Similarly Anon (1991) found no significant differences in the catches of a number of scientific nets used in the BIOMASS surveys of the South Atlantic in 1980. However, there were substantial differences between scientific nets and commercial trawls.

An interesting development to minimise the problems of net avoidance has been put forward by Sameoto *et al.* (1993) who describe the use of strobe lighting on a MOCNESS net. When a strobe light was shone ahead of the net during daytime hauls the degree of net avoidance was significantly reduced. Net density estimates with the light on were similar to those estimated with an OPC and much closer to the acoustic estimates than when the light was not used.

(c) Degree of aggregation and the representativeness of sampling

All euphausiids exhibit some degree of heterogeneous distribution or patchiness. The ultimate expression of this is the formation of discrete, high density swarms in species such as *Euphausia superba* and *Meganyctiphanes norvegica* (see reviews by Miller & Hampton, 1989; Siegel & Kalinowski, 1994). The formation of such aggregations may have a number of effects on the sampling of krill. It has been known for many years that plankton patchiness increases the sampling variability (Silliman, 1946; Anraku, 1956; Wiebe & Holland, 1968), and it has generally been determined that abundances have to differ by a factor of at least two to be considered significantly different. More recently it has been shown that the heterogeneous distribution of krill may affect more than just abundance estimates. Watkins *et al.* (1986, 1990) showed that adjacent swarms frequently contain krill which have different size or maturity status. Thus single net hauls are likely to contain krill which are totally unrepresentative of the local population. For example, it was calculated that to estimate the mean size of krill in an area with a precision similar to that which would occur if there was no heterogeneity, would require a minimum of 20 swarms to be sampled.

(d) The importance of sampling biases for different sampling objectives

In the previous paragraphs the various ways in which sampling biases may arise have been discussed. However, the relative importance of these different biases will vary according to the aims of the sampling programme. It is considered that net sampling to estimate euphausiid biomass is subject to the greatest number of biases. In Table 2.1.1 the objectives of the sampling programme have been ranked against the relevant biases. Thus availability of krill to the sampler is only a serious problem when there is no independent method of detecting where the krill are in the water column. Consequently if net sampling is used in conjunction with acoustic techniques then this is much less of a problem unless the krill are very close to the surface and above the transducer, as might happen at night. When determining the biological characteristics of the krill population it is less important whether some krill avoid the net as long as there is no significant net selection. For acoustic validation, bias caused by net selection is important if it is necessary to obtain a size classification, but if the aim is only to confirm species identification then the problem is reduced even more.

Table 2.1.1 The importance of net sampling problems for different sampling programmes ×× – significant problem or bias, × – minor problem or bias, ✓ no problem.

Net sampling problem	Net sampling programme			
	Estimates of biomass	Biological characteristics	Acoustic validation	Live animals
Availability of krill	××	✓	✓	✓
Net avoidance	××	×	×	✓
Selectivity	××	××	××	✓
Degree of aggregation	××	✓	××	✓
Representatives of sampling	××	××	××	✓

Sampler designs

Over the years the design of nets has changed in order to reduce the biases associated with net sampling and to cater for specific sampling objectives. General principles for designing nets were thoroughly discussed by Clutter & Anraku (1968). They concluded that samplers should have the following characteristics; they should be:

- as large as possible,
- propelled as fast as possible at a constant speed,
- capable of remaining at least 85% efficient throughout the haul,
- free of forward obstructions forward of the mouth such as tow-lines, and
- generally dark-coloured with no shiny metal components.

Some of these requirements are mutually exclusive, for instance net size and towing speed tend to be inversely related.

Many modern nets have the towing cable attached to a point which is not directly in front of the net (e.g. RMT – Roe & Shale, 1979; MOCNESS – Greene *et al.*, 1998; BIONESS – Herman, 1988). The plummet net described by Hovekamp (1989) is free of all obstructions forward of the mouth of the net as it has no bridles and samples by free falling through the water column.

The size of many nets has increased greatly from that of the 1 m ring net used during the Discovery expeditions and described by Kemp & Hardy (1929). Scientific nets used to sample Antarctic krill often have a mouth area of 8 m^2 (Roe & Shale, 1979). Other significant advances are the ability to take multiple samples within a single haul, either using different sized nets (e.g. Bongo and RMT8+1 M) or using multiple nets of the same size to sample different depth horizons (e.g. BIONESS, MOCNESS, RMT). For sampling at a finer spatial resolution, the LLHPR is capable of collecting up to 70 individual samples during a single haul (Bone, 1986). Such modern multi-samplers are usually deployed with integrated oceanographic instrument packages and real-time net to ship telemetry of key parameters such as net depth, flowmeter readouts and net status (e.g. mouth angle, open/closed state). With such equipment it is possible to target the net at particular targets detected on a sonar or echo-sounder (Watkins & Brierley, in press). Most recently we see that some net systems also have independent acoustic, video or particle-counter sensors attached which provide important, additional information on the quality of the net sampling (Sameoto *et al.*, 1993).

Sampling strategies to obtain representative samples of population

This section on direct sampling techniques has demonstrated that net sampling programmes are nowadays frequently used for purposes other than estimating abundance or biomass of the population. This introduces a number of new priorities when we design net sampling programmes. The plans for the CCAMLR synoptic survey (SC-CAMLR, 1999) show some of the factors that need to be considered in a modern net sampling programme which is carried out to support acoustic biomass estimates and to describe the population demography. The various sampling strategies that may be considered are discussed below.

(a) Daytime versus night-time net sampling

Evidence has already been presented on the large differences between day and night-time net avoidance. Therefore a relatively simple strategy is to restrict net sampling to the hours of darkness. Such a plan was used in the British Antarctic Survey Core Programme, designed to monitor the interannual variability of krill at South Georgia (Brierley *et al.*, 1997). Two sampling criteria needed to be satisfied for this programme. The first was to provide representative samples of krill for length frequency and maturity stage analyses. The second was to determine the species that were present in individual scattering layers. For this study station sampling with nets was undertaken at night to minimise biases due to avoidance and this was combined with acoustic surveys carried out during the day, when krill are unlikely to be found close to the surface and consequently above a hull-mounted echo-sounder transducer. The second question was addressed by undertaking a series of daytime target validation hauls with the aim of providing a qualitative indication of the dominant species that were present. A similar sampling strategy was also adopted for the recent multi-national CCAMLR Synoptic Survey of the Scotia Sea (SC-CAMLR, 1999).

(b) Station versus target fishing

Traditional net sampling programmes utilise a regular (Siegel, 1992) or random sampling grid (Everson *et al.*, 1996), however, in both cases the sampling is independent of the actual availability of krill. In such programmes the net(s) are usually deployed over the entire depth range of occurrence of the krill. In the case of

Euphausia superba, this is typically 0–250 m (Brierley *et al.*, 1997). While such sampling designs are important to enable a valid estimate of abundance they are less than optimal if this is not the main aim of the sampling programme. If the aim is to validate acoustic estimates or to assess the biological characteristics of swarms then it is most efficient to target krill aggregations. Such target fishing (Ricketts *et al.*, 1992) is relatively easily done with modern equipment (echo-sounders, net-sondes, and real-time data on net position and state).

We can see that depending on the aim it is advantageous to use either a target or 'random' net tow design. If the sampling programme has several aims we might consider using both kinds of net tow. Unless it is possible to carry out a large number of tows of each kind we must determine how representative our sampling may be. For example recruitment indices for *Euphausia superba* have been calculated from length density data collected with oblique net hauls carried out at regular stations (de la Mare, 1994; Siegel *et al.*, 1997). However, where the number of net hauls is limited it may be necessary to combine target and station net haul data sets to ensure that a sufficient number of net hauls are carried out to represent the population adequately. Watkins (1999) describes the use of a composite index which does this.

While it may be argued that targeted net hauls may give a biased estimate of the population characteristics (Pauly *et al.*, 1997; Watkins in press), a recent major survey of Antarctic krill in the Indian Ocean sector of the Southern Ocean showed that both types of net haul will tend to provide a representative description of the population providing sufficient hauls are carried out (Fig. 2.1.1) (Pauly *et al.*, 1997).

(c) How many swarms to sample?

I have already alluded to the problem of heterogeneity and sampling of krill. This problem affects the sampling design for every sampling objective apart from collecting live krill for experiments (Table 2.1.1). Watkins *et al.* (1990) analysed surveys over a number of different time- and space-scales and came up with some estimates of the number of swarms that might need to be sampled. While such figures provide a useful guide they demonstrate that it is very important that the trade off between precision and number of samples should be assessed for any particular survey. One implication that should be stressed for target hauls that are used to validate acoustic surveys is that proximity of swarms does not imply similarity. Therefore while it might appear more precise to link the results from single net hauls to small sections of survey track adjacent to the net haul, the small-scale variability present means that it is better to combine net hauls and generate an estimate of krill length distribution for a larger area. How is the size of such an area to be decided? One way may be to use ordination techniques such as cluster analysis to define the areas on the basis of a similarity matrix, such as has been done in survey analyses of population demographics of *Euphausia superba* (Siegel, 1988; Watkins *et al.* 1999).

There have been a number of papers that have considered how many krill should

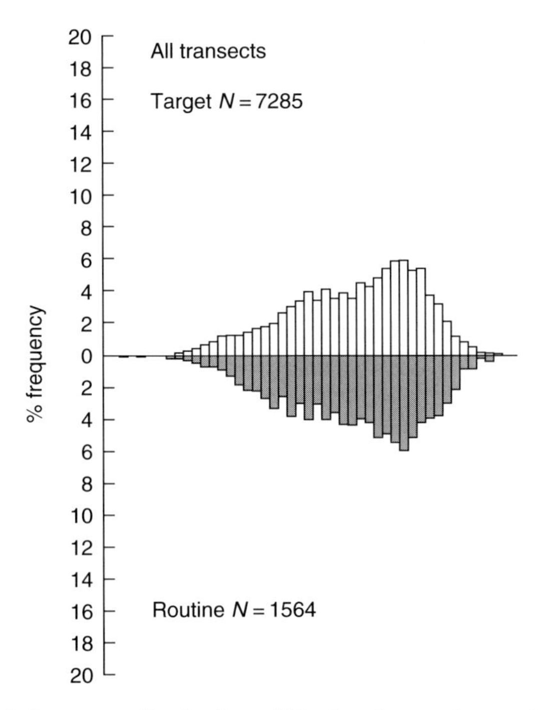

Fig. 2.1.1 The length-frequency distribution of *Euphausia superba* caught in hauls directed at acoustic targets (clear histogram) and in hauls taken at routine stations (shaded histogram). *N* refers to total number of animals measured. Figure redrawn from Nicol *et al.* (in press) with permission from Elsevier Science.

be measured in any particular net. In situations where the sample is clearly unimodal with a narrow spread, a sample site of 50 may be adequate. However, if the distribution is broad and polymodal then a sample size of 200 may be needed in order to obtain a reasonable distribution.

A further consideration with regard to the sampling design of net haul surveys concerns the trade-off between the number of hauls and duration of the individual tows. Pennington & Volstad (1991, 1994) suggest that it is more efficient to reduce the time of individual nets and carry out more net hauls than to carry out fewer, longer net hauls. This is in line with the findings of Watkins *et al.* (1986) where it is stressed that it is more important to sample more swarms of krill than to increase the number of krill sampled from within the same swarm.

(d) Can we ever estimate absolute density from nets?

It has been shown that estimating density or abundance from nets is likely to be more prone to bias than any other goal in a net sampling programme. The work of Sameoto *et al.*(1993) shows that in the future it may be possible to reduce some of the sampling biases caused by net avoidance. However, the large differences between the various methods used to estimate density using nets mean that where possible other techniques should be used to estimate the density of krill. However, it is only fair to point out that standardised net sampling programmes can have a very important role in obtaining relative density estimates that can then be used to monitor population changes (see for instance Siegel *et al.*, 1997).

(e) Alternative sampling techniques to capture krill

Net sampling has been the traditional way to capture krill for measurement to determine length frequency distributions for population demography. Depending on the location and level of effort, net sampling programmes tend to be both expensive and time consuming. It is therefore worth considering other ways in which we may obtain samples of krill. One of the more interesting techniques is to use predators to catch the krill and then sample the krill from the predators. Such techniques have been widely used in the CCAMLR Ecosystem Monitoring Programme (CEMP) (SC-CAMLR, 1999). Predators used as samplers may be prone to the same types of biases as nets. Of particular concern has been the selectivity of predators for particular sizes or types of krill (see Nicol, 1993; Hill *et al.*, 1996; Reid *et al.*, 1996). While this can undoubtably occur, Reid *et al.* (1999) showed that there was good correspondence between the combined size distribution of *Euphausia superba* taken by several different predator species (Macaroni penguins and Antarctic fur seals) and the size distribution derived from a series of net hauls conducted in the same area and at the same time (Fig. 8.8).

One of the advantages of predator sampling is that it is possible to collect data over longer temporal scales than may be possible in a ship-based sampling programme. Set against this is the problem of where predator samples may have come from and the generally more restricted spatial coverage of the sampling. In the case of sampling krill from predators at South Georgia, Reid *et al.* (1999) found a temporal change in size range of krill taken. While the temporal coverage of net-caught krill was much less, the spatial coverage was significantly greater. It was possible to see spatial trends in the data that would probably be reflected as temporal changes when sampled from a fixed point past which a stream of krill was flowing (Watkins 1999).

References

Anon. (1991) Non-acoustic krill data analysis workshop. *BIOMASS Rep. Ser.* **66**, 1–59.

Anraku, M. (1956) Some experiments on the variability of horizontal plankton hauls and on the horizontal distribution of plankton in a limited area. *Bull. Fac. Fish. Hokkaido Univ.* **7**, 1–16.

Bone, D.G. (1986) An LHPR system for adult Antarctic krill (*Euphausia superba*). *Br. Antarct. Surv. Bull.* **73**, 31–47.

Brierley, A.S., Watkins, J.L. & Murray, A.W.A. (1997) Interannual variability in krill abundance at South Georgia. *Mar. Ecol. Prog. Ser.* **150**, 87–98.

Brinton, E. (1967) Vertical migration and avoidance capability of euphausiids in the California Current. *Limnol. Oceanogr.* **12**, 451–83
Clutter, R.I. & Anraku. M. (1968) Avoidance of samplers in zooplankton sampling. In *UNESCO Monographs on Oceanographic Methodology, 2: Zooplankton sampling* (D.J. Tranter, ed.), pp. 57–76. UNESCO Press, Paris.
de la Mare, W.K. (1994) Modelling krill recruitment. *CCAMLR Sci.* **1**, 49–54.
Everson, I. (1987) Some aspects of the small scale distribution of *Euphausia crystallorophias*. *Polar Biology* **8**, 9–15.
Everson, I. & Bone, D.G. (1986a) Detection of krill (*Euphausia superba*) near the sea surface: preliminary results using a towed upward-looking echo-sounder. *British Antarctic Survey Bulletin* **72**, 61–70.
Everson, I. & Bone, D.G. (1986b) Effectiveness of the RMT8 system for sampling krill (*Euphausia superba*) swarms. *Polar Biology* **6**, 83–90.
Everson, I., Bravington, M. & Goss, C. (1996) A combined acoustic and trawl survey for efficiently estimating fish abundance. *Fisheries Research* **26**, 75–91.
Fleminger, A. & Clutter, R.I. (1965) Avoidance of towed nets by zooplankton. *Limnology and Oceanography* **10**, 96–104.
Greene, C.H., Wiebe, P.H., Pershing, A.J. *et al.* (1998) Assessing the distribution and abundance of zooplankton: a comparison of acoustic and net-sampling methods with D-BAD MOCNESS. *Deep-Sea Research II* **45**, 1219–37.
Guzman, O. & Marin, B. (1982) Hydroacoustic and photographic techniques applied to study the behavior of krill. *Mem. Natl Inst. Polar Res., Special Issue* **27**, 129–51.
Hamner, W.M., Hamner, P.P., Strand, S.W. & Gilmer, R.W. (1983) Behavior of Antarctic krill, *Euphausia superba*: chemoreception, feeding, schooling, and molting. *Science* **220**, 433–35.
Hardy, A.C. (1936) Observations on the uneven distribution of oceanic plankton. *Discovery Report* **9**, 511–38.
Herman, A.W. (1988) Simultaneous measurement of zooplankton and light attenuance with a new optical plankton counter. *Continental Shelf Research* **8**, 205–221.
Herman, A.W., Cochrane, N. A. & Sameoto, D.D. (1993) Detection and abundance estimation of euphausiids using an optical plankton counter. *Marine Ecology Progress Series* **94**, 165–73.
Hill, H.J., Trathan, P.N., Croxall, J.P. & Watkins, J.L. (1996) A comparison of Antarctic krill *Euphausia superba* caught by nets and taken by macaroni penguins *Eudyptes chrysolophus*: evidence for selection? *Marine Ecology Progress Series* **140**, 1–11.
Hovekamp, S. (1989) Avoidance of nets by *Euphausia pacifica* in Dabob Bay. *Journal of Plankton Research* **11**, 907–924.
Kemp, S. & Hardy, A.C. (1929) Discovery investigations, objects, equipment and methods. *Discovery Report* **1**, 141–232.
Mackintosh, N.A. (1934) Distribution of the macroplankton in the Atlantic sector of the Antarctic. *Discovery Rep.* **9**, 65–160.
Marschall, H.-P. (1988) The overwintering strategy of Antarctic krill under the pack-ice of the Weddell Sea. *Polar Biol.* **9**, 129–35.
Miller, D.G.M. & Hampton, I. (1989) Biomass Scientific Series, No. 9: Biology and ecology of the Antarctic Krill (*Euphausia superba Dena*). A Review SCAR and SCJR, Scott Polar Research Institute, Cambridge, England.

Nicol, S. (1986) Shape, size and density of daytime surface swarms of the euphausiid *Meganyctiphanes norvegica* in the Bay of Fundy. *J. Plankton Res.* **8**, 29–39.
Nicol, S. (1993) A comparison of Antarctic petrel (*Thalassoica antarctica*) diets with net samples of Antarctic krill (*Euphausia superba*) taken from the Prydz Bay region. *Polar Biol.* **13**, 399–403.
Nicol, T., Kitchener, J., King, R., Hosie, G.W. & de la Mare, W.K. Population structure and condition of Antarctic krill (*Euphausia superba*) off East Antarctica (80–150°E) during the Austral summer of 1995/96. *Deep-Sea Res. II*, Topical Studies in Oceanography (in press).
O'Brien, D.P. (1987) Direct observations of the behaviour of *Euphausia superba* and *Euphauisa crystallorophias* (Crustacea: Euphausiacea) under pack ice during the Antarctic spring of 1985. *J. Crustacean Biol.* **7**, 437–48.
Pauly, T., Nicol, S., de la Mare, W.K., Higginbottom, I. & Hosie, G. (1997). A comparison between the estimated density of krill from an acoustic survey with that obtained by scientific nets on the same survey. CCAMLR, Hobart, WG-EMM-97/43.
Pennington, M. & Volstad, J.H. (1991) Optimum size of sampling unit for estimating the density of marine populations. *Biometrics* **47**, 717–23.
Pennington, M. & Volstad, J.H. (1994) Assessing the effect of intra-haul correlation and variable density on estimates of population characteristics from marine surveys. *Biometrics* **50**, 725–32.
Reid, K., Trathan, P.N., Croxall, J.P. & Hill, H.J. (1996) Krill caught by predators and nets: differences between species and techniques. *Marine Ecology Progress Series* **140**, 13–20.
Reid, K.G., Watkins, J.L., Croxall, J.P. & Murphy, E.J. (1999) Krill population dynamics at South Georgia 1991-97, based on data from predators and nets. *Marine Ecology Progress Series* **177**, 103–114.
Ricketts, C., Watkins, J.L., Morris, D.J., Buchholz, F. & Priddle, J. (1992) An assessment of the biological and acoustic characteristics of swarms of Antarctic krill. *Deep-Sea Research* **39**, 359–71.
Roe, H.S.J. & Shale, D.M. (1979) A new multiple rectangular midwater trawl (RMT1+8 M) and some modification to the Institute of Oceanographic Sciences (RMT1+8). *Marine Biology* **50**, 283–88.
Sameoto, D., Cochrane, N. & Herman, A. (1993) Convergence of acoustic, optical, and net-catch estimates of euphausiid abundance: use of artificial light to reduce net avoidance. *Canadian Journal of Fisheries and Aquatic Science* **50**, 334–46.
SC-CAMLR (1999). Report of the Working Group on ecosystem monitoring and management. In: *Report of the eighteenth meeting of the Scientific Committee.* CCAMLR, Hobart.
Siegel, V. (1988). A concept of seasonal variation of krill (*Euphausia superba*) distribution and abundance west of the Antarctic peninsula. In *Antarctic Ocean and Resources Variability* (D. Sahrhage, ed.), pp. 219–30. Springer-Verlag, Berlin.
Siegel, V. (1992) Assessment of the krill (*Euphausia superba*) spawning stock off the Antarctic Peninsula. *Arch. FischWiss.* **41**, 101–130.
Siegel, V. & Kalinowski, J. (1994). Krill demography and small-scale processes: a review. In *Southern Ocean Ecology: the BIOMASS perspective* (S.Z. El-Sayed, ed.), pp. 145–63. Cambridge University Press, Cambridge.
Siegel, V., de la Mare, W.K. & Loeb, V. (1997) Long-term monitoring of krill recruitment and abundance indices in the Elephant Island area (Antarctic Peninsula). *CCAMLR Sci.* **4**, 19–35.

Silliman, R.P. (1946) A study of variability in plankton tow-net catches of Pacific Pilchard (*Sardinops caerulea*) eggs. *J. Mar. Res.* **6**, 74–83.

Tattersall, W.M. (1924) Crustacea, Pt. VIII. Euphausiacea. *British Antarctic Terra Nova Expedition 1910. Natural History Report Zoology* **8**, 1–36.

Watkins, J.L. (1999) A composite recruitment index to describe interannual changes in the population structure of Antarctic krill at South Georgia. *CCAMLR Sci.* **6**, 71–84.

Watkins, J.L. & Brierley, A.S. (in press) Verification of acoustic techniques used to identify Antarctic krill. *ICES J. mar. Sci.*

Watkins, J.L. & Murray, A.W.A. (1998) Layers of Antarctic krill, *Euphausia superba*: are they just long krill swarms? *Mar. Biol.* **131**, 237–47.

Watkins, J.L., Morris, D.J., Ricketts, C. & Priddle, J. (1986) Differences between swarms of Antarctic krill and some implications for sampling krill populations. *Mar. Biol.* **93**, 137–46.

Watkins, J.L., Morris, D.J., Ricketts, C. & Murray, A.W.A. (1990) Sampling biological characteristics of krill: effect of heterogeneous nature of swarms. *Mar. Biol.* **107**, 409–415.

Watkins, J.L., Murray, A.W.A. & Daly, H.I. (1999) Variation in the distribution of Antarctic krill *Euphausia superba* around South Georgia. *Mar. Ecol. Prog. Ser* **188**, 149–60.

Wiebe, P.H. & Holland, R.H. (1968) Plankton patchiness: effects on repeated net tows. *Limnol. Oceanogr.* **13**, 315–21.

Wiebe, P.H., Boyd, S.H., Davis, B.M. & Cox, J.L. (1982) Avoidance of towed nets by the euphausiid *Nematoscelis megalops*. *Fish. Bull.* **80**, 75–91.

2.2 Acoustic estimation of krill abundance

Michael Macaulay

Overview

Before an investigator employs acoustic methods to census krill, numerous factors must be considered. Hydroacoustics, like any remote sensing method, requires a good understanding of the sensing technology and its limitations. The investigator will need to have confidence in what is being observed and how to best deploy equipment for the task at hand. This section of the chapter will address some of the ways krill behaviour alters their acoustic signature. The changeable acoustic characteristics of krill directly influence our choice of off-the-shelf hydroacoustic instrumentation from among the different types available. Other items to be covered include: what acoustic data can tell us about the distribution of krill in relation to their environment, and how to manage the large quantities of data once we have them. All these items have been identified (GLOBEC, 1991) as problems that need consideration prior to using hydroacoustic methods to observe the highly mobile krill. This section assumes that the reader is, at least passingly, familiar with acoustic terminology. For those who want a more detailed study of acoustics, a good reference on methods is given in MacLennan & Simmonds (1992).

Tools for acoustic observation

The acoustician can choose a single-beam system, dual-beam system, or a split-beam system to conduct acoustic measurements. Single-beam systems were originally the most commonly used, but split-beam systems have become the current standard for echo integration estimation of biomass and other aspects of krill distribution. Dual-beam and split-beam systems are most frequently used to measure krill target strength and to locate targets in an ensonified volume. To obtain the highest quality volume scattering data and to make direct target strength measurements, it is now common practice (Cochrane *et al.*, 1991; Hewitt & Demer, 1993; Macaulay *et al.*, 1995) to deploy single-beam and dual-beam or split-beam systems operating at two or more frequencies. Adjunct tools to improve localisation of concentrations of krill are provided by searchlight sonar and side-scan sonar.

(a) Perfomance comparison

An investigator can measure target intensity (target strength) with single-, dual-, or split-beam methods. Figure 2.2.1 shows a comparison of single-, dual-, and split-beam systems and the possible appearance of echoes for the same set of targets for each kind of system. The small X–Y plots below each example indicate the possible intensity and/or phase signals for this set of targets with each type of system. In each case, the first target is a small fish well off the main axis of all systems (target a in

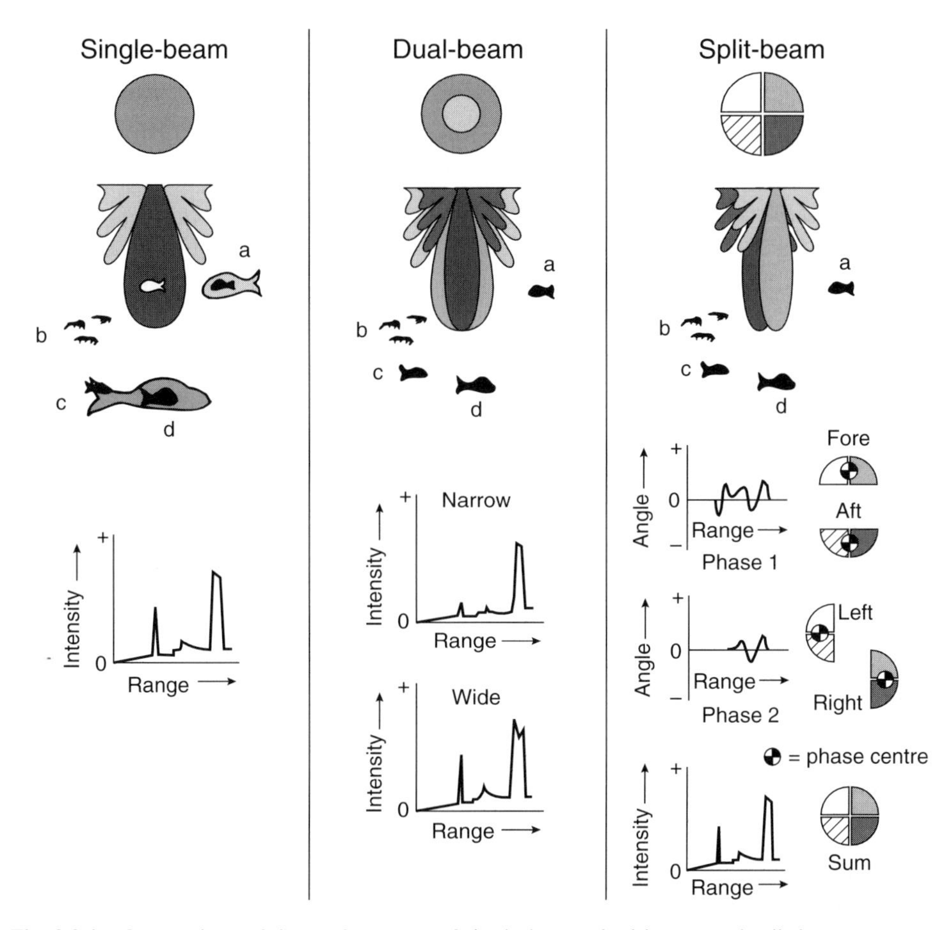

Fig. 2.2.1 Comparison of the performance of single-beam, dual-beam, and split-beam systems on a mixed population of targets. The shaded fish outlines in the single-beam illustration indicate uncertainty whether target 'a' is a small or large fish off axis or a small fish on axis, and whether targets 'c' and 'd' are possibly one larger target. The dual-beam and split-beam systems properly identify target 'a' as a small fish off axis and targets 'c' and 'd' as two separate fish.

each column). In a single-beam system, it is not clear whether this target is a small fish on axis or a larger fish off-axis (hence the shaded outlines in the single-beam illustration). This positional and size uncertainty is a common problem with single-beam systems. Dual-beam systems provide two channels of data, one from each transducer (Fig. 2.2.1, second-column of the illustration). A target's position in the acoustic beam is resolved using the signal intensity in relation to the beam patterns for the narrow and wide portions of the transducer. The angular resolution of the target's position forms an acceptance/rejection criterion, such that targets with narrow beam intensity below a pre-established level (usually 3–6 dB down from the wide beam intensity) are rejected as being too far off-axis to obtain a reliable estimate of target strength. A dual-beam system resolves the first off-axis target, (that the single-beam system could not resolve) as probably a small target off-axis (appears in wide-beam signal and poorly in the narrow-beam signal). The volume scatter from a number of small targets at mid-range (target b in all columns) is resolved as being off-axis in the dual-beam system. For the two targets at maximum range (targets c and d in all columns), a dual-beam system may resolve whether this is one larger target or two smaller targets. Split-beam systems (see the third column of Fig. 2.2.1) resolve positional uncertainties by providing a more complete definition of the spatial distribution of targets (Traynor & Ehrenberg, 1979; MacLennan & Simmonds, 1992). The split-beam system can resolve the angular bearing to each of the example targets and would probably resolve the two targets at maximum range as two separate targets, so long as the separation between the two is sufficient to provide distinct phase angle differences. The phase angles for a target directly on axis for the split-beam array would produce 0° phase angles for both axes, other positions would produce phase angles with different magnitude and sign proportional to the target's position. Range-to-a-target is measured using the sum beam signal. The two phase angles and the target range are then converted to a mechanical offset angle in two axes and expressed in Cartesian or polar coordinates to specify location of each identified target. The primary deciding factor between using dual-beam or split-beam systems for measuring target strength is one of cost (dual-beam systems are considerably less expensive). If precise positioning and wide size-range of the target are important, then a split-beam system is always preferable.

Other types of acoustic systems, which have been or are being used for acoustic observation, include: sector-scanning sonar, a means of electrically switching from transducer element to transducer element that produces a precise localisation of targets; searchlight sonar, in which a single transducer is mechanically scanned from position to position to localise targets; and, side-scan sonar which uses a long, narrow transducer with a very wide beam in one axis (often 90° or more at the narrow dimension of the transducer), and a very narrow beam for the axis at right angles to the first (often only 1–2° or less at the long dimension of the transducer). These types of systems are frequently used to localise nets and schools, or patches of fish or krill, relative to a fishing vessel for more effective catch strategies. In general, these systems are not used for quantitative measurements because they are seldom calibrated adequately to determine beam pattern, sensitivity, and source level precisely. They are, however, very effective at improving the chances of being able to locate and observe some key features of krill behaviour.

An additional type of acoustic system that aids in separating fish echoes from other signals utilises a 'chirp'. The production of a frequency band (the 'chirp') provides a way to separate individual targets from reverberation or background noise. The system relies on signal processing methods to separate portions of the returning echo into repetitive elements from those which are not (signal correlation), thereby improving recognition of discrete targets (i.e., short duration targets like fish) from more uniform, slower-varying ones like krill patches and layers or regions of acoustic noise.

Each of the types of systems described above are available as off-the-shelf systems from a number of manufacturers. The investigator may also need to obtain training on how to operate the associated data collection system. Having a clear sense of precisely what needs to be measured and why ensures selecting the proper system or systems for a particular investigation. A great deal of useful data can be collected with simple, non-quantitative echo-sounders before investing in more quantitative ones. The experience obtained may well save costly expenditure on the wrong equipment.

(b) System calibration

Acoustic systems are most often calibrated using standard targets (MacLennan, 1981; Foote, 1982; Foote *et al.*, 1987). Such targets are metal spheres of copper or tungsten carbide whose size and composition determine their target strength. Other methods of calibration have been used in the past (e.g. ANSI, 1972), but they are subject to considerable error (Blue, 1984). Single frequencies are commonly employed, though a more recent trend is to utilise more than one frequency simultaneously (Macaulay *et. al.*, 1995). Typically, frequencies from 38 kHz to 1–2 MHz have been applied in hydroacoustic investigations with biological emphasis (Holliday & Pieper, 1980). Survey sampling (where maximal depth range is desirable) requires the use of frequencies below 200 kHz, while point or profile sampling (where ranges of tens of metres or less is acceptable) can make use of frequencies above 200 kHz.

(c) Transducer design

Single-beam and dual-beam transducers are generally composed of single ceramic elements or simple arrays of elements. In a split-beam system, a bi-axial transducer is used to produce phase centres in a fore and aft plane (providing a fore and aft axis), and left and right phase centres (providing a second axis at right angles to the first one, see Fig. 2.2.1). The electrical phase difference for signals intersecting each axis from a target is converted to a mechanical angle bearing relative to a central axis perpendicular to the transducer face. These angular bearings, plus the slant range to the target (provided by the sum beam), allow precise mapping of the position of the target in Cartesian or polar coordinates, thereby removing the positional ambiguity of the target inherent in dual-beam systems. Split-beam transducers are constructed as an array of elements, especially at low (50 kHz or less) frequencies. The transducer may also be constructed from a single piece of ceramic divided into separate elements by slicing into several segments. There is no clear preference between circular or rectangular designs so long as stable phase centres can be established between the two halves of the transducer for each axis.

If the spacing and size of the transducer elements provide a distance between phase centres of less than 1/2 wavelength for each axis, the transducer will produce unambiguous phase angles over a wide range of directions away from the centre of the transducer; however, this often results in a very wide beam pattern – possibly too wide for a particular application. A more common practice in fisheries applications is to locate the phase centres a few wavelengths apart by using a narrow-beam transducer design. This minimises positional ambiguity by limiting acceptable phase angles for valid targets to those targets known to be within the scope of the main lobe of transducer sensitivity, and not detected by the side lobes. In all cases (including single-beam and dual-beam), the transducer design needs to have minimal side lobes because the angular position and level of the side lobes establishes the usable signal level and phase angle limits for determining target strength. Asymmetrical transducer arrays (wider in one axis than the other) can provide increased beam width for either axis to better fit the geometry of the environment. Transducers with asymetrical beam patterns are often used in riverine or estuarine environments where water depth may be a limiting factor.

(d) Near surface effects

The investigator should be aware of systematic limitations on the minimum range of detection in utilising hydroacoustic methods for enumerating krill populations. Because the acoustic properties of a signal near to a transducer are not well defined, we must restrict the beginning of sampling to a range where they are well defined. This range is approximately 10 transducer widths from the face of the transducer, or on the order of 1 meter. In addition, the transducer is seldom located at the surface and is usually several meters below, depending on whether it is hull-mounted or in a towed body. The presence of noise generating bubbles, either from the ship or weather-induced bubble entrainment, further limits the investigator's ability to detect krill or other targets in the immediate surface layers of the water. In practical terms, this means that the upper 10 m of the water column will be difficult or impossible to examine quantitatively for populations of krill. This acoustic 'blind-spot' creates a systematic error in estimates of krill populations whenever significant quantities of krill are located near surface. Fortunately, krill usually prefer deeper depths, at least during daylight hours, although some individuals may be present near-surface at all times. The diel migration of krill into the surface layers at night may force acoustic surveys to be done during daylight hours only. In addition, krill orientation and clumping in schools or patches is often very different between day and night, and this can alter the strength of echo returned (Greenlaw *et al.*, 1980; Sameoto, 1980; Everson, 1982). These observations are likely the result of orientation on target strength (McGehee *et al.*, 1998). Clearly, some knowledge of the local behaviour of krill populations in any given study area is required before an adequate survey design, using acoustic methods of sampling, can be developed.

Sound scatter versus reflection

Measurements of volume scattering distribution and intensity can provide estimates of biomass for many small organisms that have low individual target strength. Low target strength, ordinarily, would render the organism acoustically transparent. Because many organisms frequently aggregate in large concentrations, the effect of their combined scattering in an ensonified volume (i.e., volume scattering) produces a measurable signal. Krill are a typical case in which volume scattering measurements provide an effective way to determine distribution and abundance. Figure 2.2.2 illustrates a typical situation where the only targets are fish, while Fig. 2.2.3 shows the same situation but with the presence of a krill layer. Size distribution of krill can be extracted from multiple frequency single-beam systems (Pieper & Holliday, 1984; Holliday & Pieper, 1989).

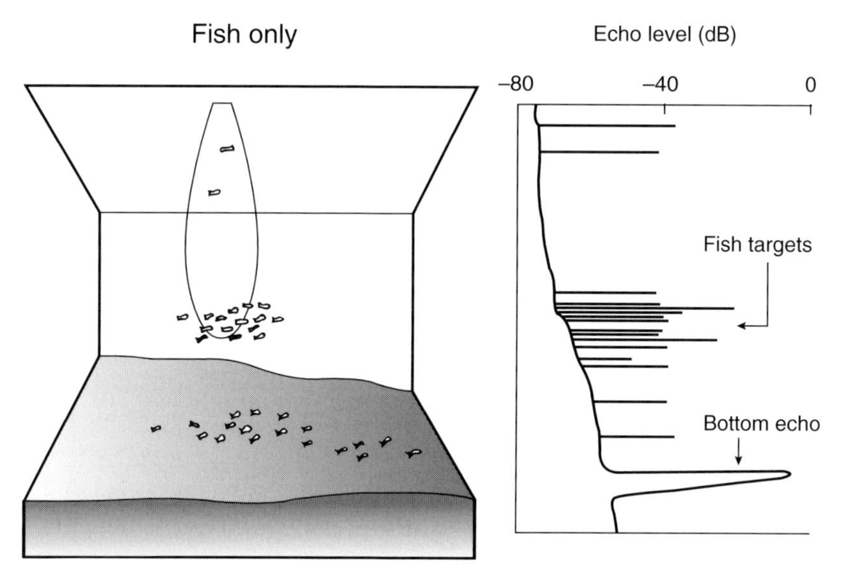

Fig. 2.2.2 Illustration of how fish targets, diagrammatically shown in the left illustration, might appear on an oscilloscope.

The estimation of krill abundance from volume scattering measurements (as is done by echointegration) requires the consideration of factors seldom used in acoustic estimation of fish abundance. Sound scatter from fish is predominantly in the geometric region of sound reflection, that is, the target is large relative to the wavelength of sound ensonifying it, and, there is little change in the strength of the echo with frequency. Sound scatter from krill, on the other hand, is often in the Rayleigh region such that the strength of the echo is strongly proportional to the size of the target, or in a resonant region where echo strength varies dramatically with changes in target size. Because krill are small and often nearly the same size as the wavelength of sound ensonifying them, there is strong frequency dependence of the level of scattering (see Fig. 2.2.4); this will create a similar frequency dependence in estimates of abundance obtained using echo integration or echo counting. Concentrations of krill (in schools or patches) produce complex modes of sound scatter

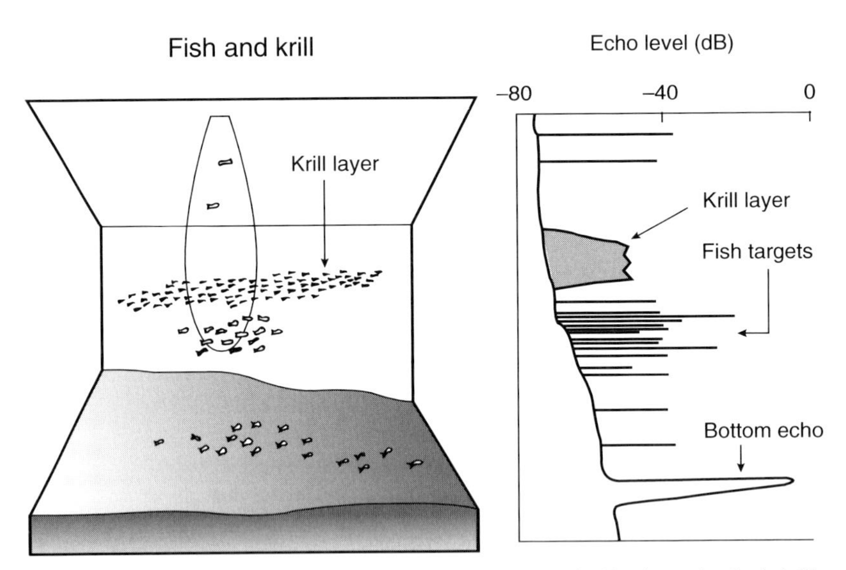

Fig. 2.2.3 Illustration of how fish targets and krill, diagrammatically shown in the left illustration, might appear on an oscilloscope. Note that the scope display shows the krill layer as a solid envelope and fish targets as distinct signal spikes. Krill will seldom appear as individual targets, rather they will be detected as a layer of volume scatter.

from the individuals comprising the layer (resonant effects). Krill, unlike fish, exhibit a much stronger orientation dependence on the levels of sound scatter they produce (MacLennan & Simmonds, 1992). The observer needs to be aware that these effects combine to produce errors in the estimate of abundance obtained acoustically. Recent papers by McGehee *et al.*, 1998, and Traykovski *et al.*, 1998, using the distorted wave Born-Oppenheimer approximation model provide explicit means of applying knowledge of the orientation of krill to adjusting target strength. Unfortunately we are seldom in possession of adequate information to utilise such a model. Currently, the linear model of Greene *et al.* (1991) and a multifrequency method by Demer & Soule (1999) are routinely applied for rejecting multiple targets when making direct measurements of target strength in the field. Because krill often actively avoid nets, estimates of abundance obtained strictly by net sampling are subject to unknown or indeterminate biases; acoustic estimation of abundance generally produces data with fewer biases and higher spatial resolution.

Target strength and identity

Target strength for zooplankton and other targets can be based on either measured values (preferred), previously established values where direct measurement is not possible, or on models like those developed by Stanton, (1989a, 1989b); Greene *et al.*, (1991); Macaulay, (1994); McGehee *et al.*, (1998), or Stanton *et al.*, (1998). A researcher can establish preliminary target identity by observing echo signals on an oscilloscope. Sound scatter produced by layers of zooplankton is often clearly

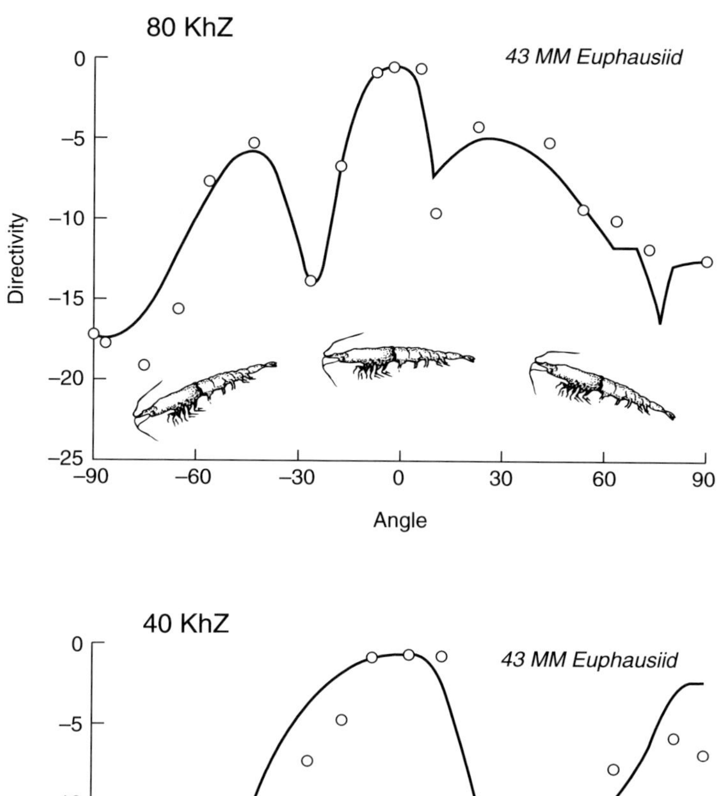

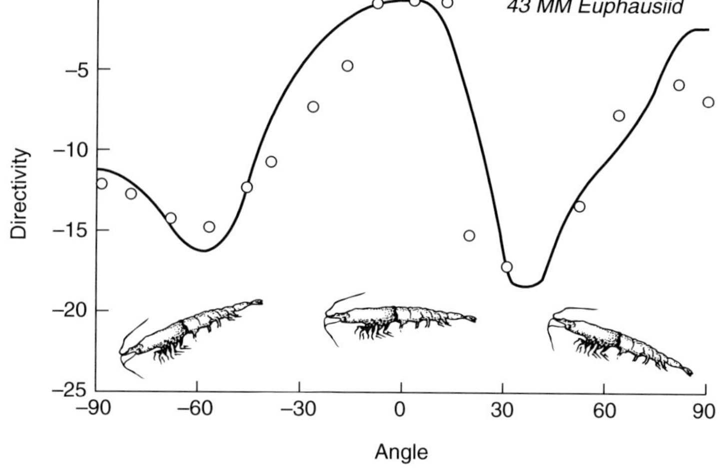

Fig. 2.2.4 Plot by frequency of normalised directivity (in dB) for a 43 mm euphausiid. Note the strong central lobe of target strength (directivity) exhibited by the target at lower frequency and the deep signal minima on either side of the peak. Individuals with a tilt angle of 30° upward orientation or with 60° of downward orientation would be nearly one tenth as good a target as one that was horizontal (i.e., −18–20 dB). Data from Kristensen & Dalen (1986) and solid curve from Macaulay (1994).

distinct from individual fish echoes. Zooplankton have a broad, low-level signal appearance as opposed to the spike-like appearance of fish targets (see Figs 2.2.3 and 2.2.4). Signal intensity is another descriptor, since fish schools, or layers, have a much higher intensity than do zooplankton layers. Copepod layers have intensities of −45 dB to −43 dB at a maximum (equivalent density of individuals from net samples was 40 000 individuals per m^3; Macaulay *et al.*, 1995), euphausiid layers are frequently 10 dB higher (i.e., −35 dB, or 3000–5000 individuals per m^3), while fish schools are often −10 dB to −20 dB (i.e., 10–100 individuals per m^3). Single echoes from fish are on the order of −30 dB to −35 dB. A factor of 6 dB in intensity is equivalent to a factor of two increase in signal strength; so, 10 dB is almost a factor of four, and 20 dB is almost a factor of ten. See Miyashita *et al.*, 1996 for an example.)

The currently accepted target strength for *Euphausia superba* at 120 kHz is

$$\text{TS(dB)} = -127.45 + 34.85 \times \log_{10} \quad \text{(length in mm)}$$

It is important that the investigator check such relationships carefully when applying them to other species of euphausiid. The relationship of target strength to frequency given in Greene *et al.* (1991) is useful as a check, bearing in mind that the inherent linearity with frequency may not always apply (cf. Traykovski *et al.*, 1998) at frequencies higher than 200 kHz.

Comparisons of echo intensity can be used to separate likely targets, so long as the types of targets are not excessively mixed. The use of multiple frequencies provides significant assistance in this task (Pieper & Holliday, 1984; Saetersdal *et al.*, 1984; Macaulay *et al.*, 1995). Miyashita *et al.* (1996) used comparisons of the scattering from 38 kHz and 120 kHz to distinguish *E. pacifica* from fish. The selection of time of day to do such comparative work (daylight hours, especially) permits separation of targets due to their diel behaviour. At night, there is considerable mixing and this necessitates using some bulk-average target strength to estimate abundance. An investigator's sampling experience in a particular area provides information on what ranges are reasonable for the resident target organisms. In general, the acoustic characteristics (receiver sensitivity, beam pattern, type of system) will determine what size of target, level of concentration, and range of observation is possible with each type of acoustic system. For krill, this usually means narrow beam transducers operating at 100–200 kHz and at ranges up to 250 m.

Krill behaviour

Krill often exhibit a strong tendency to aggregate in layers, schools, swarms, and patches (Miller & Hampton, 1989). Consequently, special consideration must be given to their behaviour in designing surveys to enumerate their populations. Because krill are highly active swimmers, they are frequently capable of avoiding many types of nets commonly used to collect biological samples. Increasing the speed of the sampler or using a downward direction for sampling have proven to increase the effectiveness of such samplers (Loeb *et al.*, 1993). Ecological interactions between populations of krill (or other zooplankton) and the environment or other organisms can be examined by spectral analysis of the spatial distributions observed hydroacoustically (Macaulay *et al.*, 1995). The spectral density of these distributions often reveals scales of patchiness that are significantly different from those for purely hydrographically distributed particles. Macaulay *et al.* (1995) suggests that there is a preferred patch size exhibited by many zooplankton and this appears to be true for krill as well (Miller & Hampton, 1989).

The analysis of acoustic data from survey transects produces estimates of biomass along the cruise track by distance. The distance covered in each integration interval will depend on the speed of the vessel. In addition, vertical profiles of distribution and abundance can be calculated for depth slices of selected thickness (usually 1–10 m). These vertical profiles can then be used for statistical comparison with hydrography and sampling by other investigations. A number of statistical methods can be used to establish relationships of acoustic data (as an estimator of zooplankton abundance) to other biological sampling and to the prevailing hydrography. These methods include cluster analysis and stepwise discriminant function analysis to develop indicators of biologically distinct areas (Macaulay *et al.*, 1995), and spectral analysis using a fast Fourier transform (FFT) to compare the spatial scales of patch structure between areas (Platt & Denman, 1975; Weber *et al.*, 1986; Macaulay, 1991; Loeb *et al.,* 1993). In addition to the collection of sound scattering measurements of biogenic origin, researchers (Wishner *et al.*, 1988; Macaulay, 1991; Loeb *et al.*, 1993; Macaulay *et al.*, 1995; Beardsley *et al.*, 1996) have shown that much ecological information can be derived from the spatial and temporal patterns observed in acoustically derived data.

Data management

To provide effective management information, survey data collection systems must provide a number of items; among these are: real-time data display including position; timely production of reports following field work; platform independence of data formats and files; and manageable data volume. Flexible data export capability allows convenient use of many commercial data analysis and display packages. Because of the rapid evolution of the tools and platforms for analysis it seems prudent not to be bound too tightly to any one system for data collection or analysis (Macaulay, 1992).

The most important component of data management is the selection of a means to store the information collected and to make it available to the user community. While many database systems may have sufficient capacity and tools to satisfy data storage and retrieval needs, there is an increasing tendency to use a geo-referenced database (or Geographical Information System, GIS) to provide these functions. There are many advantages to using regular databases and GIS systems together to maximise the ability to extract, calculate, and display the survey data in the most effective manner. There remain significant differences between display and information management tools available on different computer platforms, but generally, once they are produced the resulting data products can be utilised across platforms.

Conclusions

To make effective use of acoustics to measure the distribution and abundance of krill an investigator should use more than one frequency and often more than one type of system at the same time. Careful, regular calibration of the systems before and after use is also required. In general, frequencies in the 100–200 kHz range are the most useful for surveys, and these can be supplemented by other (higher frequency) systems where more detail is desired. Split-beam systems are preferred for target-strength measurement work and can often be used simultaneously with single-beam systems. There is a suggestion from Traykovski *et al.* (1998) that there are non-linearities in target-strength with frequency; care is needed when target-strength is determined at one frequency and applied to a different frequency using linear methods.

The most accurate results are obtained by target-strength measurements for each frequency as well as net or other sampling for target identity and size distribution. The researcher needs to examine the magnitude of near-surface concentrations either by special sampling to determine the magnitude of such concentrations, or by sampling at times when the krill are known to be less abundant. Krill behaviour will often control survey design, and krill distribution can be used to reveal environmental structure (e.g. frontal zones due to currents or hydrography). Analysis of distribution data from acoustic sampling can also be used to examine scales of patchiness related to the environment that have great influence on predator distribution. This requires the use of spectral analysis and other statistical methods to extract the complexities of the underlying distributions. Lastly, the investigator needs to consider data management before conducting acoustic surveys. Acoustic systems produce large quantities of detailed data on the abundance and distribution of krill and other organisms. These data need to be reduced and stored in ways that maximise their availability for the investigator to extract and analyse the results. To this end, it is recommended that Geographical Information Systems (preferably platform independent ones) be used to store the data. Almost all of the problems that arise from the use of acoustic methods to estimate krill abundance are well addressed by current software and hardware availability. After that, researchers must ensure that careful consideration is given to understanding the effects of the strengths and limitations that acoustic methods impose on the interpretation of results.

References

American National Standards Institute Inc. (1972) Procedures for calibration of underwater electroacoustic transducers. *N.Y., Amer. Nat. Stds. Inst., Inc.* N.Y. 10018.

Beardsley, R.C., Epstein, A.W., Chen, C., Wishner, K.F., Macaulay, M.C. & Kenney, R.D. (1996) Spatial variability in zooplankton abundance near feeding Right Whales in the Great South Channel. *Deep-Sea Res.* **45**, 7–8.

Blue, J. E. (1984) Physical calibration. *Rapp. P. -v. Reun. Cons. int. Explor. Mer.* **184**, 19–24.

Cochrane, N.A., Sameoto, D. Herman, A.W. & Neilson, J (1991) Multiple-frequency acoustic backscattering and zooplankton aggregations in the inner Scotian shelf basins. *Can. J. Fish. Aquat. Sci.* **48**, 340–55.

Demer, D. & Soule, M. (1999) Improvements to the multiple-frequency method for in-situ target strength measurements. CCAMLR WG-EMM 99/38.

Everson, I. (1982) Diurnal variations in mean volume backscattering strength of an Antarctic krill (*Euphausia superba*) patch. *J. Plank. Res.* **4**, 155–62.

Foote, K.G. (1982) Optimising copper spheres for precision calibration of hydroacoustic equipment. *J. Acous. Soc. Am.* **71**, 742–47.

Foote, K.G., Knudsen, H.P., Vestnes, G., MacLennan, D.N. & Simmonds, E.J. (1987) *Calibration of Acoustic Instruments for Fish Density Estimation: a practical guide*. ICES.

GLOBEC (1991) Workshop on acoustical technology and the integration of acoustical and optical sampling methods. *GLOBEC Report No. 4*, p. 58. Joint Oceanogr. Inst. Inc., Washington DC.

Greene, C.H., Stanton, T.K., Wiebe, P.H. & McClatchie, S. (1991) Acoustic estimates of Antarctic krill. *Nature* **349**, 110.

Greenlaw, C.F., Johnson, R.K. & Pommeranz, T. (1980) Volume scattering strength predictions for Antarctic krill (*Euphausia superba* Dana). *Meeresforsch* **28**, 48–55.

Hewitt, R.P. & Demer, D.A. (1993) Dispersion and abundance of Antarctic krill in the vicinity of Elephant Island in the 1992 austral summer. *Mar. Ecol. Prog. Ser.* **99**, 29–39.

Holliday, D.V. & Pieper, R.E. (1980) Volume scattering strengths and zooplankton distributions at acoustic frequencies between 0.5 and 3 mHz. *J. Acoust. Soc. Am.* **67**, 135–46.

Holliday, D.V. & Pieper, R.E. (1989) Determination of zooplankton size and distribution with multifrequency acoustic technology. *J. Cons. int. Explor. Mer.* **46**, 52–61.

Loeb, V.J., Amos, A., Macaulay, M.C. & Wormuth, J.H. (1993) Krill stock distribution and composition in the Elephant Island and King George Island areas, January–February, 1988. *Polar Biol.* **13**, 171–81.

Macaulay, M.C. (1991) AMLR PROGRAM: Spatial patterns in krill distribution and biomass near Elephant Island, Summer 1991. *Antarctic J.* **26**, 205–206.

Macaulay, M.C. (1992) A hybrid database management system for the collection, display, and statistical analysis of large and small scale hydrographic survey data. *IEEE Proceedings, OCEANS 92*. October 1992. pp. 91–96.

Macaulay, M.C. (1994) A generalized target strength model for euphausiids, with applications to other zooplankton. *J. Acous. Soc. Amer.* **95**, 2452–66.

Macaulay, M.C., Wishner, K.F. & Daly, K.L. (1995) Acoustic scattering from zooplankton and micronekton in relation to a whale feeding site near Georges bank, Cape Cod. *Continental Shelf Res.* **15**, 509–537.

MacLennan, D.N. (1981) The theory of solid spheres as sonar calibration targets. *Scottish Fisheries Research Report*, **22**, 17.

MacLennan, D.N. & Simmonds, E.J. (1992) Fisheries acoustics. In *Fish and Fisheries*, Series 5, Chapman and Hall, London.

McGehee, D.E., O'Driscoll, R.L., Martin, V. & Traykovski, L. (1998) Effects of orientation on acoustic scattering from Antarctic krill at 120 kHz. *Deep-sea Res.* II, **45**, 1273–94.

Miller, D.G.M. & Hampton, I. (1989) Biology and ecology of the Antarctic krill. *BIOMASS Scientific Series.*

Miyashita, K., Aoki, I., Seno, K., Taki, K. & Ogishima, T. (1996) Acoustic identification of isada krill, *Euphausia pacifica* Hansen, off the Sanriku coast, north-eastern Japan. *Fish. Oceanogr.* **6**, 266–71.

Pieper, R.E. & Holliday, D.V. (1984) Acoustic measurements of zooplankton distributions in the sea. *J. Cons. int. Explor. Mer.* **41**, 226–38.

Platt, T. & Denman, K.L. (1975) Spectral analysis in ecology. *Annual Rev. Ecol. Syst.* **61**, 189–210.

Saetersdal, G., Stromme, T. Bakken, B. & Piekutowski, L. (1984) Some observations on frequency-dependent backscattering strength. *FAO Fish. Rep.* **300** (41), 150–56.

Sameoto, D.D. (1980) Quantitative measurements of euphausiids using a 120 kHz sounder and their in situ orientation. *Can. J. Fish. Aquat. Sci.* **37**, 693–702.

Stanton, T.K. (1989a) Sound scattering by cylinders of finite length. III. Deformed cylinders. *J. Acoust. Soc. Am.* **86**, 691–705.

Stanton, T.K. (1989b) Simple approximate formulas for backscattering of sound by spherical and elongated objects. *J. Acoust. Soc. Am.* **86**, 1499–1510.

Stanton, T.K., Chu, D. & Wiebe, P.H. (1998) Sound scattering by several zooplankton groups. II: Scattering models. *J. Acoust. Soc. Am.* **103**, 236–53.

Traykovski, L., Martin, V., O'Driscoll, R.L. & McGehee, D.E. (1998) Effect of orientation on broadband acoustic scattering of Antarctic krill *Euphausia superba*: Implications for inverting zooplankton spectral acoustic signatures for angle of orientation. *J. Acoust. Soc. Am.* **104**, 2121–35.

Traynor, J.J. & Ehrenberg, J.E. (1979) Evaluation of the dual-beam acoustic fish target strength method. *J. Fish. Res. Board Can.* **36**, 1065–71.

Weber, L.H., El-Sayed, S.Z. & Hampton, I. (1986) The variance spectra of phytoplankton, krill, and water temperature in the Antarctic Ocean south of Africa. *Deep-Sea Research* **33**, 1327–43.

Wishner, K., Durbin, E., Durbin, A., Macaulay, M., Winn, H. & Kenney, R. (1988) Copepod patches and right whales in the Great South Channel off New England. *Bull. Mar. Sci.* **43**, 825–44.

2.3 Biological observations

Inigo Everson

Behind any ecological study lies a great deal of basic observation on the species of interest. For those species that are widespread in their distribution or that have been studied by researchers from differing organisations it is quite common for different criteria to be applied to the observations. Common cases where this has happened in the past have been for length, maturity stage and feeding status. All of these are important in population ecology and their value is enhanced when they can be easily compared with results from other studies.

Length measurement

The morphology of euphausiids, with their long antennae and antennulae extending forward, means that several definitions of length are possible. Even restricting the definition to the anterior end of the main body poses difficulties because krill are fragile and break easily, particularly in fixatives. Furthermore, they do not always lie 'straight' and although possessing a rigid exoskeleton, they can be compressed or extended. Because it is so fundamental to many population studies length measurement is worth some consideration here.

Mauchline (1980) when considering the Antarctic krill gave a reference measurement of total body length which was defined as:

> The lateral distance between the base of the eye-stalk and the posterior end of the uropods, excluding their terminal setae.

In this context, the base of the eye-stalk is taken to mean the 'anterior leading edge of the carapace'. This is a measurement that is unfortunately quite tedious and requires a low-power binocular microscope, consequently it is not surprising that it has not been used for observations on fresh krill at sea.

A 'straw poll' of colleagues known to the author and who have been involved in measuring Antarctic krill at sea provided several approaches to what on the face of it appears to be a relatively simple and standardised procedure. The standard reference publication used by many workers is Mauchline (1980) who considered the most practical total length measurement for *E. superba*, specified as Standard Measure 1, to be:

> The lateral or dorsal distance between the anterior tip of the rostrum and the posterior end of the uropods, excluding their terminal setae.

The strict definition specified above is used by Australian workers (Nicol, pers. comm.) although others take the posterior end as the telson tip rather than the uropods. This is probably of minor importance in *E. superba* because when the uropods are laid lengthwise they extend to approximately the end of the telson. This may not necessarily be true for all species.

The Standard 1 measurement, taking either the uropods or telson as the posterior end, has been used by researchers from Australia, Japan and the United States (personal communications from S. Nicol, T. Ichii and R. Ross). For the anterior end such a definition is satisfactory for those species where the rostrum is small but if it is of any significant size and likely to be prone to damage it may not be ideal. An alternative that was used on the 'Discovery Investigations' is:

> from the front of the eye to the tip of the telson excluding the setae.

This has been used by researchers from Germany (V. Siegel, pers. comm.), South Africa (D. Miller, pers comm) and the United Kingdom.

The question of length measurement was reviewed by CCAMLR for application in its Observer Programme (CCAMLR, 1997). The requirement in this case was for a method of measurement that could be undertaken at sea on commercial fishing vessels. The definition given is from the front of the eye to the tip of the telson, the 'Discovery' standard. At the anterior end, the eye is more clearly visible than the rostrum so that the two ends of the animal are thus clearly discernible. In making the measurement it is probably best to hold the animal by the antennae and lay it on a ruler or flat surface. Care needs to be taken due to the movement of the eyes on their stalks as the animal is laid down. The different measurements are indicated in Fig. 2.3.1.

Krill measurements have also been reported from samples obtained from predators. In such cases the individuals are often damaged and the options for measurement are restricted to the carapace, the largest hard part of the body. Mauchline (1980) gives three measures which can be applied but favours the following definition of carapace length:

> the dorsal distance between the anterior tip of the rostrum and mid-dorsal posterior edge of the carapace.

This measurement, indicated in Fig. 2.3.2. has been used by Australian and UK workers (Nicol, pers. comm., Hill, 1990).

Even though the differences between results using the different methods are small, of the order of one or two millimetres in the case of Antarctic krill, they do introduce biases which could compromise the results based on comparisons between locations or between years at the same locality. Precise standardisation would therefore be an advantage for future studies.

A further point of similar concern is that of observer bias. Where groups of workers are engaged in a series of measurements it is important to include an identifier for the observer within the database and then to test for biases in the measurements before drawing conclusions that rely on comparisons of size frequency distributions.

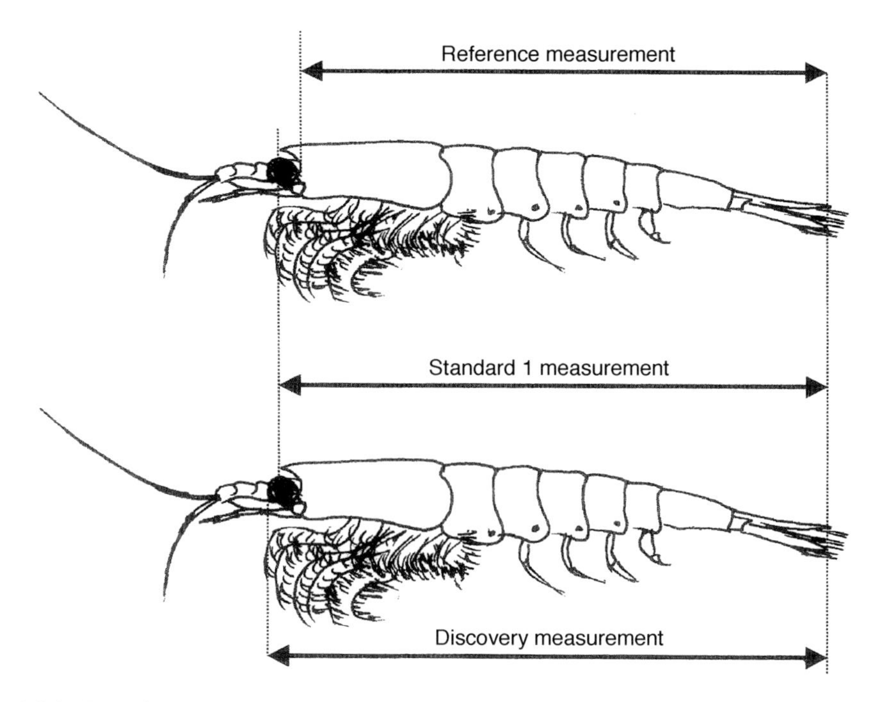

Fig. 2.3.1 Length measurements proposed for euphausiids. For explanation see text.

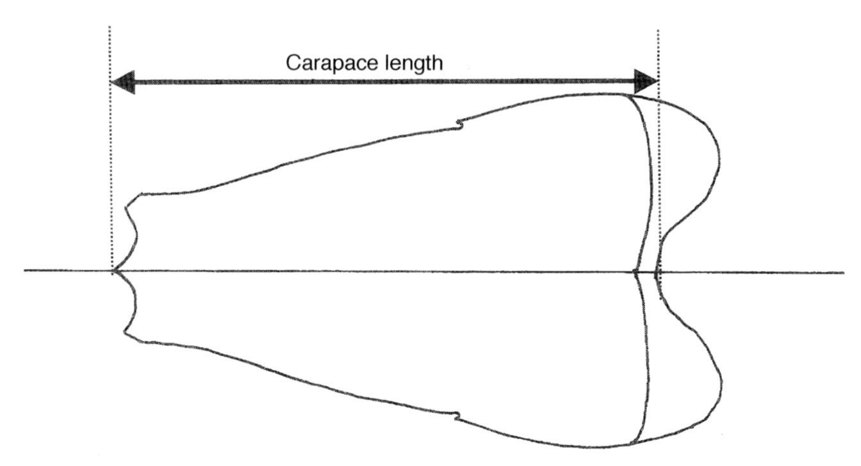

Fig. 2.3.2 Dorsal view of carapace to indicate carapace length measurement. (With permission from and after Hill, 1990.)

Maturity stages

The secondary sexual organs of mature krill are reasonably easy to distinguish. Mature male krill have petasmae present as modified endopods of the first pair of pleopods (Fig. 2.3.3). They are swollen structures having a complex shape and are used to transfer the spermatophores to the female. Paired vasa deferentia open

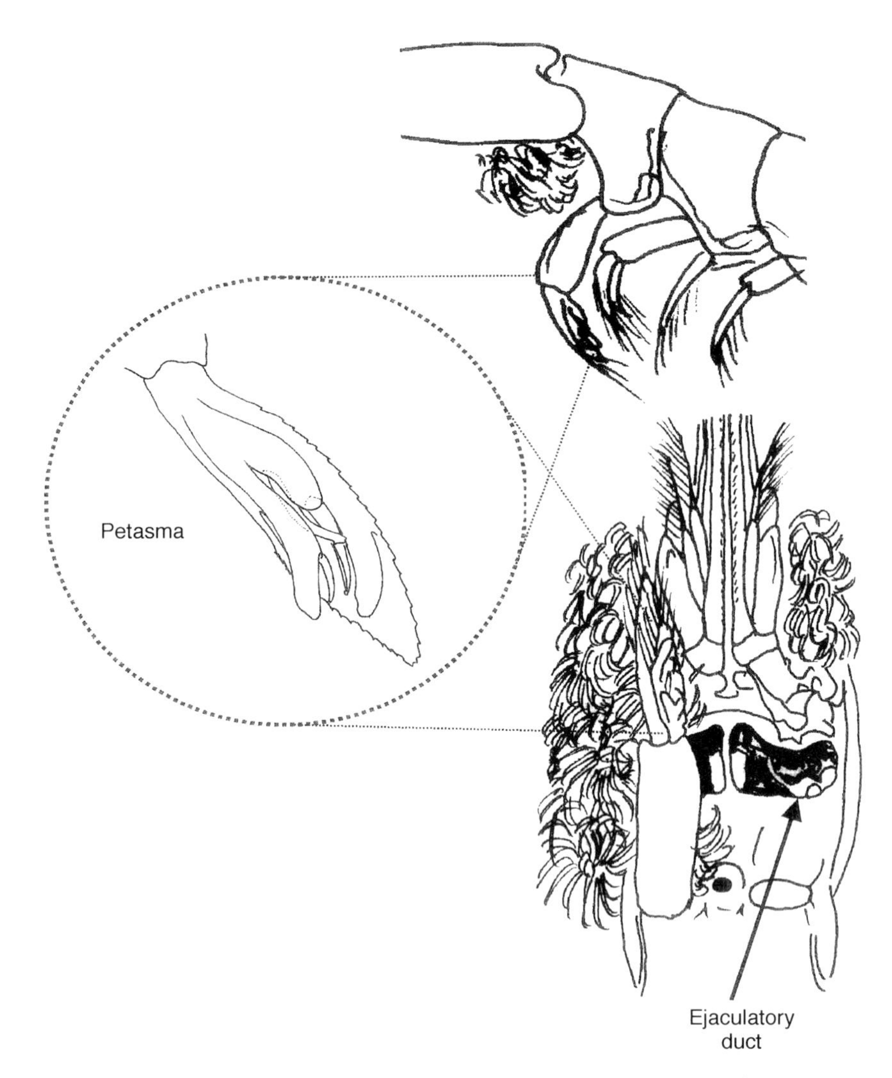

Fig. 2.3.3 Mature male krill. Lateral view, upper diagram, to show location of petasma on anteriormost pleopod and ventral view, lower diagram, to show location of ejaculatory duct at posterior end of cephalothorax. With permission from SCAR. (After Makarov & Denys, 1981.)

externally on the ventral side of the thorax, their distal regions being swollen to form ejaculatory ducts which are seen as paired red patches in the ventral integument.

In female krill the first pair of pleopods are not modified as sexual appendages. On the ventral posterior region of the female cephalothorax there is a median tri-lobed structure, the thelycum. This is shown in Fig. 2.3.4. and is red in colour and often has small white vesicles, spermatophores, attached to it. In unmated females the red colour of the thelycum is often particularly bright.

The ability to determine whether krill are juvenile (in which case they would have no external sexual characteristics visible) male or female is sufficient for many

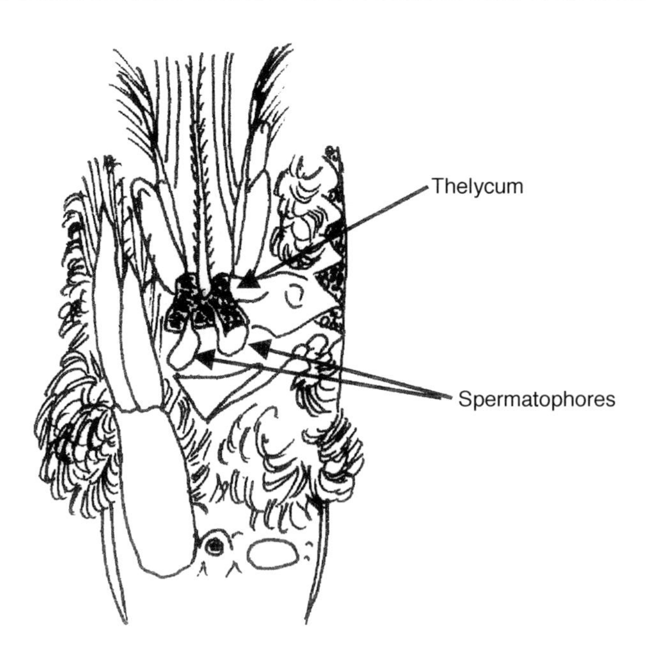

Fig. 2.3.4 Ventral view of a mature female krill to show location of thelycum at posterior end of cephalothorax. With permission from SCAR. (After Makarov & Denys, 1981.)

studies. When one is dealing with small species this is probably the limit which can be achieved on fresh material at sea. For those species which grow to five or six centimetres in length it is possible to categorise the krill into a series of maturity stages with either the naked eye or else with the aid of a simple low-power magnifier. A scheme that has been used for many years for Antarctic krill was devised by Makarov & Denys (1981) and this is set out below. This has been further developed by the CCAMLR Scientific Observer Programme for observations on Antarctic krill (CCAMLR, 1997). In common with the length measurements mentioned above the system relies on freshly caught specimens and the manual provides the following general guidelines. The basic Makarov & Denys (1981) descriptions are given below.

Juveniles

Stage I

- Secondary sexual characteristics, petasma and thelycum are not visible.

Males

Stage II Sub-Adults

- Developing petasmae visible but not fully developed. Ejaculatory ducts are small and not coloured.

Stage III Adults

- Petasmae fully developed; ejaculatory ducts red in colour and clearly visible through the gills.

- *IIIA*. Fully formed spermatophores are not present within the ejaculatory ducts.
- *IIIB*. Fully formed spermatophores are present within the ejaculatory ducts. Such spermatophores are easily ejected by exerting pressure on the ducts.

Females

Stage II Sub-Adults

- Developing thelycum is present but its colour is pale or colourless.

Stage III Adults

- The thelycum is fully developed, red in colour, and is clearly visible through the gills.
  - *Stage IIIA*. The thelycum bears no spermatophores and as a result, is very brightly coloured. The body is not swollen.
  - *Stage IIIB*. The thelycum is dirty red in colour owing to the presence of empty spermatophores or, more rarely, has two opaque white points – filled spermatophores – present. There is a free space between the ovary and the walls of the body which is not swollen.
  - *Stage IIIC*. Ovary fills whole thoracic space as well as the dorsal regions of the first and second abdominal segments but the body is not noticeably swollen.
  - *Stage IIID*. Thorax and first two abdominal segments are swollen owing to the presence of the enlarged ovary which is clearly visible through the transparent integument.
  - *Stage IIIE*. Thorax and first two abdominal segments are swollen but no enlarged ovary is visible through the integument. Instead, the internal body cavities are empty, the eggs having been laid. A small ovary is situated in the middle of this space.

Systems of classifications such as this one where a continuous process is described as a series of steps, always raises difficulties when 'intermediate' forms are found. However, since the key is being applied generally to determine the proportion of krill of a given maturity stage that is present in the population then this problem is of minor consequence. It is important, however, to operate a consistent policy with regard to the 'intermediates'. More detailed information on the gonad development processes will be found in Ross & Quetin (Chapter 6).

Krill coloration

The feeding status of Antarctic krill, and other species that are filter feeding on phytoplankton, can be investigated by examining the colour of the carapace and hepatopancreas. As well as having an application in understanding ecosystem interactions this analysis can be important in determining the quality of krill for commercial processing, as discussed by Ichii, (Chapter 9) and Nicol *et al.*, (Chapter 10).

The CCAMLR Scheme requires that the krill are compared against the coloured standard shown in Plate 2 (facing p. 182). It is stressed that the krill should be absolutely fresh and undamaged and that the analysis should take place in a cold well-lit area. These criteria are important as autodigestion can be rapid and is hastened in a warm environment. The krill themselves should be arranged into colour groups and viewed against a white background. Four categories are described for each of the carapace colour and the hepatopancreas colour and transparency.

References

CCAMLR (1997) *Scientific Observers Manual.* Commission for the Conservation of Antarctic Marine Living Resources, Hobart, Australia.

Hill, H. (1990) A new method for the measurement of Antarctic krill *Euphausia superba* Dana from predator food samples. *Polar Biol.* **10**, 317–20.

Makarov, R.R. & Denys, C. (1981) Stages of sexual maturity of *Euphausia superba* Dana. *BIOMASS Handbook* No 11. 11 pp.

Mauchline, J.R. (1980) Measurement of body length of *Euphausia superba* Dana. *BIOMASS Handbook* No 4. 9 pp.

Chapter 3
Distribution and Standing Stock

3.1 Japanese waters
Yoshi Endo

Euphausia pacifica is widely distributed in the North Pacific Ocean. In the Japanese waters of the Pacific Ocean, it occurs as far south as Suruga Bay, 34°50′N (Sawamoto, 1992) and extends northwards as far as the southwestern area of the Okhotsk Sea (Ponomareva, 1963; Ohtsuki, 1975). It is present over almost the entire Sea of Japan extending northwards as far as the southern part of the Gulf of Tartary (Komaki & Matsue, 1958; Ponomareva, 1963).

Oceanographic conditions and the distribution of Euphausia pacifica

A part of the Oyashio Current flows southward along the east coast of Honshu Island as the first and second Oyashio branches. Another part returns northeastward as the Oyashio Extension, forming the Oyashio Front with the high salinity waters originating mostly from the subtropical gyre (Fig. 3.1.1) (Yasuda *et al.*, 1996). The first and second Oyashio branches, modified through contact with Tsugaru Warm Current water and warm-core rings cut off from the Kuroshio Extension, meet with the Kuroshio Extension and form the Kuroshio Front. The region between the Oyashio and Kuroshio waters is referred to as the mixed water region, perturbed area or transition zone (Yasuda *et al.*, 1996).

The relationship between oceanographic conditions and fishable *E. pacifica* aggregations in Sanriku waters is well documented by Odate (1991) and Kodama & Izumi (1994). Using information on fishing conditions, Kodama & Izumi (1994) reported that the fishing grounds are formed when the water mass whose temperature is in the range 7° to 8°C is pushed to the shore by the approach of the first Oyashio branch, which has a temperature of <5°C from the surface to the bottom. The authors speculated that the optimal temperature for *E. pacifica* in the fishing season is 7° to 8°C and that this species avoids water colder than 5°C or warmer than 10°C. Odate (1991) also concluded that the main distributional area of the species in the Sanriku and Joban regions is confined to waters where the temperature is between 5° and 10°C. The lower temperature limit for the species, based on the fishing condition data in the area, is around 4°C.

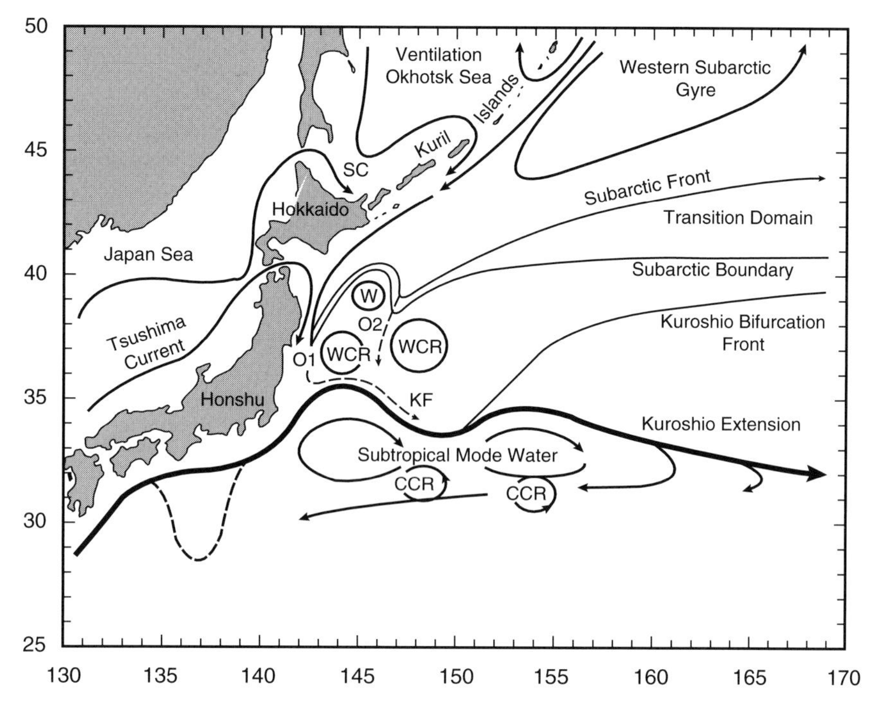

Fig. 3.1.1 Schematic representation of the current and frontal systems in Japanese waters (Yasuda *et al.*, 1996). TW: Tsugaru Warm Current; O1: Oyashio first branch (Coastal Oyashio Intrusion); O2: Oyashio second branch (Offshore Oyashio Intrusion); WCR: warm-core ring, CCR: cold-core ring; KF: Kuroshio Front.

More recently, the distribution of the species has been investigated throughout the year in the Sanriku and Joban areas (Taki *et al.*, 1996; Taki & Ogishima, 1997). Based on samples collected with a beam trawl and an Ocean Research Institute net (Omori, 1965), with a mouth diameter of 1.6 m and a mesh size of 330 μm, both of which were fished along a transect line off Onagawa by day and night every two months from April 1993 to February 1994, Taki *et al.* (1996) reported seasonal change in horizontal distribution of *E. pacifica*. In April, the krill were distributed pelagically mainly in coastal areas from 100 and 150 m depth. In other seasons the main distributional area shifted offshore within the depth range from 200 to 300 m. During the period from August to December, adults were mostly benthopelagic during daytime, but immature animals had a more pelagic distribution. Furcilia larvae, which occurred throughout the year, but were most abundant in April, were entirely pelagic. The bottom temperature was 2° to 8°C when benthopelagic krill appeared. Benthopelagic concentrations rarely occurred when the bottom temperature exceeded 10°C.

Taki & Ogishima (1997) reported the seasonal distribution of all developmental stages as well as growth of *E. pacifica* using data from monthly samples, over the period from February to November 1994. These were obtained using Norpac nets

(Motoda, 1957), with a mouth diameter of 45 cm and a mesh size of 330 μm, hauled vertically from 150 m to the surface from wider areas of Sanriku and Joban waters. Although this was the most extensive krill survey conducted in the area, the net was not large enough to collect adults effectively and the sampling depth did not cover the whole daytime distributional range of the species. Taki & Ogishima (1997) found that all the developmental stages appeared throughout the survey period and were most abundant in the transition zone where the water temperature was in the range from 5° to 10°C at 100 m depth.

In February, eggs and early larvae did not occur in the area under the influence of the Oyashio Current where the water temperature is <5°C at a depth of 100 m. However, they were present in water of 5° to 10°C at that depth. They occurred in colder waters later in the season at which time they were concentrated in water of about 5°C at 100 m depth. Most spawning activity appears to have occurred in April–June, as has been suggested by previous workers (see Nicol & Endo, 1997).

In 1996, the North Pacific Krill Resources Research Group undertook a co-operative winter–spring survey of the horizontal distribution of euphausiids in Sanriku waters. The improved larva net with a mouth diameter of 1.3 m and a mesh size of 0.45 mm (Watanabe, 1988) was hauled obliquely from 150 m to the surface at night. During January and February, the abundance of adult *E. pacifica* was higher in Oyashio than in the waters related to Kuroshio (including the Tsugaru Warm Current water and warm water related to the Kuroshio Extension). It increased by an order of magnitude in the Kuroshio area in late March to late April. This suggests that there was a shift in horizontal distribution of the species from a cold to a warmer regime just before the onset of the fishing season (Taki *et al.* in preparation).

E. pacifica is known to form dense swarms during the fishing season (Komaki, 1967; Terazaki, 1980; Endo, 1984; Hanamura *et al.*, 1984). Juveniles of the species also swarm in the Kuroshio Front at night (Nishikawa *et al.*, 1995). Nishikawa *et al.* (1995) also investigated the distribution of euphausiids in the Kuroshio Front and warm water tongue in the Kashima-nada Sea, the southernmost distributional area of *E. pacifica* (Fig. 3.1.2). This warm tongue of Kuroshio water separated from the main current and stretched to the north along the continental shelf in the Kashima-nada Sea. In the Kuroshio area, *E. similis* and *Nematoscelis microps* were dominant. *E. pacifica* was the dominant species in the front area and the warm water tongue, comprising more than 95% of the total number of euphausiids in the frontal area. They were present in numerical densities up to 238 individuals m^{-3} in the frontal area. Surface aggregations of juvenile *E. pacifica*, of around 8 mm body length, have also been observed with the naked eye at night in the frontal area. Nishikawa *et al.* (1995) concluded that the convergent flow pattern at the front may be an important factor leading to the formation of surface aggregations of juvenile *E. pacifica.*

Taki *et al.* (1996) reported that both juvenile and adult *E. pacifica* were abundant in the water column when the average temperature was 7° to 14°C (Fig. 3.1.3). If the surface water temperature does not exceed 20°C, they migrate to the surface layer at

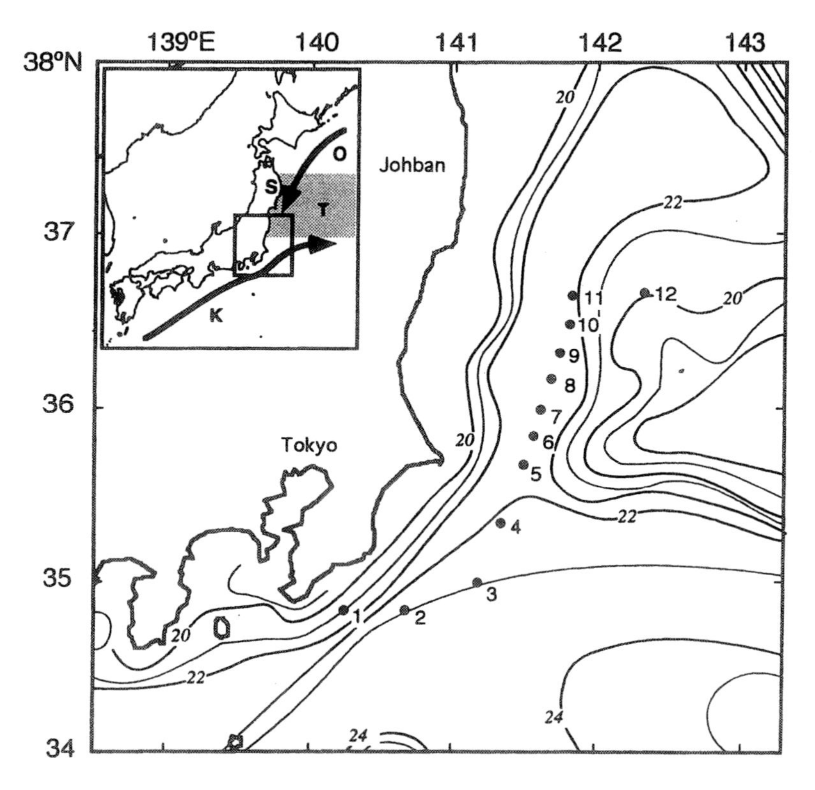

Fig. 3.1.2 Locations of sampling stations and surface water temperature gradients in the Kashima-nada Sea (Nishikawa *et al.*, 1995). A warm water tongue is delineated by 22°C contour line. O: Oyashio Current; S: Sanriku coast; T: Transition zone; K: Kuroshio Current.

night (Iguchi *et al.*, 1993). The occurrence of juvenile *E. pacifica* in the Kuroshio front and the warm water tongue is not surprising in view of the water temperature.

Warm core rings and Euphausia pacifica *distribution*

Warm-core rings shed from the Kuroshio Extension are reported to have profound effects on the distribution of copepods (Hattori, 1991, 1993) and chaetognaths (Terazaki, 1992) and the same has proved to be true for the euphausiid population. Taki (1998a) investigated the euphausiid distribution in and around warm-core ring 93A which is believed to have been formed in 1993. He used MOCNESS-1 (Wiebe *et al.*, 1985) to collect euphausiids from 5 depth layers down to 1000 m at 6 stations along 39°30′N. Five warm-water euphausiid species, *E. gibboides*, *E. similis*, *E. hemigibba*, *Nematobrachion flexipes* and *Nematoscelis gracilis*, were present in warm-core ring 93A; the temperature of the mixed layer formed by the winter convection was 12° to 13°C in July 1995. At that time *E. pacifica* were scarce, representing only 0 to 8% numerically of the total adult euphausiids in the centre of

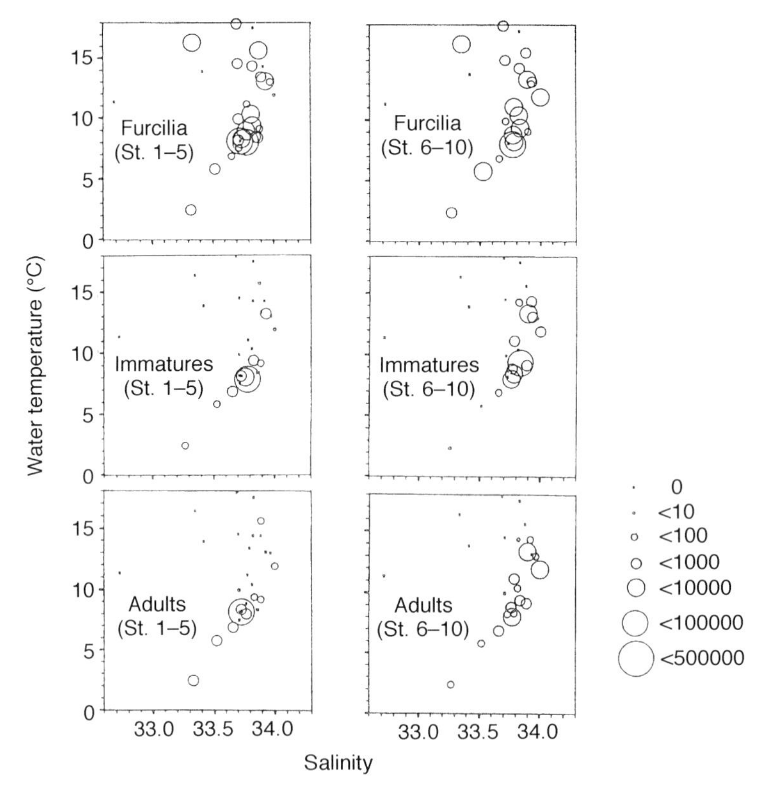

Fig. 3.1.3 Occurrence of furcilia, immature and adult of *Euphausia pacifica* during the day (left panels) and night (right panels) at stations off Onagawa plotted on T–S coordinates (Taki *et al.*, 1996).

the ring, but they were abundant outside and beneath the ring accounting for up to 90% of juvenile and adult euphausiids. In the next year, *E. pacifica* were still scarce in 93A, but abundant outside of the ring, although in January 1996 the mixed-layer temperature had decreased to 10° to 11°C (Taki *et al.* in preparation). In an old warm-core ring with an observed temperature of 4° to 5°C, Endo (1993) reported that *E. pacifica* dominated the euphausiid population accounting for more than 95% numerically of the catches; these included very small numbers of *E. gibboides*, the sole warm-water species. Further studies are needed to determine the processes associated with regulating the incursion of *E. pacifica* into warm-core rings.

Benthopelagic aggregation

At the end of the fishing season, surface aggregations of *E. pacifica*, which earlier in the season are present in the coastal waters of Ibaraki Prefecture, disappear as the surface water temperature increases. At that time, aggregations were distributed in the cooler mid-water zone (Suzuki, 1986). Since this discovery, mid-water swarms as well as benthopelagic and surface swarms have been targeted by the fishery (Nakamura, 1992). According to Nakamura (1992), surface aggregations were formed in years when cold water (5° to 7°C) prevailed from the surface to the bottom, whereas benthopelagic aggregations were formed when the water column was stratified with warm water >10°C at the surface overlying a thick cold (6° to 9°C) deeper layer. When the water column was stratified and the cold water layer was thin, neither surface aggregations nor benthopelagic aggregations appeared.

Monthly investigations of the benthopelagic *E. pacifica* have been made since September 1989 at five stations off Nakaminato (36°20′N) (Nakamura, 1991; Ebisawa, 1994). A beam trawl with a mouth area of 1 × 2 m and a mesh size of 3 mm at the cod end was used to sample *E. pacifica*. The benthopelagic population appeared throughout the year in a bottom layer at 280 m depth, with higher catches in winter than in summer. A significant catch was made when the thickness of the cold water layer, with the temperature less than 9°C, was more than 55 m in vertical extent.

Observations from submersibles have been made on benthopelagic *E. pacifica* over the period from 1994 to 1996 in the sea area near Kinkasan Island, around 38°30′N, the central fishing grounds of the species. Additional observations have been made off Kamaishi, around 39°20′N, which is at the northern part of these fishing grounds (Fig. 3.1.4). Surveys were undertaken using video cameras in daylight during the summer.

For most of the time, the submersibles cruised 0.5 to 1.0 m above the bottom, but moved away from the bottom in order to determine the numerical density and dimensions of benthopelagic aggregations. The results are still being analysed (Endo and Kodama in preparation; videotapes have been prepared from observations made off Kamaishi and have been used for krill analysis, courtesy of Dr. T. Fujita of National Science Museum, see Stancyk *et al.*, 1998). From these observations, spread over a period of three years, it is evident that benthopelagic *E. pacifica* are present over the whole surveyed depth range from 160 to 400 m. The east to west continuity of some benthopelagic aggregations extended as far as 7 km. These observations have been limited, however, to two localities, in the sea areas near Kinkasan and off Kamaishi. Additional observations, made with a beam trawl by other workers off Ibaraki Prefecture support these findings (Nakamura, 1991; Ebisawa, 1994).

Observations on the stomach contents of demersal fish over the whole region, discussed in Chapter 7, indicate that it is very likely that the distribution of benthopelagic *E. pacifica* extends about 600 km from Aomori to Ibaraki Prefectures along the Pacific coast of northern Japan (Fig. 3.1.4).

The water temperature near the bottom varied from year to year and from area to area, with minima of 2° to 4°C in 1994 off Onagawa and 3° to 4°C off Kamaishi in 1995 and 1996, and a maximum of 7° to 8°C in 1996 off Onagawa. Aggregations of *E. pacifica* are rarely encountered during the fishing season, when the water temperature is less than 5°C (Odate, 1991; Kodama & Izumi, 1994). Results from a series of monthly surveys mentioned above over several years off Ibaraki Prefecture

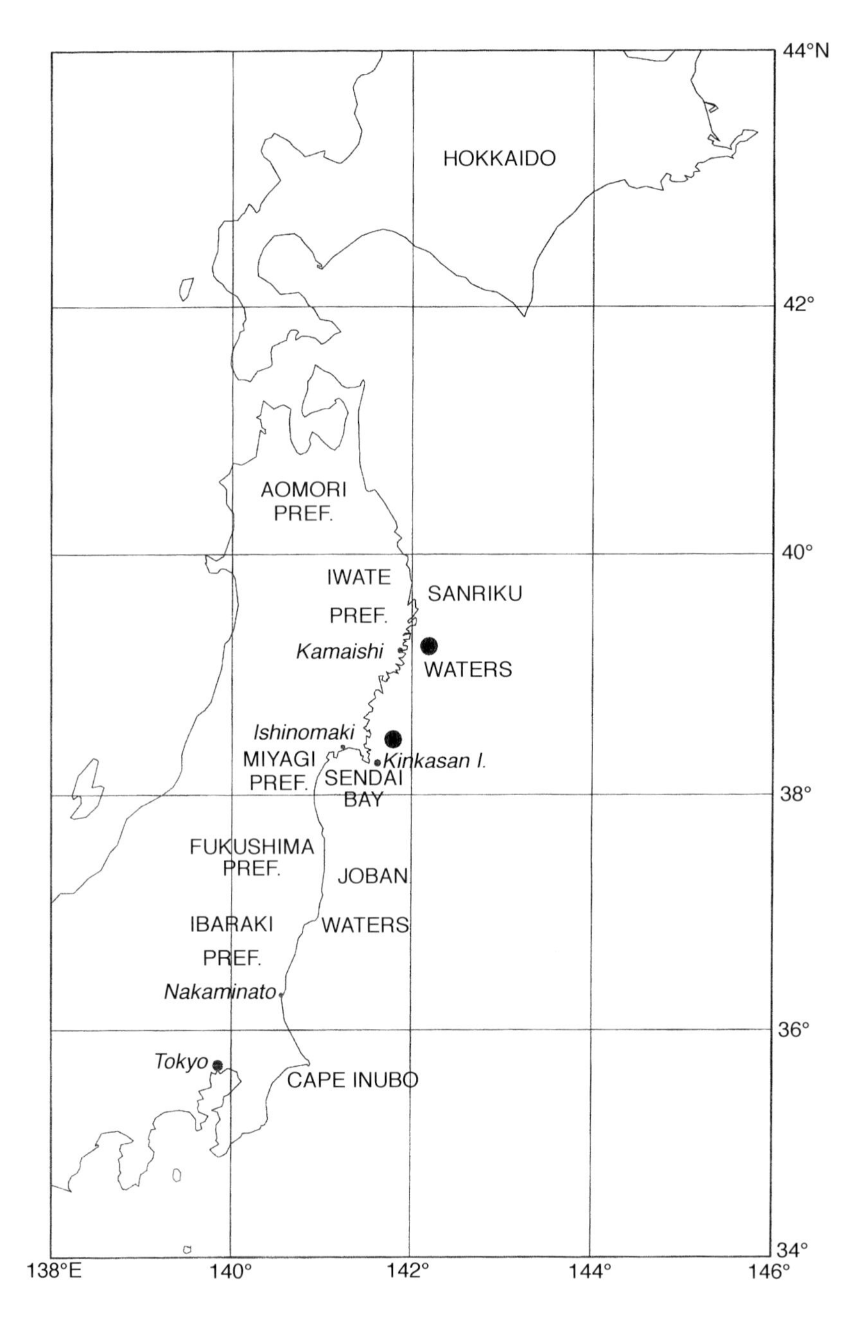

Fig. 3.1.4 Map of northern Japan indicating regions mentioned in text. Submersible observations were made on benthopelagic population of *Euphausia pacifica* at two locations shown by black circles.

(Nakamura, 1991; Ebisawa, 1994) indicate that benthopelagic concentrations do not occur unless the water temperature is less than 9°C. The near-bottom water temperature in this region rarely drops below 5°C, the lower limit, as noted above, for such concentrations to occur.

The maximal krill density from these submersible observations was estimated to be up to 30 000 individuals m^{-3}. As krill did not show parallel orientation, the dense part of the benthopelagic aggregation corresponds to Mauchline's (1980) definition of a swarm as opposed to a school in which animals have uniform orientation. The densest part was restricted to the several metres closest to the bottom.

E. pacifica performs diel vertical migration with daytime depth of 200 to 400 m and night-time depth of less than 150 m in Sanriku waters (Endo, 1981). Benthopelagic *E. pacifica* may occur as a result of encountering the sea bed during vertical migration as suggested for *Meganyctiphanes norvegica* by Greene *et al.* (1988). If this hypothesis is correct, benthopelagic *E. pacifica* should appear as deep as 400 m; this corresponds to the upper continental slope. Further offshore, *E. pacifica* may be pelagic with a lower population density. Demersal fishes are reported to be most abundant along the upper slope in Sanriku waters (Fujita *et al.*, 1993), the dominant species being: walleye pollack, *Theragra chalcogramma*, and Pacific cod, *Gadus macrocephalus*, both of which feed on zooplankton in general, and *E. pacifica* in particular. This indicates that benthopelagic krill are probably supporting a very high abundance of demersal fish in the upper slope region.

Taki *et al.* (1996) reported the seasonal occurrence of benthopelagic *E. pacifica* sampled by a beam-trawl net, which was able to collect plankton from 50 cm to 1 m above the seabed. The biomass was highest in June (2 g m^{-2} at 300 m depth) and lowest in February and April (0.02 g m^{-2} at 300 m depth). In April, benthopelagic krill occurred at the depth of 150 m, while they occurred in offshore areas at depths of 200 to 300 m in other seasons. This seasonal shift of near bottom biomass, estimated from beam-trawl surveys coincided with the distribution of pelagic biomass in the water column, estimated from net tows (Taki *et al.*, 1996). Stomach contents analysis of demersal fishes shows a general trend indicating that the importance of *E. pacifica* is highest in the period from May to July and lowest in the winter season (Fujita, 1994). The seasonal shift in benthopelagic krill distribution may be related to seasonal changes in the temperature in Sanriku waters, with a minimum in spring around 2°C and a maximum in autumn of around 22°C at the surface.

Benthopelagic populations are also known for other species of krill such as Antarctic krill, *Euphausia superba* (Gutt & Siegel, 1994; Takahashi & Iwami, 1997), North Atlantic krill, *Meganyctiphanes norvegica* (Simard *et al.*, 1986; Greene *et al.*, 1988) and *Thysanoessa raschi* (Simard *et al.*, 1986). Takahashi & Iwami (1997) reported that benthopelagic *E. superba* is distributed down to depths of around 800 m in the vicinity of the South Shetland Islands and constitutes one of the main food resources for demersal fishes in summer.

Quantitative estimates of the biomass of benthopelagic *E. pacifica* are needed over the whole distributional range in the Japanese waters to elucidate better the importance of the species in the pelagic as well as benthopelagic food webs in Sanriku and Joban waters.

Abundance estimation using acoustic and spawning survey methods

A feasibility study of species identification and abundance estimation of *E. pacifica* was undertaken using a dual-frequency (38 and 120 kHz) echo-sounder in April 1996 (Miyashita *et al.*, 1997). *E. pacifica* form dense aggregations and are the dominant zooplankters in the fishing season. These characteristics are ideal for the estimation of standing stock acoustically. Miyashita *et al.* (1997) adopted the straight-cylinder model taking account of swimming angle to estimate the acoustic target strength. This was then used to estimate the size and abundance of krill. Assuming a swimming angle of 15°, the observed value when actively swimming (Miyashita *et al.*, 1996), the TS difference between the two frequencies ($\Delta TS = TS_{120} - TS_{38}$) of *E. pacifica* of body length 16.5 mm was estimated to be between 10 and 15 dB. Those echoes which fell within this ΔTS range were assumed to be *E. pacifica* (Plate 3, facing p. 182) and thus distinguishable from those echoes due to walleye pollack, the ΔTS of which, using the swim bladder resonance model, would be less than 2 dB. The estimated size and numerical density of *E. pacifica* at one point on the echogram, for example, were 17.0 mm and 8.3 individuals m^{-3}, respectively.

In April 1997, using the same criteria, Miyashita *et al.* (1998) estimated the numerical density of the species along Sanriku and Joban coasts in the depth range from 150 m to 300 m in April 1997. In this case, the theoretical scattering model (high-pass sphere model) of Johnson (1977) was used to estimate TS. The highest density, 9519.5 individuals m^{-2}, was found in the sea area off Fukushima Prefecture where commercial fishing had been in progress during the survey period. In the southernmost survey area, where the water temperature had increased rapidly from 9° to 15°C, *E. pacifica* were absent.

Another promising method to assess krill biomass might be the spawning survey method, in which adult biomass is estimated from egg abundance. This method has been used for sardines and mackerel since the 1940s in Japan and USA (Watanabe, 1983). Egg abundance is estimated more effectively than is possible with adult krill because eggs do not form dense aggregations and are not able to undertake any avoidance reaction. Even so, several assumptions must be made to estimate adult biomass from egg abundance. Taki (1998b) adopted this method to estimate *E. pacifica* biomass using samples collected with Norpac nets hauled vertically from 150 m depth to the surface at 17 stations allocated alongshore in both Sanriku and Joban waters (Fig. 3.1.5). The acoustic survey mentioned above was conducted simultaneously along the cruise track (Miyashita *et al.*, 1998). The estimated biomass (W_i) of *E. pacifica* at station i can be obtained from equation (3.1.1).

$$W_i = \varpi N_i \tag{3.1.1}$$

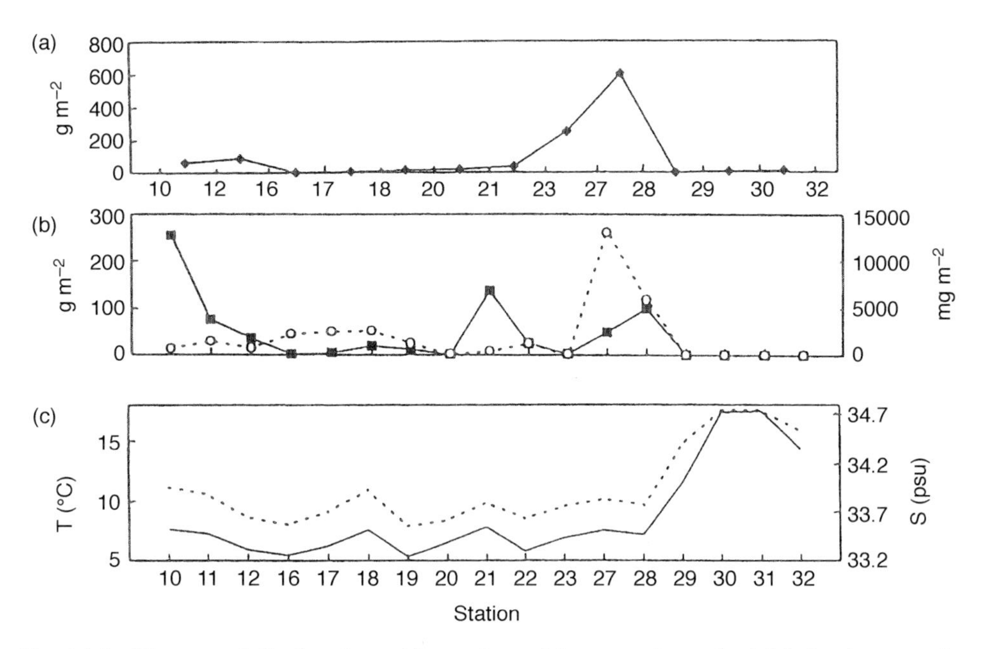

Fig. 3.1.5 Biomass of *Euphausia pacifica* estimated by acoustic method (a); by the spawning survey method (solid line) and by net samplings with the improved larva net (broken line) (b); and water temperature (solid line) and salinity (broken line) at 50 m depth (c); at 17 stations allocated alongshore from 36°50′N to 40°20′N (Redrawn from Taki, 1998b.)

where ϖ is the mean wet weight of *E. pacifica*, and N_i is the individual number under unit area of sea surface at station *i*.

The number of eggs spawned in a day at station i (P_i) is given in equation (3.1.2).

$$P_i = \alpha p_i = sN_i \sum_{j=k}^{m} q_j \gamma_j \delta_j \Phi_j \tag{3.1.2}$$

where α is the turnover rate of the time needed for hatching in a day, p_i is the number of eggs per unit surface area (1 m^2), s is the survival rate of eggs, q_j proportion of individuals of the body length class j, γ_j sex ratio of females, δ_j the number of spawning episodes in a day, and Φ_j the number of eggs spawned by one female per episode, respectively. If we assume all the spawned eggs hatch in the upper 150 m, then

$$N_i = \frac{\alpha p_i}{s \sum_{j=k}^{m} q_j \gamma_j \delta_j \Phi_j} \tag{3.1.3}$$

Substituting equation (3.1.3) into equation (3.1.1), we can get

$$W_i = \frac{\varpi \alpha p_i}{s \sum_{j=k}^{m} q_j \gamma_j \delta_j \Phi_j}$$

Taki (1998b) reported high abundance of krill at Stn 28 off Fukushima Prefecture by this method, which was confirmed by acoustic results by Miyashita *et al.* (1998) (Fig. 3.1.5). In the spawning survey method, however, two other abundance peaks appeared in the northern area at Stns 10 and 21, which were not supported by the acoustic method. The discrepancy may be attributable to the time lag between spawning and net sampling. The abundance trend estimated by the acoustic method and the net sampling with the improved larva net showed good coincidence (Fig. 3.1.5).

Both of these methods are likely to provide estimates of standing stock, the accurate estimation of which, along with estimates of production of *E. pacifica* are necessary for the management of this fishery.

References

Ebisawa, Y. (1994) On *Euphausia pacifica* in the southern Joban area. *Gekkan Kaiyo (Kaiyo Monthly)* **26**, 255–59 (in Japanese).

Endo, Y. (1981) Ecological Studies on the Euphausiids Occurring in the Sanriku waters with Special Reference to Their Life History and Aggregated Distribution. PhD. thesis, Tohoku University, Sendai, 166 pp. (in Japanese with English abstract).

Endo, Y. (1984) Daytime surface swarming of *Euphausia pacifica* (Crustacea: Euphausiacea) in the Sanriku coastal waters off northeastern Japan. *Mar. Biol.* **79**, 269–76.

Endo, Y. (1993) Distribution of zooplankton in relation to warm core rings and warm streamers. *Gekkan Kaiyo (Kaiyo Monthly)* **25**, 94–99 (in Japanese).

Fujita, T. (1994) Importance of planktonic organisms as food of demersal fish population. *Gekkan Kaiyo (Kaiyo Monthly)* **26**, 236–41 (in Japanese).

Fujita, T., Inada, T. & Ishito, Y. (1993) Density, biomass and community structure of demersal fishes off the Pacific coast of northeastern Japan. *J. Oceanogr.* **49**, 211–29.

Greene, C.H., Wiebe, P.H., Burczynski, J. & Youngbluth, M.J. (1988) Acoustical detection of high-density demersal krill layers in the submarine canyons off Georges Bank. *Science* **241**, 359–61.

Gutt, J. & Siegel, V. (1994) Benthopelagic aggregations of krill (*Euphausia superba*) on the deeper shelf of the Weddell Sea. *Deep-Sea Res.* **41**, 169–78.

Hanamura, Y., Endo, Y. & Taniguchi, A. (1984) Underwater observations on the surface swarm of a euphausiid, *Euphausia pacifica* in Sendai Bay, northeastern Japan. *La mer* **22**, 63–68.

Hattori, H. (1991) Vertical distribution of zooplankton in the warm core off Sanriku (86B) and adjacent Oyashio water, with special reference to copepods record. *Bull. Hokkaido Natl. Fish. Res. Inst.* **55**, 59–77 (in Japanese with English abstract).

Hattori, H. (1993) Copepod distribution in a warm-core ring. *Gekkan Kaiyo (Kaiyo Monthly)* **25**, 99–104 (in Japanese).

Iguchi, N., Ikeda, T. & Imamura, A. (1993) Growth and life cycle of a euphausiid crustacean (*Euphausia pacifica* Hansen) in Toyama Bay, southern Japan Sea. *Bull. Japan Sea Natl.*

Fish. Res. Inst. **43**, 69–81.

Johnson, R.K. (1977) Sound scattering from a fluid sphere revisited. *J. Acoust. Soc. Am.* **61**, 375–77.

Kodama, J. & Izumi, Y. (1994) Factors relevant to the fishing ground formation of *Euphausia pacifica* and the relation to the demersal fish resources. *Gekkan Kaiyo (Kaiyo Monthly)* **26**, 228–35 (in Japanese).

Komaki, Y. (1967) On the surface swarming of euphausiid crustaceans. *Pacif. Sci.* **21**, 433–48.

Komaki, Y. & Matsue, Y. (1958) Ecological studies on the Euphausiacea distributed in the Japan Sea. *Report of the Cooperative Survey of the Warm Tsushima Current and Related Waters* **2**, 146–59 (in Japanese).

Mauchline, J. (1980) The biology of mysids and euphausiids. *Adv. Mar. Biol.* **18**, 1–681.

Miyashita, K., Aoki, I. & Inagaki, T. (1996) Swimming behaviour and target strength of isada krill (*Euphausia pacifica*). *ICES J. Mar. Sci.* **53**, 303–308.

Miyashita, K., Aoki, I., Seno, K., Taki, K. & Ogishima, T. (1997) Acoustic identification of isada krill, *Euphausia pacifica* Hansen, off the Sanriku coast, north-eastern Japan. *Fish. Oceanogr.* **6**, 266–71.

Miyashita, K., Aoki, I., Asami, N., Mori, H. & Taki, K. (1998) A feasibility study on the distribution and standing stock of *Euphausia pacifica* in Sanriku coastal waters in spring. *Rep. Res. Meeting of Krill Res. and Oceanogr. Conditions in the northwest. Pacific*, pp. 110–124 (in Japanese).

Motoda, S. (1957) North Pacific standard plankton net. *Inform. Bull. Planktol. Japan* **4**, 13–15 (in Japanese).

Nakamura, T. (1991) Distribution and fishing ground formation of *Euphausia pacifica* in the southern Joban area. *Bull. Tohoku Branch Jap. Soc. Sci. Fish.* **41**, 44–46 (in Japanese).

Nakamura, T. (1992) Recent aspect of krill fishing grounds off Joban–Kashima area in relation to warming tendency. *Bull. Jap. Soc. Fish. Oceanogr.* **56**, 155–57 (in Japanese).

Nicol, S. & Endo, Y. (1997) Krill fisheries of the world. *FAO Fish. Tech. Paper* **367**, 1–100.

Nishikawa, J., Tsuda, A., Ishigaki, T. & Terazaki, M. (1995) Distribution of euphausiids in the Kuroshio front and warm water tongue with special reference to the surface aggregation of *Euphausia pacifica*. *J. Plankt. Res.* **17**, 611–29.

Odate, K. (1991) *Fishery Biology of the Krill, Euphausia pacifica, in the Northeastern Coasts of Japan.* Suisan Kenkyu Sosho (Library of Fisheries Study) No 40, pp. 1–100 (in Japanese with English abstract).

Ohtsuki, T. (1975) Distribution of euphausiids and physical environment in the south-west area of the Okhotsk Sea. *Monthly Rep. Hokkaido Fish. Exp. St.* **32**, 1–10 (in Japanese).

Omori, M. (1965) 160 cm opening-closing plankton net-I. Description of the gear. *J. Oceanogr. Soc. Jap.* **21**, 212–20.

Ponomareva, L. A. (1963) *The Euphausiids of the North Pacific, their Distribution and Ecology*, 142 pp., Dokl. Akad. Nauk. SSSR Israel programme for Sci. Transl. 1966.

Sawamoto, S. (1992) Species composition of euphausiids in Suruga Bay, central Japan. *Bull. Inst. Oceanic Res. & Develop., Tokai Univ.* **13**, 85–96.

Simard, Y., de Ladurantaye, R. & Therriault, J.-C. (1986) Aggregation of euphausiids along a coastal shelf in an upwelling environment. *Mar. Ecol. Prog. Ser.* **32**, 203–215.

Stancyk, S. E., Fujita, T. & Muir, C. (1998) Predation behavior on swimming organisms by *Ophiura sarsii*. In *Echinoderms: San Francisco* (R. Mooi and M. Telford, eds.), pp. 425–29.

Balkema, Rotterdam.

Suzuki, M. (1986) On the distribution of *Euphausia pacifica* in the southern Joban and Kashima-nada areas. *Bull. Tohoku Branch Jap. Soc. Sci. Fish.* **37**, 30–31 (in Japanese).

Takahashi, M. & Iwami, T. (1997) The summer diet of demersal fish at the South Shetland Islands. *Antarct. Sci.* **9**, 407–413.

Taki, K. (1998a) Horizontal distribution and diel vertical migration of *Euphausia pacifica* Hansen in summer in and around a warm-core ring off Sanriku, northwestern Pacific. *Bull. Tohoku Natl. Fish. Res. Inst.* **60**, 49–61.

Taki, K. (1998b) A feasibility study to estimate standing stock of *Euphausia pacifica* using a dual-frequency echosounder and egg production method. *T.N.F.R.I. News* **53**, 4–6 (in Japanese).

Taki, K. & Ogishima, T. (1997) Distribution of some developmental stages and growth of *Euphausia pacifica* Hansen in the northwestern Pacific on the basis of Norpac net samples. *Bull. Tohoku Natl. Fish. Res. Inst.* **59**, 95–117 (in Japanese with English abstract).

Taki, K., Kotani, Y. & Endo, Y. (1996) Ecological studies of *Euphausia pacifica* Hansen and seasonal change of its environment off Onagawa, Miyagi Prefecture III. Distribution and diel vertical migration. *Bull. Tohoku Natl. Fish. Res. Inst.* **58**, 89–104 (in Japanese with English abstract).

Terazaki, M. (1980) Surface swarms of a euphausiid *Euphausia pacifica* in Otsuchi Bay, northern Japan. *Bull. Plankt. Soc. Jap.* **27**, 19–25.

Terazaki, M. (1992) Horizontal and vertical distribution of chaetognaths in a Kuroshio warm core ring. *Deep-Sea Res.* **39**, Supplementary Issue No. 1A, S231–S245.

Watanabe, T. (1983) Spawning survey method. In *Population Dynamics of Fishery Resources* (T. Ishii, ed.), pp. 9–29. Fisheries Science Series No 46. Koseisha-Koseikaku, Tokyo (in Japanese).

Watanabe, Y. (1988) Design of surface towing nets and analysis method of sampling data. In *Manual on Sampling and analysis of Eggs and Larvae of Pelagic Fishes.* Natl. Res. Inst. Fish. Sci., pp. 15–22 (in Japanese).

Wiebe, P.H., Morton, A. W., Bradley, A. M. *et al.* (1985) New developments in the MOCNESS, an apparatus for sampling zooplankton and micronekton. *Mar. Biol.* **87**, 313–23.

Yasuda, I., Okuda, K. & Shimizu, Y. (1996) Distribution and Modification of North Pacific Intermediate Water in the Kuroshio-Oyashio interfrontal zone. *J. Phys. Oceanogr.* **26**, 448–65.

3.2 Canadian waters

Inigo Everson

Although there has been some commercial interest shown in the krill that are to be found on both the eastern and western Canadian seaboards, this has been restricted to two relatively small regions. These are the Nova Scotian shelf and specifically the St Lawrence estuary on the Atlantic Ocean side and the Vancouver and Queen Charlotte Islands area on the Pacific Ocean side. Due to their geographical separation these regions are considered separately.

Pacific Ocean seaboard of Canada

Of the many species of euphausiid known to be present in the area only three, *Euphausia pacifica, Thysannoëssa spinifera* and *T. longipes,* figure greatly in ecological studies in the region. Fulton and LeBrasseur (1984) estimated that 90% of the euphausiid biomass around the Queen Charlotte Islands was composed of these three species and although they were present on the shelf, the distribution of all three species extended into deep water. Fulton *et al.* (1982), when describing results from winter and spring surveys in 1980 around the Queen Charlotte Islands, found that the seasonal mean euphausiid biomass was similar in Hecate Strait and Dixon Entrance but was 2–4 times greater off the west coast of the southern islands in the group, a region where the continental shelf is very narrow. In the Dixon Entrance, situated at the northern end of the Queen Charlotte Group *T. longipes* made up the greater proportion of the biomass but at the southern end *T. spinifera* was the dominant species. These areas of concentration are shown in Fig. 3.2.1.

Further south the trend continues such that *E. pacifica* and *T. spinifera* are the dominant species and account for 90% of the euphausiid biomass around Vancouver Island. Using the results from a series of surveys from 1991 to 1997 (Tanasichuk, 1998a , 1998b) showed the extent of inter-annual variation in standing stock. During that study, sea temperatures had been anomalously warm in 1992, a period coincident with an El Niño Southern Oscillation (ENSO) event. In that season the abundance of *E. pacifica* larvae was found to be six times the level it had been in 1991. In the case of adults, due to the strong recruitment of the 1992 larvae, the abundance was twice as high in 1992 and 1993 as it had been in 1991. Subsequent to the ENSO event larval abundance has continued to fluctuate although the abundance of adults returned to the pre-ENSO level.

A slightly different story emerges from an examination of the results for *T. spinifera* (Tanasichuk, 1998b). Larval abundance was 2.4 times higher in 1992 than it had been in 1991 and became low for 1993 and 1994 when it declined to 0.6 and 0.4 respectively of the 1992 value. It was, however, higher again in 1995 at 6.3 and 1996 at 1.9. In spite of this the abundance of adult *T. spinifera* tells a different story since there had been a steady decline from 1991 through to 1996 by which time it was only

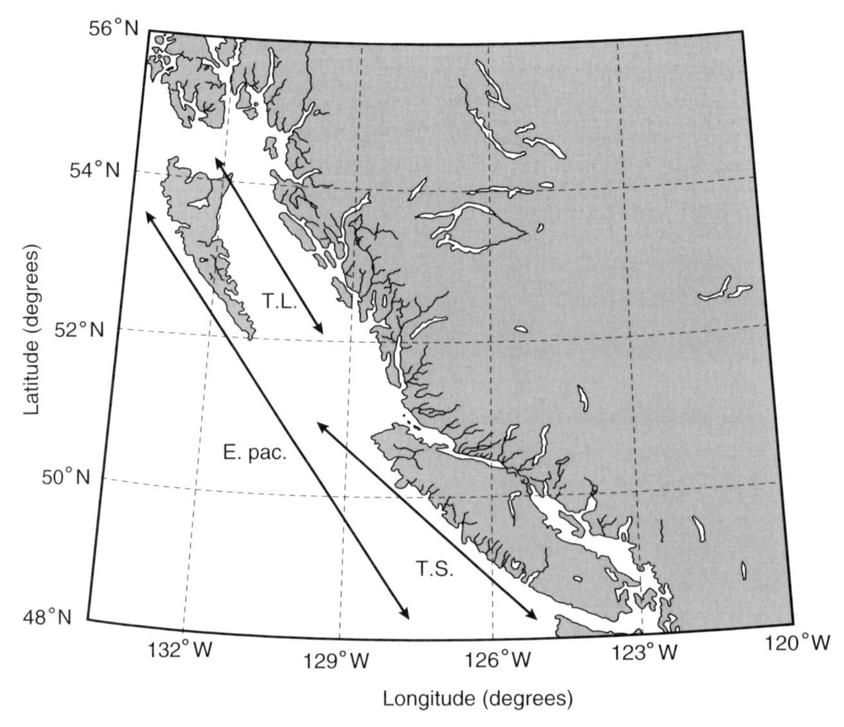

Fig. 3.2.1 Coastline of British Columbia showing the latitudinal distribution of the three dominant euphausiids in the region. T.L. = *Thysanoessa longipes*; T.S. = *T. spinifera*; E. pac = *Euphausia pacifica*.

about 0.14 times the pre-ENSO levels. Tanasichuk (1995) suggests that the reason for this decline may not be directly related to the environment but might be a result of changes in predation pressure from fish. Euphausiids figure significantly in the diet of hake and it appears that these have increased over the same time frame.

Off Vancouver Island in June and August 1986, an acoustic and net sampling programme found that the macrozooplankton layers were dominated by the euphausiids *E. pacifica* and *T. spinifera*. During both the surveys of this study by Simard & Mackas (1989) dense aggregations were found in the same two general regions and shown in Fig. 3.2.2. The first was along the shelf break, where the highest biomass was observed. A second broader and more diffuse aggregation, covering about 200 km^2, occupied the deeper south-east end of the shelf area with a density of 1–2 g/m^2. The scattering layers undertook a consistent diel vertical migration with daytime depth at the shelf break varying from around 150 to 200 m while on the shelf it was shallower (~100–150 m) although occasionally extending to the bottom. Simard & Mackas (1989) noted that these day-depths corresponded to the location of California Undercurrent waters with a temperature of 6–7°C and a salinity of 33.75 ppt. These studies all indicate that while *E. pacifica* is present throughout most of the region there is a transition from *T. longipes* in the north to *T. spinifera* in the south.

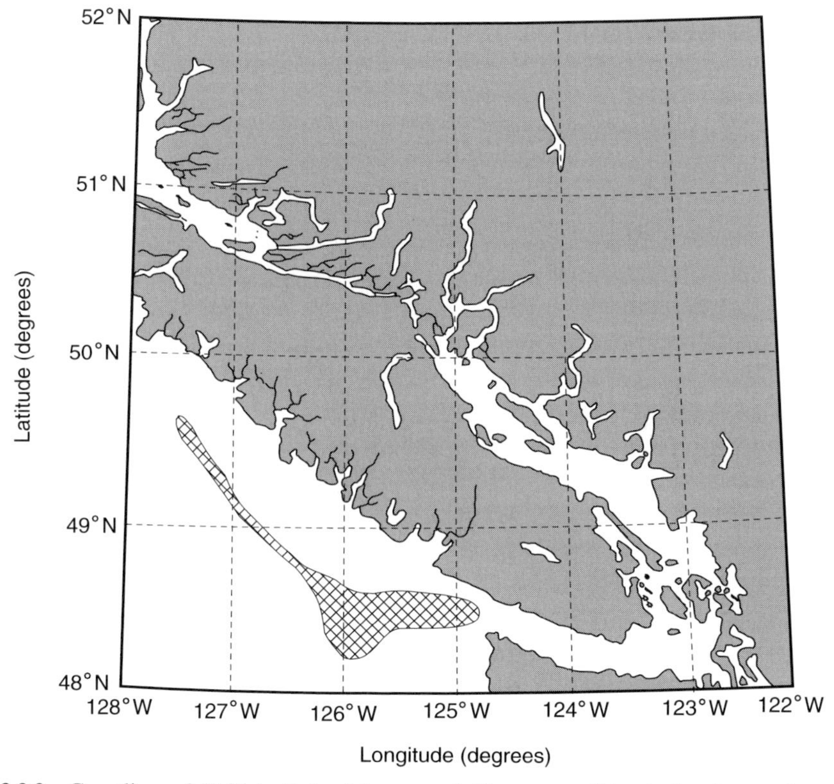

Fig. 3.2.2 Coastline of British Columbia around Vancouver Island showing region of high euphausiid biomass, *Euphausia pacifica* and *Thysanoessa spinifera*. (From information in Simard & Mackas, 1989.)

Even though there is an annual total allowable catch (TAC) of 500 t for the entire British Columbia coast, the main centre of commercial fishing has been restricted to the Strait of Georgia and concentrated in the Jervis Inlet and adjacent Malaspina Strait (Romaine *et al.*, 1995). This is a small area relative to the western Canadian seaboard. Jervis inlet, shown in Fig. 13.2, is about 316 km^2 and since it is mostly enclosed short-term changes due to physical transport are unlikely to be of major significance. A series of acoustic surveys have been undertaken in the region to provide standing stock estimates. Because the entrance to the inlet is relatively narrow, with an assumed very low level of interchange with the Strait of Georgia, it has been possible to follow the seasonal changes at the population level. Two cohorts with modal lengths 9 and 20 mm were present in June 1994 and the smaller cohort (9 mm) grew over the sampling period to reach 20 mm by May 1995. The larger (20 mm) cohort seen in June 1994 was not present in samples after October 1994. The loss of this larger cohort at that time probably explains the decline in biomass up to January 1995 shown in Fig. 3.2.3.

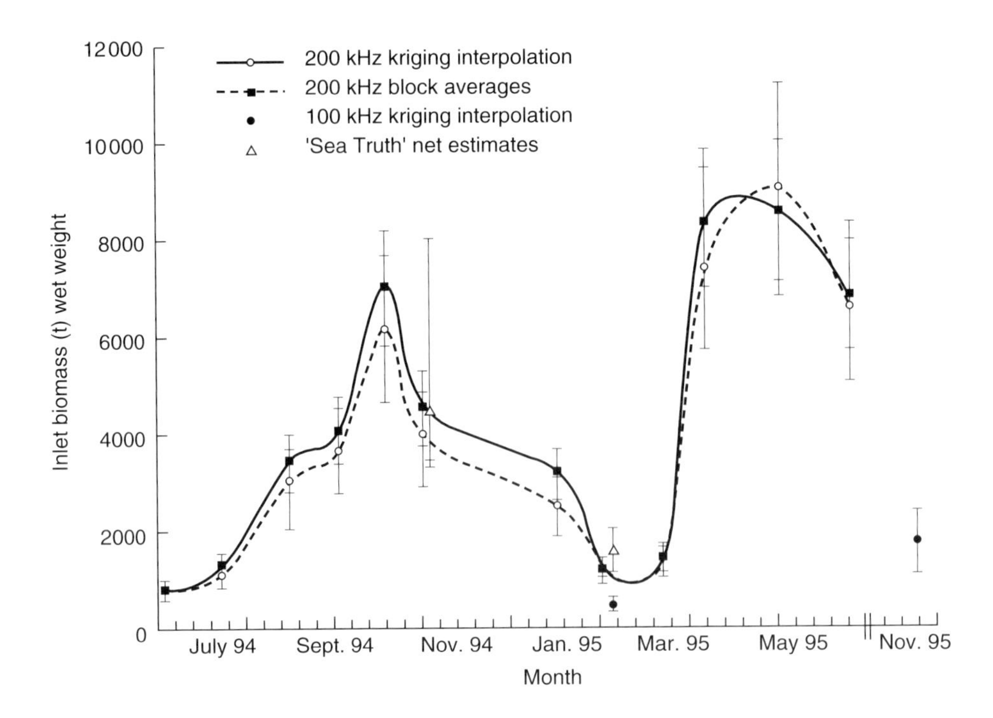

Fig. 3.2.3 Euphausiid biomass within Jervis Inlet determined from a series of acoustic surveys supported by net sampling. The pronounced within season variation in biomass is clearly evident. (With permission from Romaine *et al.*, 1996.)

Eastern Canadian seaboard

The main focus of interest on the Atlantic seaboard of Canada has been in the St Lawrence estuary, Bay of Fundy and the Scotian shelf where four species are of significance in this coastal ecosystem, *Thysannoëssa lomgicaudata*, *T. raschii*, *T. inermis* and *Meganyctiphanes norvegica.*

Sameoto (1982) showed that although *T. longicaudata* was present on the Scotian shelf, where the water is relatively colder than offshore, its abundance was much lower than in the deeper oceanic water at and beyond the shelf break. This largely confirmed the conclusions of Berkes (1976) who only found this species in the deeper eastern part of the Gulf of St Lawrence. The distribution of the Scotian shelf species is indicated in Fig. 3.2.4.

From a study examining surface swarms in the Bay of Fundy using aerial photography, Nicol (1986) noted that the density of *M. norvegica* within these swarms varied from 77.8 to 778 g /m^2. Further studies, specifically related to surveys to estimate standing stock, do not appear to have been made for this locality. Arising from this a conservative estimate of the density of krill within the bay has been inferred from estimates coming from the Scotian shelf (DFO, 1996).

The Gulf of St Lawrence is a semi-enclosed sea connected to the Atlantic Ocean

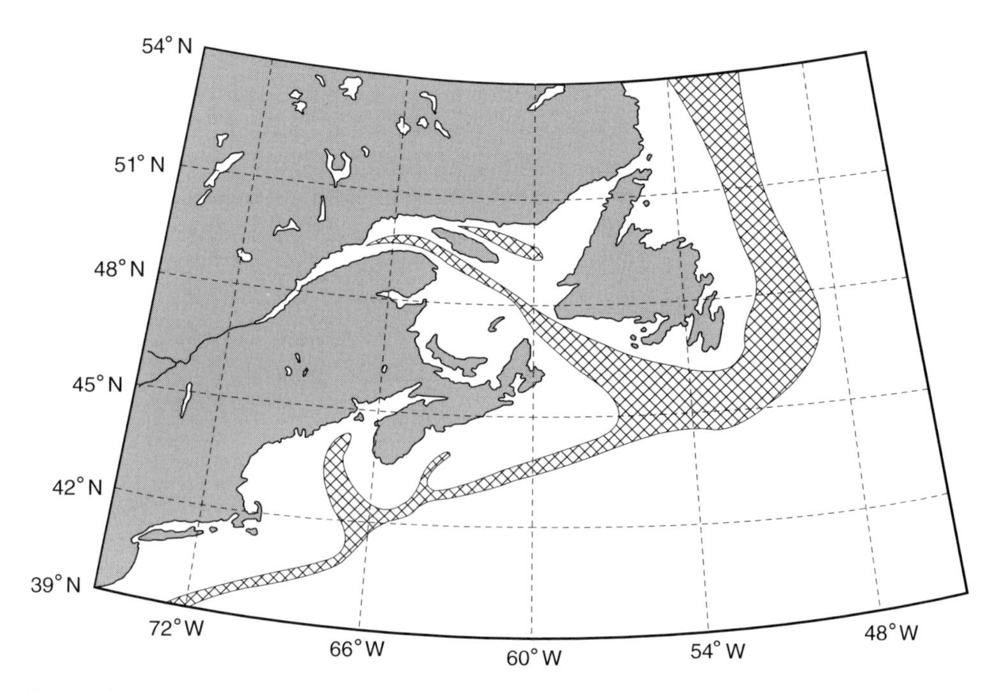

Fig. 3.2.4 Areas of concentration, shown shaded, of euphausiid biomass on the Nova Scotia shelf. The main species in the area is *Meganyctiphanes norvegica*. (DFO, 1996.)

by two straits, the southern Cabot Strait about 450 m deep and the shallower northeastern Belle-Isle Strait which has a maximum depth of around 70 m. The Gulf is divided by a network of three channels about 350 m deep which merge in the Cabot Strait and extend up to the continental slope between the Scotian Shelf and the Grand Banks. These main features are shown in Fig. 3.2.5.

The circulation in the region is complex and is the result of different forces operating. The main system is of a two-layer estuarine flow, driven by the large freshwater discharge of the St Lawrence which strongly influences the mean circulation pattern (Koutitonsky & Bugden, 1991). The freshwater discharge has a clear annual signature, with a strong peak in Spring following the thaw. Both the water mass turbidity, which is always higher in the St Lawrence estuary than in the Gulf, and the phytoplankton production have Spring peaks, which are spatially modulated (Therriault & Levasseur, 1985). These features which have a major effect on the local environment appear to be important in controlling the abundance and spatial structure of euphausiid populations in the region.

Within the Gulf of St Lawrence there have been a series of studies over a number of years. Analysing results from a series of ten cruises over two years, Berkes (1976, 1977) noted that adults of *M. norvegica*, *T. raschii* and *T. inermis* occurred most abundantly in the western Gulf while their larvae were present in the Magdalen Shallows. These three species, along with *T. longicaudata,* reach reproductive maturity at one year although *T. longicaudata* is the only one for which the samples

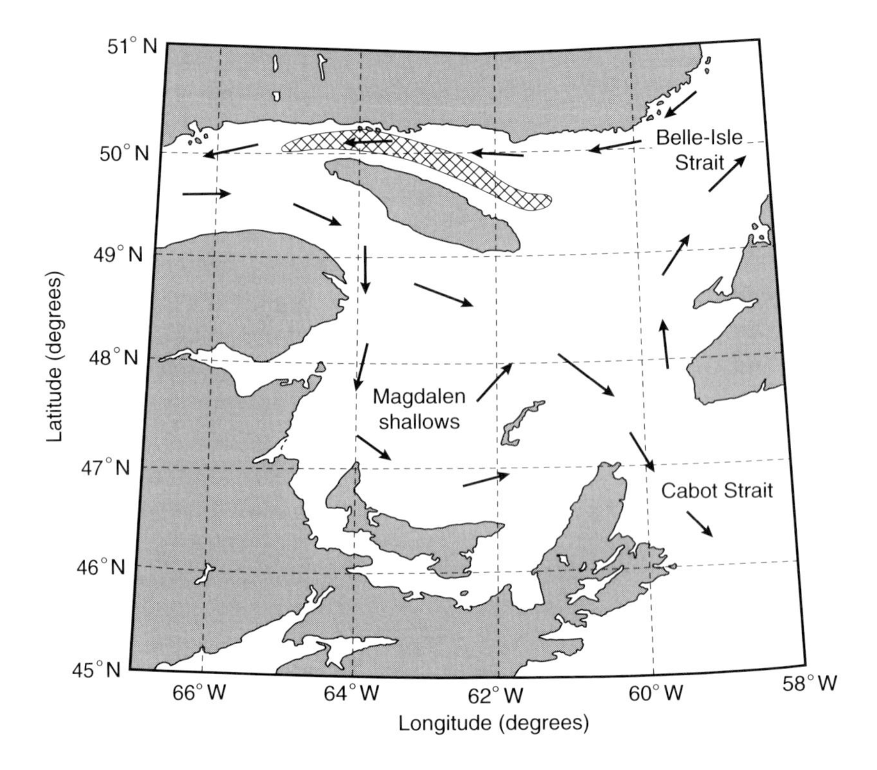

Fig. 3.2.5 Water circulation pattern in the Gulf of St Lawrence shown by arrows (Berkes, 1976) and main concentration, shown shaded, of *Meganyctiphanes norvegica*.

indicated that there were never more than two breeding-year classes present in the population. Spawning in all the three *Thysannoëssa* species takes place in April, a month that is coincident with the phytoplankton bloom.

A three-year study in the Gulf of St Lawrence estuary, reported by Sameoto (1976), using a 120 kHz echo-sounder demonstrated the existence of large populations of *T. raschii*, *T. inermis* and *Meganyctiphanes norvegica*. Chlorophyll a concentrations, measured within 5 m of the surface at the same time as the sound scattering layers were sampled, showed significant correlation with the numerical density of euphausiids in the area.

In the June 1973 survey *M. norvegica* was the most abundant species with aggregations concentrated in the region of the Laurentian channel between the tip of Gaspé and Anticosti Islands. The same species was also very common and numerous on the north side of the St Lawrence estuary between Sanuenay River and Pte des Monts. *T. raschii* was the dominant species in the area along the north shore and in the Anticosti Channel as well as along the Gaspé coast whereas *T. inermis* was dominant only on the eastern tip of Anticosti Island.

During the September 1973 survey the dominant species was *T. raschii* which was concentrated in dense scattering layers along Gaspé coast. That same survey showed there to be small concentrations of *M. norvegica* in the Laurentian channel at that time. Continuing the sequence, the December 1973 and May 1974 surveys indicated that the scattering layers sampled along the coast of Gaspé and in the mouth of the St Lawrence River west of Pte des Monts contained predominantly *T. raschii.*

There was a difference in the relative importance of the species from the different years. Numerically *T. raschii* was the dominant species over much of the area in 1973 and increased greatly over the period from September to December 1973 as shown in Table 3.2.1. All three species were concentrated along the Gaspé coast during May and June, a period when primary production in the area was at a peak.

Table 3.2.1 Ratio of the numerical abundance of each species in each survey relative to the lowest density for that species.

Date	Ratio		
	T. raschii	*T. inermis*	*M. norvegica*
June 1972	1	7.5	1.5
June 1973	5.8	2.1	1
Sept 1973	416	1	1.9
Dec 1973	165	12	1
May 1974	10	1	1

The results of a study reported by Simard *et al.* (1986) noted that most of the euphausiids were aggregated in a patch greater than 100 km in length, but only one to seven kilometres wide, along the northern edge of the Laurentian Channel. Within this patch the biomass was greater than 1 g dry wt/m^2 and on a scale of kilometres showed well-defined higher density zones where the biomass was found to be up to 22.2 g dry wt/m^2. Overall the mean density within the patch was around 60 mg dry wt/m^3.

In that study euphausiids were found at depths varying from 50 to 175 m with some of these large concentrations extending down and coming close to the bottom. The krill were present in two distinct layers with modal depths of 75 and 150 m, the shallower layer being dominated by *T. raschii* while the deeper layer contained mostly *M. norvegica*. Vertical and horizontal distribution patterns of temperature and salinity indicated that upwelling was occurring during the survey. The authors suggest that the aggregations of euphausiids along one edge of the channel resulted from the interaction of the negative phototaxis of the animals with the upwelling and mean circulation in the area. It is suggested that the euphausiids were transported towards the north shore by the circulation outlined above. However, they were not carried up to the surface but remained at a level governed by the depth to which daylight penetrated underwater. The authors suggest that this resulted in the formation of a long narrow patch running along the bathymetric contours on the upwelling side of the Laurentian Channel in the St Lawrence estuary.

Topographic effects on the circulation are also thought to affect the density along the aggregations so that the channel heads, in particular the head of the Laurentian channel, are the major sites where the aggregations occur. These same sites are also known to be the traditional feeding grounds for baleen whales (Simard *et al.*, 1986; Simard & Lavoie, 1999).

The varying degree of intensity of sampling over the region has made it difficult to derive an overall estimate of standing stock for the Maritimes Region. Standing stock estimates of *M. norvegica* per km^2 within two depth strata have been made by taking the results from the different studies and analysing them together. The two depth strata selected have been the shelf, where the water is less than 200 m, and deep water to the continental slope where the depth is between 200 and 1000 m. Density estimates have been determined acoustically during the spring and summer as noted above. These results were used along with known areas of distribution derived from net surveys in order to estimate krill biomass. To allow for uncertainties in the estimation procedure conservative values were taken for the estimates. These were 10 g m^{-2} as the estimated total density in both the Emerald and LaHave Basins while a much lower value of 2 g m^{-2} was used for all other localities (DFO, 1996). The estimated values are shown in Table 3.2.2.

Table 3.2.2 Krill biomass estimates for the eastern Canadian seaboard. Data from DFO (1996).

Location	Area (km^2)	Krill biomass (t)
Emerald and LaHave Basin	5296	52 960*
Basins in eastern half of Shelf	2399	4800**
Fundy Channel and Jordan Basin	7714	15 000**
Continental shelf edge including Laurentian Channel (200–1000 m)	25 406	51 000**
Total Shelf		123 760

* Based on an average biomass of 10 g m^{-2}
** Estimates are based on the conservative assumption of a biomass of 2 g m^{-2}

These results are substantially lower than those given by Runge & Joly (1995) from a study using net tows in association with information from historical sources. They estimated the biomass to be 500 000 t for an area of 21 000 km^2 in the western Gulf of St Lawrence. There is no way with the information available to indicate which of these estimates is the more reliable. They are both estimating standing stock but using different methods; some difference is unsurprising. What is important is that they demonstrate the level of uncertainty associated with the estimates, a factor that should be incorporated into any model taking the information forward in order to provide scientific advice for the management of any fishery that might develop in the region.

References

Berkes, F. (1976) Ecology of euphausiids in the Gulf of St Lawrence. *J. Fish. Res. Board Can.* **33**, 1894–1905.

Berkes, F. (1977) Production of the euphausiid *Thysanoëssa raschii* in the Gulf of St Lawrence. *J. Fish. Res. Board Can.* **34**, 443–46.

DFO (1996) Krill on the Scotian Shelf. DFO Atlantic Fisheries Stock Status Report **96/106E**.

Fulton, J. & LeBrasseur, R. (1984) Euphausiids of the continental shelf and slope of the Pacific coast of Canada. *La mer* **22**, 268–76.

Fulton, J., Arai, M.N. & Mason, J.C. (1982) Euphasuiids, coelenterates, ctenophores and other zooplankton from the Canadian Pacific coast ichthyoplankton survey, 1980. *Can. Tech. Rep. Fish. Aquat. Sci.*, **1125**, 75 pp.

Koutitonsky, V.G. & Bugden, G.L. (1991) The physical oceanography of the Gulf of St Lawrence: a review with emphasis on the synoptic variability of the motion. *Can. Spec. Publ. Fish. Aquat Sci.* **113**, 57–90.

Nicol, S. (1986) Shape, size and density of daytime surface swarms of the euphausiid *Meganyctiphanes norvegica* in the Bay of Fundy. *J. Plank. Res.* **8**, 29–39.

Romaine, S.J., Mackas, D.L., Macaulay, M.C. & Saxby, D.J. (1995) Comparisons of repeat acoustic surveys in Jervis inlet, British Columbia, 1994–1995. In *Harvesting Krill: Ecological Impact, Assessment Products and Markets.* Fisheries Centre Research Reports, University of British Columbia, Canada. 1995 Vol. 3 No. 3. pp. 48–52.

Runge, J.A. & Joly, P. (1995). Zooplancton (euphausiacés et calanus) de l'estuaire et du golfe du Saint-Laurent. In *Rapport sur l'état des invertébrés en 1994: crustacés et mollusques des côtes du Québec, crevette nordique et zooplancton de l'estuaire et du golfe du Saint-Laurent* (L. Savard, ed.), Chap. 7, pp. 129–37. *Rapp. Manus. Can. Sci. halieut. Aquat.* **2323**.

Sameoto, D.D. (1976) Distribution of sound scattering layers caused by euphausiids and their relationship to chlorophyll a concentrations in the Gulf of St Lawrence estuary. *J. Fish. Res. Board Can.* **33**, 681–87.

Sameoto, D.D. (1982) Zooplankton and micronekton abundance in acoustic scattering layers on the Nova Scotian slope. *Can. J. Fish. Aquat. Sci.* **39**, 760–77.

Simard, Y. & Lavoie, D. (1999) The rich krill aggregation of the Saguenay-St Lawrence Marine Park: hydroacoustic and geostatistical biomass estimates, structure, variability and significance for whales. *Can. J. Fish. Aquat. Sci.* **56**, 1182–97.

Simard, Y. & Mackas, D. (1989) Mesoscale aggregations of euphausiid sound scattering layers on the continental shelf of Vancouver Island. *Can. J. Fish. Aquat. Sci.* **46**, 1238–49.

Simard, Y., de Ladurantaye, R. & Therriault, J.-C. (1986) Aggregation of euphausiids along a coastal shelf in an upwelling environment. *Mar. Ecol. Prog. Ser.* **32**, 203–215.

Tanasichuk, R.W. (1995) The influence of inter-annual variations in sea temperature on the population biology and productivity of euphausiids in Barkley Sound, Canada. In *Harvesting Krill: Ecological Impact, Assessment Products and Markets.* Fisheries Centre Research Reports, University of British Columbia, Canada. 1995 Vol. **3** No. 3. pp. 62–64.

Tanasichuk, R.W. (1998a) Interannual variations in the population biology and productivity of *Euphausia pacifica* in Barkley Sound, Canada, with special reference to the 1992 and 1993 warm ocean years. *Mar. Ecol. Prog. Ser.* **173**, 163–80.

Tanasichuk, R.W. (1998b) Interannual variations in the population biology and productivity of *Thysnaoessa spinifera* in Barkley Sound, Canada, with special reference to the 1992 and 1993 warm ocean years. *Mar. Ecol. Prog. Ser.* **173**, 181–95.

Therriault, J.C. & Levasseur, M. (1985) Control of phytoplankton production in the lower St Lawrence estuary: light and freshwater runoff. *Naturaliste Can.* **112**, 77–96.

3.3 The Southern Ocean
Inigo Everson

Large-scale distribution

The circumpolar nature of the Southern Ocean with its dominant West to East water movement suggests that many planktonic species may be circumpolar in distribution. That this hypothesis is largely correct was shown by Baker (1954) studying the zooplankton and Hart (1942) the phytoplankton samples of 'Discovery Investigations' both of whom found that the species composition did not change markedly around the continent.

There are in the Southern Ocean the following commonly occurring species of euphausiid crustacean: *Euphausia superba*, *E. vallentini*, *E. triacantha*, *E. frigida*, *E. crystallorophias* and *Thysanoessa vicina* and *T. macrura*. Due to the general circulation pattern of the Southern Ocean they are found all around the continent although they are separated reasonably well latitudinally as is shown in Fig. 3.3.1. Of the species which commonly occur in the Southern Ocean there are only two, *E. superba* and *E. crystallorophias*, which regularly occur in dense swarms and are likely to figure in commercial catches.

E. crystallorophias is restricted more or less entirely to the shelf region of the Antarctic continent. In the south-west Indian Ocean sector, it has been shown by Hosie (1994) and Hosie & Cochran (1994) that it is predominant in the neritic zone and conversely that substantial catches of *E. superba* are unlikely to be made in localities close to the Antarctic continent. During a recent survey extending from 80° to 150° East in the Indian Ocean sector Pauly *et al.* (1996) found *E. superba* in deep oceanic water but as soon as the survey extended on to the continental shelf the catches were almost entirely composed of *E. crystallorophias*. In the Pacific sector a recent series of observations reported by Azzali & Kalinowski (1999) found *E. superba* extending to around 74° South in the eastern part of the Ross Sea and only south of that latitude was *E. crystallorophias* the dominant species.

The high latitude neritic distribution for *E. crystallorophias* is also typical of the Atlantic sector (Mauchline & Fisher 1969) although there have been records away from the Antarctic coastal region. This species has been reported from the caldera of Deception Island, South Shetlands by Everson (1987), another shallow locality, while Brierley & Brandon (1999) reported a concentration of *E. crystallorophias* from deep oceanic water in the Scotia Sea which appeared to be within a tongue of cold continental shelf water.

By far the most important species commercially in the Southern Ocean is the Antarctic krill, *E. superba*, and the remainder of this part of the chapter will be restricted to its consideration. The broad scale distribution of *E. superba* has been described by Marr (1962) from a large series of net hauls during the 'Discovery

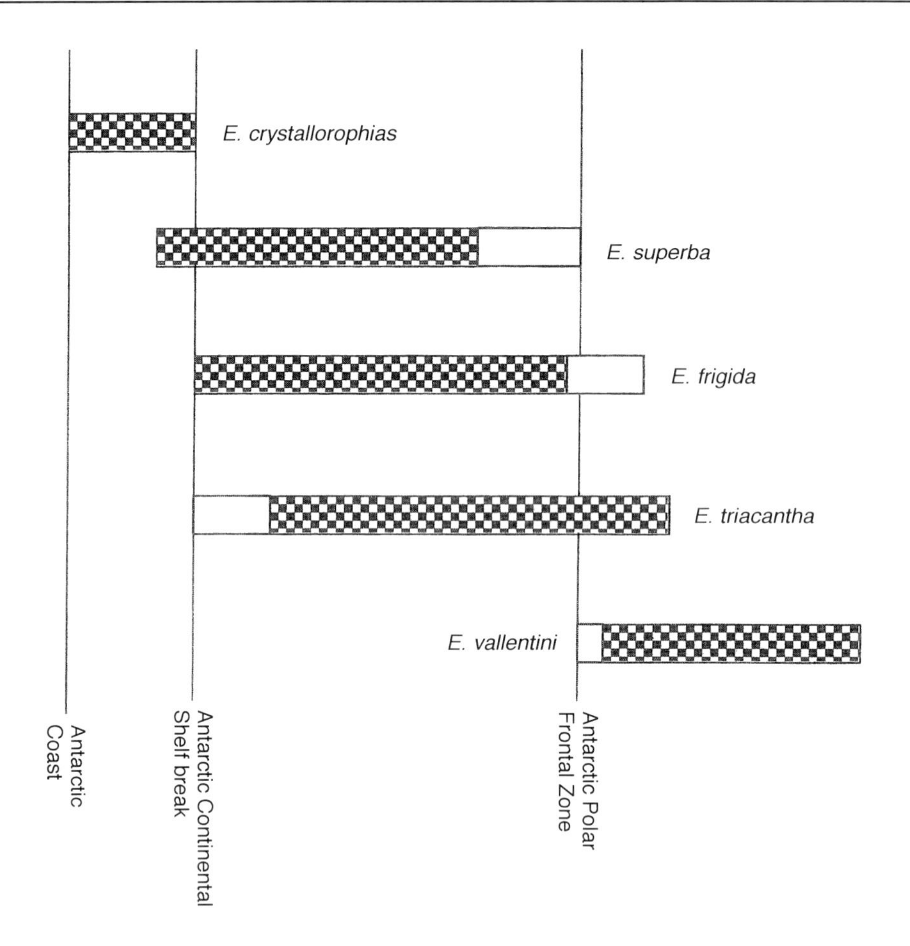

Fig. 3.3.1 Latitudinal distribution of the species of the genus Euphausia in the Southern Ocean. The shaded portion represents the main range, the open portion the extreme range. With permission from Cambridge University Press. (After John, 1937.)

Investigations.' Although there have been many large-scale surveys subsequently, that picture of the general distribution extending from the high Antarctic continental shelf north as far as the Antarctic Polar Front Zone (APFZ) and shown in Fig. 3.3.2. is still accepted. It was this distribution which was behind the designation of the management area for the Commission for the Conservation of Antarctic Marine Living Resources (CCAMLR). The CCAMLR area, shown in Fig. 12.1, was designed to cover the distribution of all species harvested within the Southern Ocean and for these the APFZ provides a very convenient northern limit as proposed by Everson (1977). The initial plan, with the exception of the part between 55° and 60° South and 50° to 60° West which was excluded for political reasons, in the Atlantic sector was accepted. This locality, within FAO Division 41.3.2, is the only area outside of the CCAMLR region from which Antarctic krill catches have been reported (Anon, 1998).

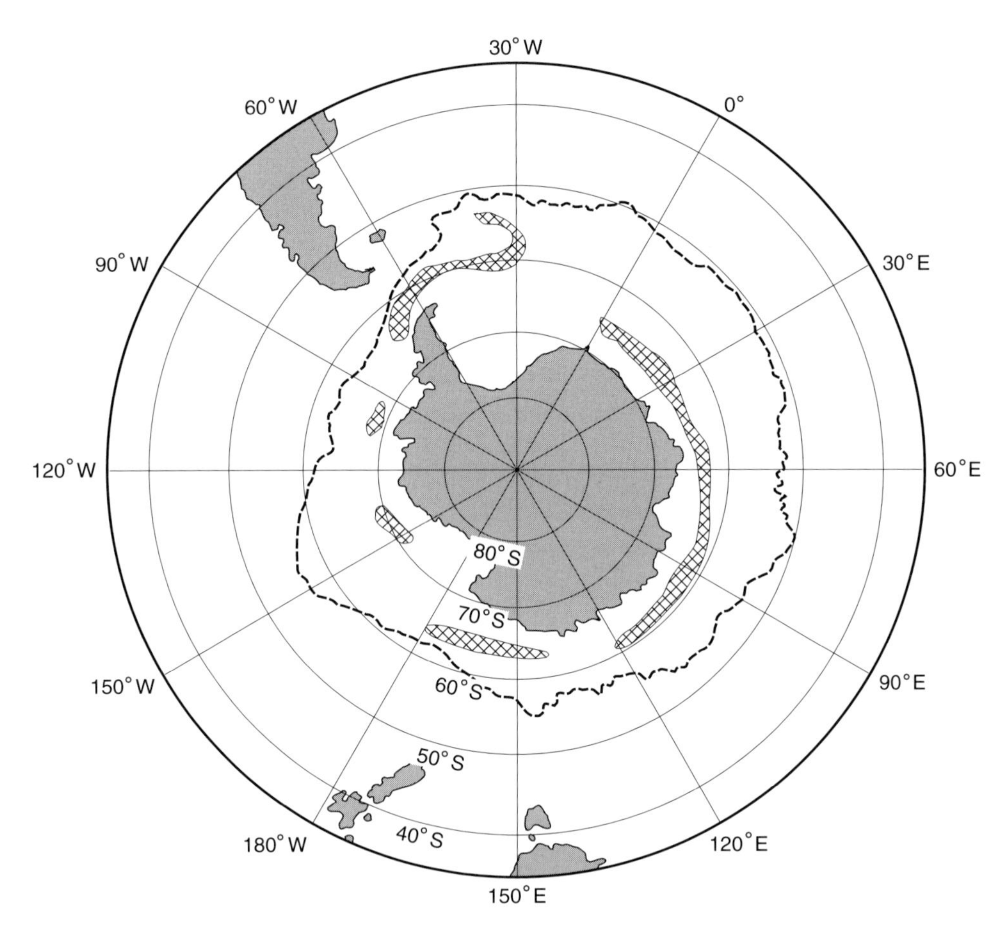

Fig. 3.3.2 Large-scale distribution of krill. Based on surveys and commercial fishing activity. (Based on Marr, 1962; Everson & Miller, 1994; Nicol & Endo, 1997.)

Large-scale estimation of standing stock

With a distribution extending over much of the 36 million square kilometres which make up the Southern Ocean, estimation of the standing stock of *E. superba* presents enormous logistical difficulties. A broad-scale survey with transects spaced even as close as 30 mile intervals would take about 30 vessels a month to complete, clearly an unrealistic commitment. Added to this is the logistical problem caused by a significant part of the area being covered by sea-ice. It is no surprise therefore that there has never been a single synoptic survey for this species.

Three large-scale surveys have been undertaken. The first to estimate the standing stock of krill in the Southern Ocean, formed a major part of the Biological Investigations of Marine Antarctic Systems and Stocks (BIOMASS) programme and as the First International BIOMASS EXperiment (FIBEX) covered a large part of the Southwest Atlantic and Western Indian Ocean sectors. This was a major undertaking involving 11 vessels from 10 nations (Hempel, 1983) and involved a flexible survey design to take account of time lost due to adverse weather (Everson *et al.*, 1989). That design was developed further and used for other acoustic surveys (Jolly & Hampton, 1990). The results from the survey are summarised in Table 3.3.1.

Table 3.3.1 Krill (*Euphausia superba*) standing stock estimates from the FIBEX survey which took place between 16 January and 12 March 1981. Data from Trathan *et al.*, 1992.

CCAMLR or FAO Area of subarea	Area surveyed (km^2)	Standing stock million tonnes	CV (%)
41	75 000	3.66	29.6
48.1	283 000	10.54	35.0
48.2	242 000	15.61	22.2
48.3	25 000	1.51	37.9
48.6	576 000	4.63	22.9
58.4.2	1 711 000	3.93	32.0

During January and February 2000 there was a further survey in the Atlantic sector to estimate standing stock over a larger area than had been covered by the FIBEX survey. All four participating vessels were equipped with identical multi-frequency echosounder and data management systems. Preliminary analyses of the data indicate that the estimated standing stock of 44.3 million tonnes (SC-CAMLR 2000) is similar to that estimated from the FIBEX results.

The 1996 survey in the Southeast Indian Ocean sector by the Australian Antarctic Division was undertaken by a single vessel over a period of two months and extended from 80° to 150° East (Pauly *et al.*, 1996). In order to cover the area designated for the survey, the transects were widely spaced, depending on the sampling regime along a particular transect the spacing was 150 or 80 nautical miles. Density was not uniform over the area and krill were more abundant in the west of the surveyed area, between 80° and 120° East. The results are summarised in Table 3.3.2.

Although on a much smaller scale, the surveys undertaken by Italian scientists working in the Ross Sea provide further insight into the distribution of krill (Azzali & Kalinowski, 1999). These surveys in 1989–90 and in 1994 demonstrated that *E. superba* were present as far as 76° South in an area previously thought to contain few krill (Marr, 1962).

These large-scale acoustic surveys although covering less than half of the Southern Ocean have provided confirmation of the pattern of distribution determined from nets in these areas. The swarming nature of krill, discussed by Watkins in Chapter 4, means that the distribution is decidedly non-uniform. This component

Table 3.3.2 Results of an Australian krill survey in the Indian Ocean sector of the Southern Ocean from 29th January to 21st March 1996. Data from Pauly *et al.* (2000).

Longitudinal range	80–115°E	115–150°E
Area (million km^2)	2.34	2.34
Number density (#/m^2)	10.6	6.7
Weight density (g/m^2)	6.7	4.2
Standing stock (million of tonnes)	3.0	1.8
Coefficient of Variation (%)	19	30

of the spatial distribution must also be taken into account in order to gain an understanding of the system.

Small-scale distribution

Turning to a smaller scale of distribution, of the order of one to ten kilometres, we need to look at situations where intensive sampling has been taking place within limited areas and a good starting point for this is through an analysis of commercial fishing activity.

Several studies have been undertaken which look at the spatial distribution of krill fishing. From an analysis of data in 10-day intervals at a scale of half a degree of latitude by one degree of longitude Everson & Goss (1992) showed that in the Atlantic sector fishing tended to be restricted to the coastal or shelf region. On a finer scale, haul by haul data have been analysed by Endo & Ichii (1989), Ichii *et al.* (1992), Kasatkina & Latogursky (1991), Vagin *et al.* (1992), Sushin (1998) and Trathan *et al.* (1998) and all these studies indicate that in the Atlantic sector the commercial fishery concentrates on the continental shelf close to the shelf break around the main island groups. Elsewhere there have been few studies although in the Indian Ocean sector similar analyses have shown that the fishery there is concentrated at the shelf break (Ichii, 1990).

There are two reasons why the fishery concentrates in coastal regions in the Atlantic sector. The obvious one is that this is where the densest concentrations of krill are likely to be found. The other explanation is that the coastal region provides shelter for fishing vessels in the event of adverse weather. Additionally, in the case of the South Shetlands, the fishing grounds are in closer proximity to port facilities in South America than anywhere else in the Southern Ocean. These points are discussed further by Ichii, in Chapter 9.

Examination of the distribution of fishing activity might lead to the conclusion that the shelf area and shelf break zone are the only major sites of krill abundance. In terms of abundance estimation it would be very convenient if that were the case since with much lower concentrations offshore this would greatly reduce the area of a high density stratum for abundance estimation surveys. Consideration needs to be given to this situation by assessing what proportion of the krill stock is present in the oceanic deep water between the main shelf regions. This in turn raises further questions concerning the relationship between the krill that are present around one island group with those present in locations further 'downstream.'

Interactions between areas

Surface water in the West Wind drift flows in a manner analogous to a conveyer belt travelling from west to east on which krill can be carried over the deep ocean. Everson & Murphy (1987) demonstrated that krill swarms can be carried within the surface current in the coastal waters of the Bransfield Strait. From an analysis of the locations of sequential hauls by commercial fishing vessels Endo & Ichii (1989) demonstrated that over a period of a few days individual trawlers were targeting on the same krill concentration as it was carried along the north coast of the South Shetland Islands in the eastward flowing current. There is therefore good evidence that krill are likely to be carried on the ocean currents and probably at the same speed.

Extending this idea to the Scotia Sea and using the Fine Resolution Antarctic Model (FRAM) to determine the main flow fields, Murphy *et al.* (1998) and Hofmann *et al.* (1998) demonstrated that krill from around Elephant Island could be carried across the Scotia Sea to arrive off South Georgia several months later. Thus although the largest concentrations of krill may be found on the shelf and at the shelf break it is highly likely that these concentrations are being carried at some time over the deep ocean between the island groups. Looked at it in this light it appears that the circumpolar current is effectively a conveyer belt on which the krill are carried. In this simple conceptual framework it is right to expect krill concentrations to be found almost anywhere within the Antarctic Circum-Polar Current (ACC).

Commercial fishing, as already noted, tends to be concentrated in a few restricted localities. At the start of the fishing season at South Georgia the fleets generally begin operations to the north-east of Cumberland Bay (Trathan *et al.*, 1998), while in the South Shetland Islands the fleet appears to be operating most regularly to the north of King George Island (Endo & Ichii, 1989). Although krill concentrations have been encountered during research surveys in the deep ocean their location appears to be far less predictable than those found on the shelf or shelf-break. Thus although the commercial fleets provide some insight into the location of krill patches, this is only part of the story and because the current krill catch is very much less than the estimated production that proportion on the shelf may be only a small part of the overall standing stock.

Several studies have looked at the mesoscale distribution of krill and a good example is that reported by Sushin & Shulgovsky (1999). The results, summarised on a subarea scale, are shown in Table 3.3.3. These show that in the three study periods there was reasonable consistency between the South Shetlands and South Orkneys but that the pattern at South Georgia was totally different. Various authors

Table 3.3.3 Krill standing stock estimates from research surveys in the Southwest Atlantic sector. The values are the mean (g/1000 m^3) along with upper and lower boundary of the 95% confidence intervals for the estimates. (Data from Sushin & Shulgovsky, 1999.)

Dates of surveys	Subarea 48.1 (South Shetlands)	Subarea 48.2 (South Orkneys)	Subarea 48.3 (South Georgia)
27 Jan–16 Mar 1984	119 (38–231)	105 (26–205)	1.5 (0.15–3.14)
11 Oct–14 Dec 1984	177 (56–95)	131 (65–208)	0.6 (0.29–0.88)
20 Jan–9 Mar 1988	48 (15–91)	71 (26–132)	147.5 (78–238)

have noted this phenomenon and concluded that it is a result of the oceanic circulation pattern across the Scotia Sea (e.g. Mackintosh, 1973; Priddle *et al.*, 1988, Maslennikov & Solyankin, 1988; and Sushin & Shulgovsky, 1999). Proximity to each other and their location within the circumpolar current means that links between the South Shetlands and South Orkneys are likely to be strong. By contrast South Georgia lying to the northern side of the ACC is likely to be more greatly affected by any latitudinal variation of the main flow as it passes through the Drake Passage.

As well as highlighting differences at the statistical subarea scale (subareas are defined by Miller & Agnew, Chapter 12), the study by Sushin & Shulgovsky (1999) also demonstrated differences within these localities. One feature observed was the existence of zones of high biomass to the west of about 40° West. These appear to be similar in location to some favoured fishing grounds although there are important differences particularly in the Bransfield Strait. The authors note a much greater degree of variation in the eastern part of the area which they attribute to variation in the geostrophic currents within the ACC already noted.

Arising from this we can conclude that a main factor controlling the distribution of krill is the broad-scale geostrophic circulation. This brings the krill to the shelf regions within which there are further oceanic components affecting the distribution. In this region we have again components of scale which need to be considered, a shelf and shelf-break scale covering perhaps 50 to 100 km and the finer-scale effects which appear to be topographically controlled.

Models of local krill distribution

From an analysis of large numbers of net hauls in the vicinity of the South Shetlands, Siegel (1988), Siegel *et al.* (1997, 1998) noted a seasonal pattern of distribution of juvenile, subadult and adult stages, shown diagrammatically in Fig. 3.3.3. The initial concepts are set out in Siegel (1988). In summer, during the spawning season, there is a spatial separation of developmental stages along the Antarctic Peninsula, from the inner shelf where the juveniles are found, to oceanic waters where the adults are

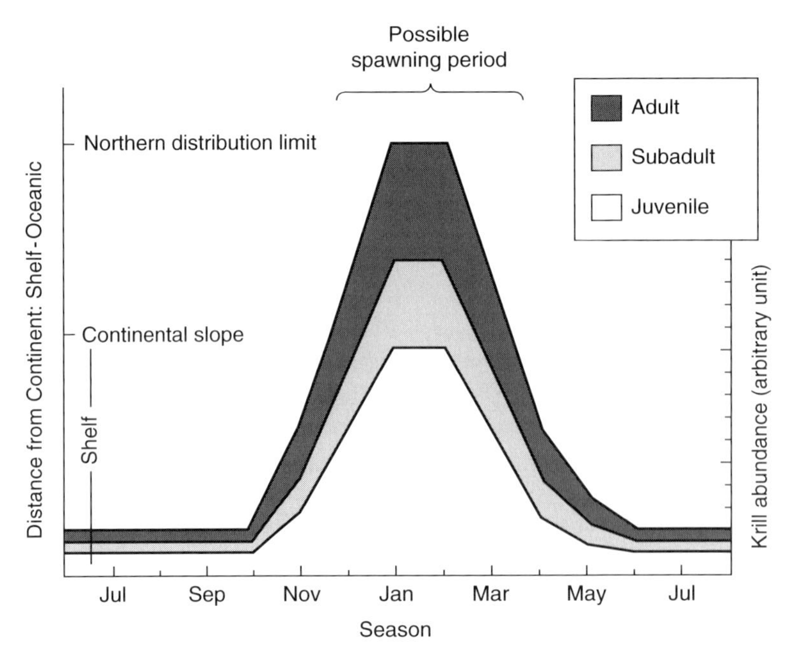

Fig. 3.3.3 Generalised picture of the seasonal fluctuations in the abundance and distribution of krill in waters of the Antarctic Peninsula and the spatial succession of developmental stages between coastal and oceanic waters. (With permission from Siegel *et al.*, 1997.)

present. Overlying this is a temporal succession of abundance from early spring when juveniles predominate through to the summer when all stages are present. Siegel and his colleagues develop this concept by postulating an offshore migration in summer and onshore migration during autumn and winter which is represented diagrammatically in Fig. 3.3.4. Support for these ideas comes from the study by Ichii *et al.* (1998) working in the vicinity of the South Shetland Island. They found that over the period from early to mid-summer the krill density increased and showed distinct offshore–onshore differences in abundance and maturity stages. In mid-summer the krill density was low, 8 g m^{-2}, in the oceanic region and higher, 36 g m^{-2} in the frontal zone on the slope. The highest values, 131 g m^{-2}, occurred along the shelf break in the inshore zone.

The annual cycle proposed by Siegel came as a result of the analysis of data from several calender months over a number of years. Working in an area further south of Elephant Island in the Palmer Long Term Ecological Research Programme (LTER), Quetin *et al.* (1996) undertook an extensive sampling programme over a number of seasons. Their results from winter sampling led them to develop hypotheses linking krill distribution and production to the extent of sea-ice cover. They suggest that when the winter sea-ice is extensive the ice-associated biota provide favourable feeding conditions for juvenile krill with the result that standing stock in the following season will be higher. A similar conclusion was reached by Kawaguchi & Satake (1994) who noted that heavy sea-ice in the austral winter leads

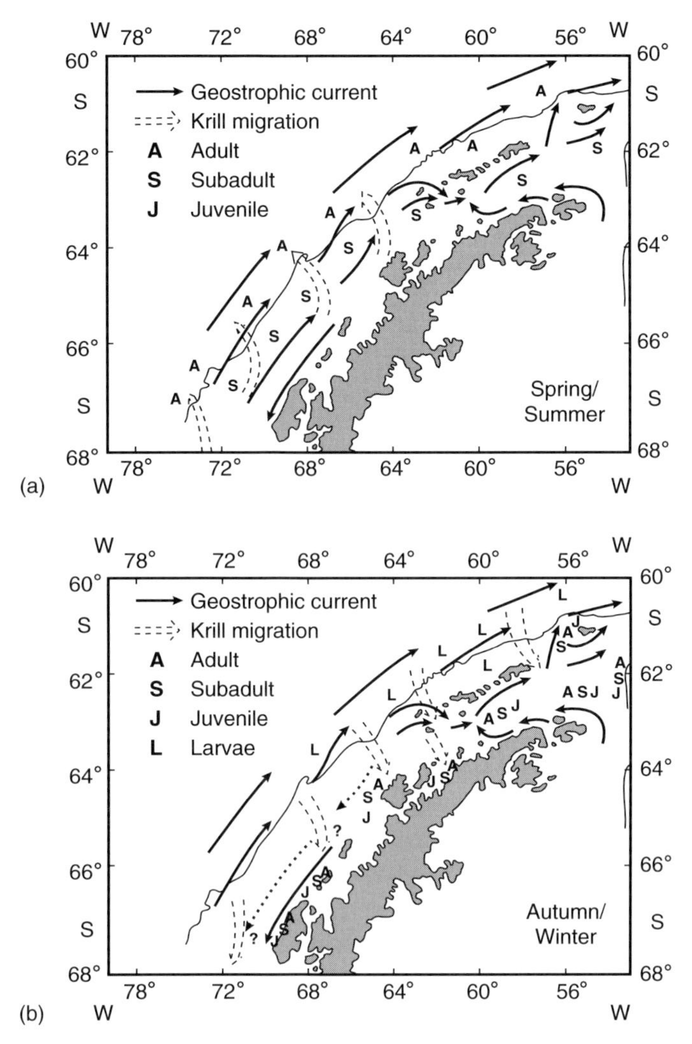

Fig. 3.3.4 Seasonal distribution pattern of juvenile, subadult and adult krill developmental stages off the Antarctic Peninsula. (a) Krill migration into oceanic waters during the spring/summer spawning season. (b) Post-spawning movement into the neritic zone. (With permission from Siegel *et al.* 1988.)

to a relatively high abundance of small krill in the population in the following spring. This is in line with Quetin *et al.* (1996) who suggest that years of greatest sea-ice extent will result in a higher reproductive output, as measured by egg production, in the following season. These two criteria, however, are not always in phase because the recruitment potential reflects environmental conditions prior to the spawning season and winter survival, a result of high production during the season, reflects conditions in the following season (Ross & Quetin, 1991).

All of these studies have been developed as a result of extensive research activity taking place in one sector of the Southern Ocean. The types of model developed there might also have application in other regions such as in the Indian Ocean sector where krill have been the subject of commercial fishing at the shelf break (Endo & Ichii, 1989).

The Siegel and the Quetin *et al.* models incorporate information from direct observations on the krill and relate them to the water circulation. They do not explicitly take account of predation on krill during the summer period, a topic discussed by Everson, Chapter 8. Thus a reduction in standing stock towards the end of the summer period could be due to predation on krill by land-based predators during the breeding season. If that were the case then the simpler model of the conveyer outlined earlier in this chapter would apply. Without such detailed information it is impossible to develop these ideas further.

The data derived from the series of surveys in the vicinity of Elephant Island highlight a further point with regard to the interpretation of the data. Sampling was undertaken by two methods and the results presented separately. These are brought together in Table 3.3.4. When comparing these results there are some key points that need to be borne in mind. The first is that the density units are different, acoustics are presented relative to sea surface area whereas the net hauls are presented as relative to the volume of water filtered. Providing both methods are sampling the same depth horizon this will not affect the trend, only the absolute values. The second point is that there is a different pattern in the seasonal values by the two methods. The reason for this difference is because the net haul data have been restricted to observations made during January and February, the period during which, according to the Siegel hypothesis, the abundance should be at its greatest. The acoustic data cover a much wider period of the summer season.

Small scale distribution

Moving now to a finer scale of the order of kilometres there are krill patches. A good example is the 'super-swarm' described by Mathisen & Macaulay (1983) which was present on the shelf close to Elephant Island, South Shetlands. This large krill patch persisted for several weeks and was targeted by commercial fishing vessels. Similar patches have been noted by other workers (see Watkins, Chapter 4) although in these cases no commercial fishing was taking place in the vicinity. A detailed discussion of these features is outside the scope of the present chapter but the key point to be noted is that they persisted for days or even weeks. In order for this to happen the localities must have been sites where the krill, for some reason, became concentrated. This may appear to be contrary to the conveyer belt analogy developed for the deep ocean situation described above.

In their study in the Bransfield Strait, Everson & Murphy (1987) noted that in the strongly flowing coastal current to the south of King George Island krill swarms appeared to be moving with the current. Close by in the channel between King George and Elephant Islands the water movements were more confused and a clear

Table 3.3.4 Estimates of krill density in the vicinity of Elephant Island from acoustics (Hewitt & Demer, 1994) and net hauls (Siegel *et al.*, 1997).

Acoustic		Net hauls	
Date	Density (g/m^2)	Season	Density (N/1000 m^3)
		1977/78	348.3
Mar 1981	68.5	1980/81	161.4
		1981/82	324.87
		1982/83	276.7
Nov 1983	13.4		
Mar 1984	22.5	1983/84	241.7
Nov 1984	63.7	1984/85	85.85
Mar 1985	8.4		
Jan 1987	57.2		
Jan 1988	41.6	1987/88	27.64
Feb 1989	82.4	1988/89	79.84
Early Jan 1990	17.1	1989/90	15.39
Late Jan	46.9		
Early Feb	78.4		
Late Feb	90.9		
Late Jan 1991	23.8	1990/91	6.46
Late Feb	28.8		
Late Jan 1992	61.2	1991/92	19.86
Early Mar	29.6		
Jan 1993	134.5	1992/93	29.17
Feb	88.2		
		1993/94	32.96
		1994/95	73.31
		1995/96	119.97
		1996/97	212.51

pattern was not present; the krill swarms in that region did not appear to be moving nearly as fast. That study used the set and drift information derived from successive satellite passes, recorded by a transit satellite navigator, at intervals of an hour or two. In the study it was assumed that the vessel was being set by an amount indicative of the surface current.

Recently the use of Acoustic Doppler Current Profilers (ADCP) has meant that information at a much finer horizontal resolution is available and furthermore that the movements through the water column can also be quantified (Brandon *et al.*, 2000). Although there have not been studies whereby individual patches have been monitored for a period of time using ADCP the information on water circulation on the shelf indicates that the local eddying effects are such that krill patches could be retained at a location for quite some time.

Further evidence for this can be inferred from the locations of commercial fishing around South Georgia. Analyses of haul by haul data have demonstrated that fishing operations commence at more or less the same location each season (Murphy *et al.*, 1997; Trathan *et al.*, 1998). This demonstrates that some locations, such as the shelf break north-east of Cumberland Bay, are ones where good concentrations are likely to be encountered regularly.

While this information is supportive of the general picture of the krill, with localised areas of concentration, this needs to be incorporated into a model which includes population dynamics. Thus the age-density of the krill present at a given location must have arisen as a result of population recruitment, growth and mortality. Where part of the population is sampled in one location the age composition should reflect these demographic changes at sites sampled downstream at a later date. Watkins (1999) has shown that in some seasons this is the case but there are, even so, some disparities which remain unexplained. Clearly there is more scope for research into understanding the distribution of *E. superba* at all time and space scales.

Application of information on Southern Ocean krill distribution

Management units

Having considered the distribution of *E. superba* in the Southern Ocean we can now consider applications of this information. Because of its central role in the Southern Ocean food web knowledge of the ecology of krill is central to understanding the dynamic functioning of the ecosystem. As an extension of this concept, it is important in the development of advice for management of the fishery.

The foregoing part of this chapter has shown the broad geographical range of *E. superba* from the continental shelf zone northwards as far as the APFZ. Locally there are concentrations some of which occur in predictable localities but others are dependent on oceanic circulation. At the Southern Ocean scale we have a very large management unit which due to its enormous area is of limited meaningful value for management. Subdivision of the area in some form is therefore essential and it is advisable that any such breakdown should have a strong ecological justification.

The CCAMLR region is divided into three major Areas, 48 the Atlantic Ocean sector, 58 Indian Ocean sector and 88 the Pacific Ocean sector. This division was made by extending the FAO Statistical Areas southwards into the Southern Ocean. While such a division outside the Southern Ocean can be seen to have an ecological justification, the circumpolar continuity of the Southern Ocean means that such a division is arbitrary in nature and of minor ecological significance.

Within these areas there have been further subdivisions most of which were designated so as to separate areas of continental shelf which harboured different finfish stocks (Everson, 1977). Due to the localisation of the commercial krill fishery on the shelf and at the shelf break these same subareas are also reasonable descriptors of the limits of the fishery. However the circumpolar current means that there are clear and recognisable interactions between these subareas.

Although the Southern Ocean circulation is basically circumpolar it is subject to a large degree of variation due to internal variation, atmospheric circulation and bottom topography. Arising from this there are regions of gyres, some of which have been thought of as containing local populations. With the recent developments in molecular biology it has been considered possible to identify such 'populations'. Unfortunately the results of such studies have been unable to determine the extent of isolation or mixing. In the absence of a clear molecular biology signal alternative approaches have been considered.

Krill flux

Central to these have been a consideration of the oceanographic circulation as a key to the transport of krill to and through a region. This concept, referred to as 'krill flux', is being actively considered within CCAMLR.

Studies of krill flux can be made at a variety of different scales. At the large oceanic scale, where the flow fields are well described by the geostrophic circulation, the situation can be modelled as an extension of the work described by Murphy *et al.* (1998)and Hofmann *et al.* (1998). Greater problems arise in the vicinity of the shelf break and on the shelf where the circulation becomes less predictable. These same regions, as outlined above, are the sites where the denser krill patches have been found and also where commercial fishing takes place.

To address this problem consideration has been given to dividing the region into small rectangles (Anon. 1994). Estimation of the flow field across and the krill density along the rectangle boundaries can be used to provide an indication of the krill flux into and out of the rectangle. The theoretical basis for such a model to estimate the flux between two adjacent boxes shown in Fig. 3.3.5 was developed in Anon. (1994 Attachment D) and is set out below. For this the residence time in box 1 (r_1) is given by V_1/f_{O1} and that for box 2 (r_2) by $V_2/(f_{12}+f_{O2})$. Note that the subscript convention used is 'O' Outside with the numbers referring to individual boxes. The overall residence time R is:

$$R = \left(\frac{V_1}{f_{O1}}\right)\left(\frac{f_{O1}}{f_{O1}+f_{O2}}\right) + \left(\frac{V_2}{f_{12}+f_{O2}}\right)\left(\frac{f_{12}+f_{O2}}{f_{O1}+f_{O2}}\right)$$

This equation can be reorganised since the first and third factors are the residence times r_1 and r_2 described above. The second and fourth factors, termed 'pooling factors' in Anon. (1994) and designated w_1 and w_2, lead to a simplified version of the equation as:

1-BOX SYSTEM – Example

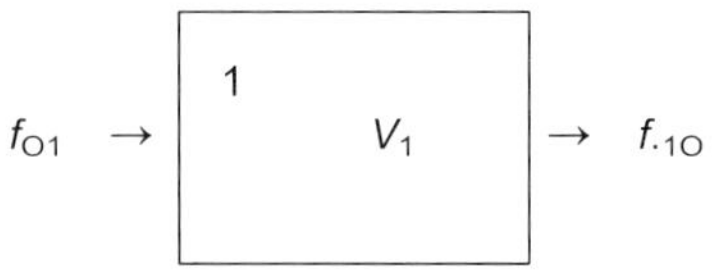

f_{O1} → [1 V_1] → $f_{.1O}$

V_1 = volume (e.g., water volume) in box 1 (e.g., km^3)

f_{O1} = input from 'outside' into box 1 (e.g., in km^3/day)

f_{10} = outflow from box 1 to the 'outside' (e.g., in km^3/day)

The subscript 'O' refers to 'outside'

T_1 = turnover for box 1 = $\frac{f_{O1}}{V_1}$

r_1 = residence time in box 1 = $\frac{V_1}{f_{O1}}$ (e.g., in days)

2-BOX SYSTEM – Example

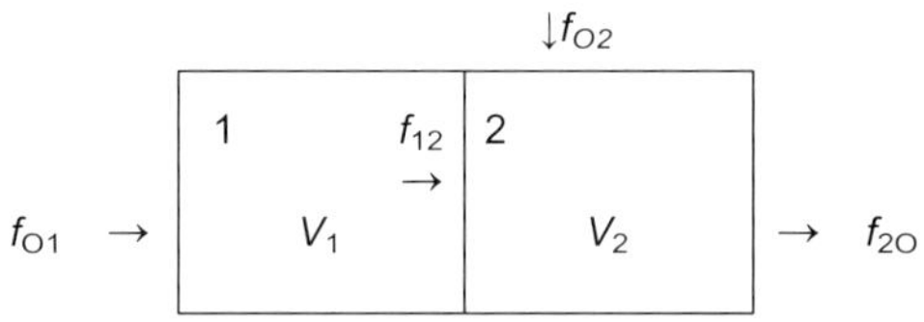

$\downarrow f_{O2}$

f_{O1} → [1 V_1 f_{12} → | 2 V_2] → f_{2O}

Fig. 3.3.5 Schematic representation of the krill flux model. With permission from the SC-CAMLK. (Anon., 1994.)

$$R = r_1 w_1 + r_2 w_2$$

This can then be taken to the generalised case for a large number of boxes:

$$R = \sum_{i=1}^{N} r_i w_i$$

Estimates of residence times have been made using the data from FRAM and the FIBEX study and these fall into line with expectation. Those analyses were undertaken to demonstrate the general validity of the process. They are not reproduced here because for them to have any real purpose it is important that the flux estimates are calculated so as to reflect a specific area of interaction. Such areas might be for management purposes but equally can reflect the overlap areas of commercial fishing and predator foraging. The importance of this information can be seen when the implications of the management issues that are raised in Chapters 8 and 12 are considered. The ecosystem approach to management requires that consideration be given to the target species, in this case krill, and the species dependent upon it. Thus the management units need to cater for the spatio-temporal variation in krill, water circulation as well as those of the dependent species.

References

Anon. (1994) Report of the workshop on evaluating krill flux factors. Report of the thirteenth meeting of the Scientific Committee for the Conservation of Antarctic Marine Living Resources. *SC-CAMLR-XIII*, Annex 5, Appendix D.

Anon. (1998) Report of the seventeenth meeting of the Scientific Committee for the Conservation of Antarctic Marine Living Resources. *SC-CAMLR-XVII.*

Azzali, M. & Kalinowski, J. (1999) Spatial and temporal distribution of krill *Euphausia superba* biomass in the Ross Sea (1989–1990 and 1994). In *Ross Sea Ecology Italiantartide Expeditions (1987–1995)* (F.M. Faranda, L. Guillielmo & A. Ianora, eds), Chapter 31, pp. 433–55. Springer-Verlag, Berlin.

Baker, A. de C. (1954) The circumpolar continuity of Antarctic plankton species. *Discovery Rep.* **27**, 201–218.

Brandon, M., Murphy, E.J., Trathan, P.N. & Bone, D.G. (2000) Physical oceanographic conditions to the northwest of the sub-Antarctic Island of South Georgia. *J. Geophys. Res.* (In press.)

Brierley, A.S. & Brandon, M.A. (1999) Potential for long-distance dispersal of *Euphausia crystallorophias* in fast current jets. *Mar. Biol.* **135**, 77–82.

Endo, Y. & Ichii, T. (1989) CPUEs body length and greenness of Antarctic krill during 1987/88 season in the fishing ground north of Livingston Island. *Scientific Committee for the Conservation of Antarctic Marine Living Resources, Selected Scientific Papers*. SC-CAMLR-SSP/6, pp. 323–45.

Everson, I. (1977) *The Southern Ocean. The living resources of the Southern Ocean.* FAO Southern Ocean Fisheries Survey Programme GLO/SO/77/1, 156 pp.

Everson, I. (1987) Some aspects of the small scale distribution of *Euphausia crystallorophias*. *Polar Biol* **8**, 9–15.

Everson, I. & Goss, C. (1992) Krill fishing activity in the southwest Atlantic. *Antarctic Sci.* **3**, 351–58.

Everson, I. & Murphy, E. J. (1987) Mesoscale variability in the distribution of krill *Euphausia superba*. *Mar. Ecol. Prog. Ser.* **40**, 53–60.

Everson, I., Hampton, I. & Jolly, G.M. (1989) Survey design to estimate krill abundance during FIBEX. *Scientific Committee for the Conservation of Antarctic Marine Living Resources, Selected Scientific Papers.* SC-CAMLR-SSP/5, pp. 253–63.

Hart, T.J. (1942) Phytoplankton periodicity in Antarctic surface waters. *Discovery Rep.* **21**, 261–356.

Hempel, G. (1983) FIBEX – An international survey in the Southern Ocean: Review and outlook. *Memoirs of the National Institute of Polar Research*, Special Issue No 27, 1–15.

Hewitt, R.P. & Demer, D.A. (1994) Acoustic estimates of krill biomass in the Elephant Island Area: 1981–1993. *CCAMLR Sci.* **1**, 1–6.

Hofmann, E., Klinck, J.M., Locarnini, R.A., Fach, B. & Murphy, E. (1998) Krill transport in the Scotia Sea and environs. *Antarctic Sci.* **10**, 406–415.

Hosie, G.W. (1994) The macrozooplankton communities in the Prydz Bay region Antarctic.

In *Southern Ocean Ecology: The BIOMASS Perspective* (S.Z. El-Sayed, ed.), pp. 93–123. Cambridge University Press, Cambridge.

Hosie, G.W. & Cochran, T.G. (1994) Mesoscale distribution patterns of macrozooplankton communities in Prydz Bay, Antarctica – January to February 1991. *Mar. Ecol. Prog. Ser.* **106**, 21–39.

Ichii, T. (1990) Distribution of Antarctic krill concentrations exploited by Japanese krill trawlers and minke whales. In *Proceedings of the National Institute of Polar Research Symposium in* Polar Biology **3**, 36–56.

Ichii, T., Ishii, H. & Naganobu, M. (1992) Abundance size and maturity of krill (*Euphausia superba*) in the krill fishing ground of Subarea 48.1 during 1990/91 austral summer. *Scientific Committee for the Conservation of Antarctic Marine Living Resources, Selected Scientific Papers.* SC-CAMLR-SSP/9, pp. 183–99.

Ichii, T., Katayama, K., Obitsu, N., Ishii, H. & Naganobu, M. (1998). Occurrence of Antarctic krill (*Euphausia superba*) concentrations in the vicinity of the South Shetland Islands: relationship to environmental parameters. *Deep-Sea Res. I* **45**, 1235–62.

John, D.D. (1937) The southern species of the genus *Euphausia. Discovery Rep.* **14**, 195–324.

Jolly, G.M. & Hampton, I. (1990) Some problems in the statistical design and analysis of acoustic surveys to asses fish biomass. *Rapp. P.-v. Réun. Cons. Perm. Int. Explor. Mer* **189**, 415–20.

Kasatkina, S. M. & Latogursky, V.I. (1991) Distribution characteristics of krill aggregations in the fishing ground off Coronation Island in the 1989/90 season. *Scientific Committee for the Conservation of Antarctic Marine Living Resources, Selected Scientific Papers.* SC-CAMLR-SSP/7, pp. 49–73.

Kawaguchi, S. & Satake, M. (1994) Relationship between recruitment of the Antarctic krill and the degree of ice cover near the South Shetland Islands. *Fish. Sci.* **60**, 123–24.

Mackintosh, N.A. (1973) Distribution of post larval krill in the Antarctic. *Discovery Rep.* **36**, 95–156.

Marr, J.W.S. (1962) The natural history and geography of the Antarctic krill *(Euphausia superba* Dana). *Discovery Rep.* **32**, 33–464.

Maslennikov, V.V. & Solyankin, E.V. (1988) Patterns of fluctuations in the hydrological conditions of the Antarctic and their effect on the distribution of Antarctic krill. In *Antarctic Ocean and Resources Variability* D. Sahrhage, ed.), pp. 209–213. Springer-Verlag, Berlin.

Mathisen, O.A. & Macaulay, M.C. (1983) The morphological features of a super swarm of krill, *Euphausia superba. Memoirs of the National Institute of Polar Research*, Special Issue No 27, 153–64.

Mauchline, J.R. & Fisher, L.R. (1969) The biology of euphausiids. *Adv. Mar. Biol.* **7**, 1–454.

Murphy, E.J., Everson, I., Parkes, G. & Daunt, F.H.J. (1997) Detailed distribution of krill fishing around South Georgia. *CCAMLR Sci.* **4**, 1–17.

Murphy, E.J., Watkins, J.L., Reid, K., *et al.* (1998) Interannual variability of the South Georgia marine ecosystem: biological and physical sources of variation in the abundance of krill. *Fish. Oceanogr.* **7**, 381–90.

Pauly, T., Higginbottom, I., Nicol, S. & de la Mare, W. (1996) Results of a hydroacoustic survey of Antarctic krill populations in CCAMLR Division 58.4.1 carried out in January to April 1996. SC-CAMLR WG-EMM-96/28.

Pauly, T., Nicol, S., Higginbottom, I., Hosie, G. & Kitchener, J. (2000) Distribution and abundance of Antarctic krill (*Euphausia Superba*) off East Antarctic (80–150°E) during the Austral summer of 1995/96. *Deep-Sea Research* II. (In press).

Priddle, J., Croxall, J.P., Everson, I., *et al.* (1988) Large scale fluctuations in distribution and abundance of krill – a discussion of possible causes. In *Antarctic Ocean and Resources Variability* (D. Sahrhage, ed.), pp. 169–82. Springer-Verlag, Berlin.

Quetin, L.B., Ross, R.M., Frazer, T.K. & Haberman, K.L. (1996) Factors affecting distribution and abundance of zooplankton, with an emphasis on Antarctic Krill, *Euphausia superba*. In *Foundations for Ecological Research West of the Antarctic Peninsular*, Vol. 70, (R.M. Ross, E.E. Hofmann & L.B. Quetin, eds), pp. 357–71. American Geophysical Union, Washington, D.C.

Ross, R.M. & Quetin, L.B. (1991) Ecological physiology of larval euphausiids, *Euphausia superba* (Euphausiacea). Memoirs of the Queensland Museum **31**, 321–33.

SC-CAMLR (2000) Report of the Nineteenth Meeting of the Scientific Committee. *CCAMLR*, Hobart. (In preparation.)

Siegel, V. (1988) A concept of seasonal variation of krill (*Euphausia superba*) distribution and abundance west of the Antarctic Peninsula. In *Antarctic Ocean and Resources Variability* (D. Sahrhage, ed.), pp. 219–30. Springer-Verlag, Berlin.

Siegel, V., de la Mare, W. & Loeb, V. (1997) Long term monitoring of krill recruitment and abundance indices in the Elephant Island area (Antarctic Peninsula). *CCAMLR Sci.* **4**, 19–35.

Siegel, V., Loeb, V. & Groger, J. (1998) Krill (*Euphausia superba*) density, proportional and absolute recruitment and biomass in the Elephant Island region (Antarctic Peninsula) during the period 1977 to 1997. *Polar Biol.* **19**, 393–98.

Sushin, V.A. (1998) Distribution of the Soviet krill fishing fleet in the South Orkneys area (Subarea 48.2) during 1989/1990. *CCAMLR Sci.* **5**, 51–62.

Sushin, V.A. & Shulgovsky, K.E. (1999) Krill distribution in the western Atlantic sector of the Southern Ocean during 1983/84 and 1987/88 based on the results of Soviet mesoscale surveys conducted using an Isaacs-Kidd midwater trawl. *CCAMLR Sci.* **6**, 59–70.

Trathan, P.N., Everson, I., Murphy, E.J. & Parkes, G.B. (1998) Analysis of haul by haul data from the South Georgia krill fishery. *CCAMLR Sci.* **5**, 9–30.

Trathan, P.N., Agnew, D.J., Miller, D.G.M., Watkins, J.L., Everson, I., Thorley, M.R., Murphy, E.J., Murray, A.W.A. & Goss, C. (1992) Krill biomass in Area 48 and Area 58: recalculation of FIBEX data. In *Selected Scientific Papers, 1992* (SC-CAMLR-SSP/9) pp. 157–81. CCAMLR, Hobart, Australia.

Vagin, A.V., Makarov, R.R. & Menshinina, L.L. (1992) Dirunal variations in biological characteristics of krill *Euphausia superba* Dana to the west of the South Orkney Islands 24 March to 18 June 1990 – based on data reported by a biologist-observer. *Scientific Committee for the Conservation of Antarctic Marine Living Resources, Selected Scientific Papers.* SC-CAMLR-SSP/9, pp. 201–222.

Watkins, J.L. (1999) A composite recruitment index to describe interannual changes in the population structure of Antarctic krill at South Georgia. *CCAMLR Sci.* **6**, 71–84.

Chapter 4
Aggregation and Vertical Migration

Jon Watkins

4.1 Introduction

Patchiness, or heterogeneity in the distribution of marine organisms is extremely common across many spatial scales (Haury *et al.*, 1978; Murphy *et al.*, 1988; Mangel, 1994). In euphausiids patchiness is frequently manifested through the formation of aggregations and according to Mauchline (1980) it is likely that all species of euphausiids aggregate to some extent at certain times or locations. In species such as *Euphausia superba* some of the most extreme examples of aggregation occur with a substantial proportion of the population occurring in swarms where densities may reach thousands of krill m^{-3}. The formation of aggregations affects many aspects of the euphausiid life cycle. Thus feeding, and hence growth, are affected by the interaction between the patchy distributions of the krill and the food supply; similarly predation and hence mortality of euphausiids are influenced markedly by the aggregation or diurnal migration of the krill. The phenomenon of aggregation in euphausiids is also of great importance to commerical fisheries, for instance off Tasmania if *Nyctiphanes australis* do not swarm then the jack mackerel do not school in commercial quantities and that fishery fails (Harris *et al.*, 1992). As a result of these important interactions considerable work has been undertaken on the phenomenon of euphausiid aggregation over the years which has been summarised in a series of very detailed reviews; the first review discussing aggregation and diurnal migration for all species of euphausiids was produced by Mauchline & Fisher (1969) and this was then extended and updated by Mauchline (1980). In addition to these works dealing with euphausiids in general there have been reviews concentrating on *E. superba* (Miller and Hampton, 1989a; Siegel and Kalinowski, 1994). Most recently the Second International Krill Symposium held in Santa Cruz, USA, in August 1999 held a session on krill swarming and aggregation.

Acoustic records show that krill aggregations vary from swarms or schools that are dense and compact (only metres in any dimension) through to large layers, aggregations or deep scattering layers that may be kilometres in horizontal extent. The schools tend have a structure like fish schools with individuals showing parallel orientation within the school while in swarms the orientation of individuals is irregular (Ritz, 1994). Frequently the animals with the swarm or school are much more similar to each other than they are to krill in neighbouring swarms, so animals within a swarm may be of a very similar size or maturity stage or at the same stage in the moult cycle (Watkins *et al.*, 1992; Buchholz *et al.*, 1996). The densest aggregations are likely to be formed and maintained through social interactions (Ritz, 1994). These krill aggregations are dynamic and this is reflected in the dispersal or concentration of aggregations over various temporal scales. The aggregations are capable of directed movements which in the vertical plane are often expressed as a diurnal vertical migration, while large-scale horizontal movements have also been suggested (Kanda *et al.*, 1982). Previously diurnal migration and aggregation have been considered quite separately (Mauchline & Fisher, 1969; Mauchline, 1980; Miller & Hampton, 1989), indeed largely in different sections of these publications. However, as Ritz (1994) notes it is quite likely that an ultimate benefit of both swarming and diurnal vertical migration is protection from predation. Here, therefore, they are considered together as differing behavioural strategies influenced by the same biotic and abiotic factors and having similar costs and benefits.

A major focus of this book is considering the processes of ecosystem assessment, and here the focus is on aspects of aggregation and diurnal migration that are likely to be relevant to that process. Therefore a major aim is to describe patterns of aggregation and how these vary over different time and space scales. The comprehensive review on *Euphausia superba* by Miller & Hampton (1989a) contains considerable information on aggregation in krill and, in addition, highlighted priority issues for future work in the study of aggregations. Therefore how the study of aggregation and vertical migration has progressed in the last ten years is considered, particularly with reference to key issues of describing aggregations, the continuing problem of estimating density in aggregations and understanding the spatial and temporal scales of aggregation, and areas where further work needs to be carried out as a priority.

4.2 Physical and biological descriptions of krill aggregations: classification or not?

Aggregation of euphausiids occurs over a range of temporal and spatial scales, creating what is often considered to be a hierarchy of different kinds of aggregation, from the dense school at the scale of metres to the patch or the concentration at the scale of up to hundreds of kilometres (Murphy *et al.*, 1988). Our ability to sample these euphausiid aggregations over a range of temporal and spatial scales using a variety of sampling devices, such as acoustics, direct visual observations and nets (see Chapter 2), has led to various ways to summarise the vast amount of data. Frequently this has been done by producing a classification scheme for different types of aggregations (Mauchline, 1980; Kalinowski & Witek, 1985; Murphy *et al.*, 1988; Miller & Hampton, 1989a; Siegel & Kalinowski, 1994). The original scheme used by Mauchline (1980) was developed for mysids and euphausiids, and utilised physical parameters of size and density coupled with observations of internal structure, in particular the degree of parallel orientation between individuals within the aggregation. With the increasing use of acoustic observation techniques this scheme was modified by Kalinowski & Witek (1985) just to include parameters that could be determined from an echo-chart produced by an echo-sounder. Their swarm classification has been modified several times and the most recent manifestation is that given by Siegel & Kalinowski (1994) and reproduced here in Fig. 4.1. The scheme is hierarchical and typically there have been three levels of organisation corresponding to characteristic spatial (and hence temporal) scales which were originally defined by Haury *et al.* (1978). At the micro-scale (distances < 10 km) aggregations consist of swarms, layers and scattered forms, irregular forms, non-aggregated forms and super-swarms. Each of these different types has a different combination of size, shape and density (expressed as number of individuals or mass m^{-3}). Groups of swarms, layers or other micro-scale forms may be grouped together at the meso-scale (distances 10–1000 km) to form patches and similarly groups of patches may be grouped into a concentration of krill which covers spatial scales of more than 1000 km.

In the case of *E. superba*, swarms are the most frequent type of aggregation observed and the physical size of swarms from a number of studies has been summarised by Miller & Hampton (1989a, b), and Siegel & Kalinowski (1994). Typically aggregations classified as swarms are small, for example the mean horizontal extent of swarms recorded on nine summer cruises to the west of the Antarctic Peninsula varied between 54 and 104 m and the mean vertical extent between 11 and 17 m (Ross *et al.*, 1996).

The acoustic record is very good at providing information on the physical dimensions of krill swarms; however it cannot, as yet, provide any information on the biological composition of the individual krill in the swarm or the behaviour of the krill. Our information on these factors has come from two different kinds of studies; high resolution net sampling and direct *in situ* observations. High resolution net sampling in conjunction with acoustics has been used to obtain discrete samples from within individual swarms that were in close proximity (Ricketts *et al.*, 1992). This leads to the conclusion that the krill within a swarm are more similar than krill in adjacent swarms – thus it might be said that proximity does not imply similarity (Watkins, 1986; Watkins *et al.*, 1986; Fig. 4.2). The relationship between the biological characteristics of krill in swarms, that is whether the krill are large or small, moulting or not moulting, reproductively mature or immature, did not bear any relationship to the acoustic description of the swarms (Ricketts *et al.*, 1992). According to Ritz (1994) this is not surprising given that social behaviour is likely to be very important in determining what happens at the scale of individual krill swarms (see also Hamner, 1984). For instance the description of the physical and biological characteristics of krill swarms usually does not take account of the costs and benefits to individuals that are in a swarm. The calculation of such costs and benefits is not only complex but also very dynamic. In the case of searching for food, groups appear to be more successful at finding patchily distributed food than do

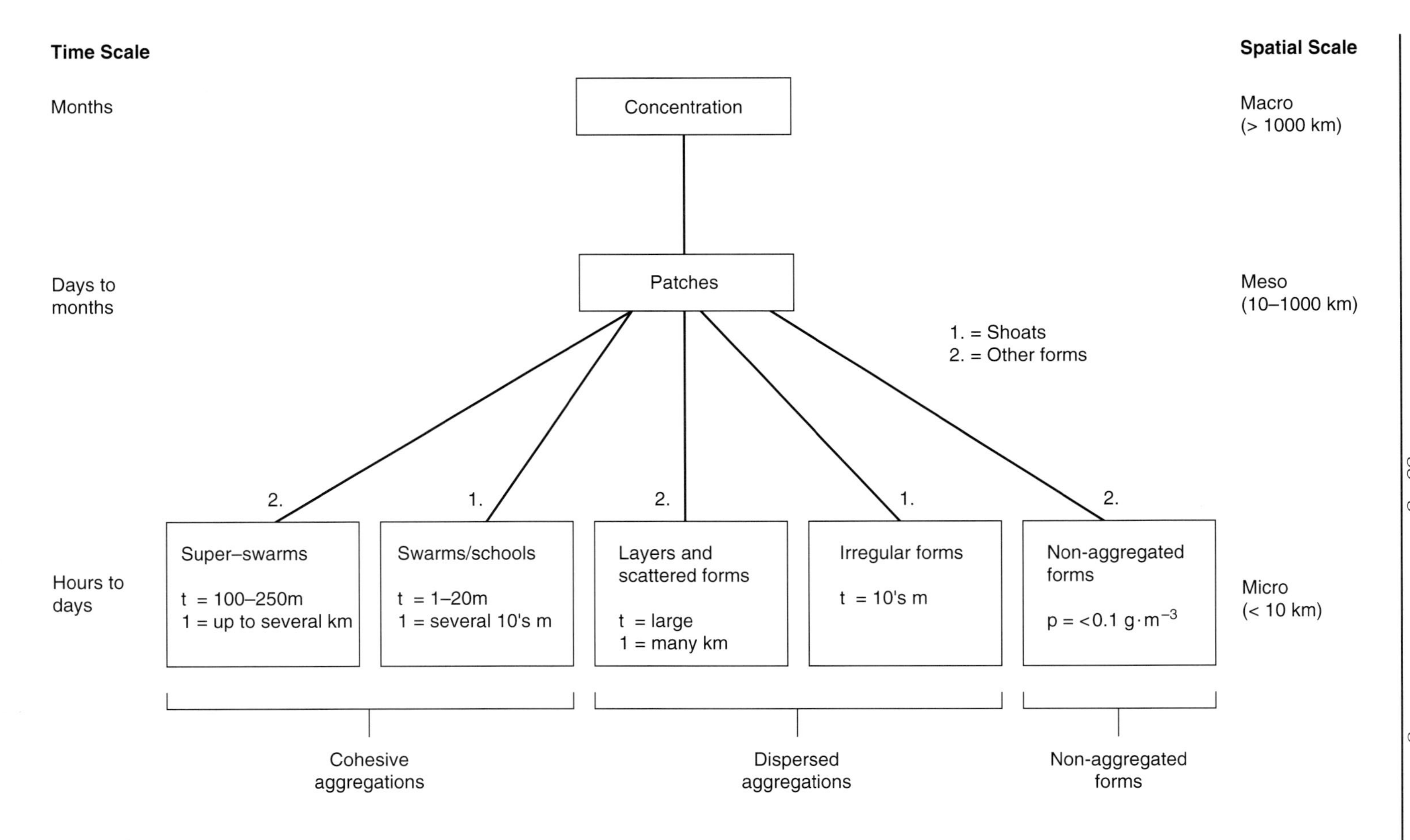

Fig. 4.1 Classification scheme for krill aggregations as modified by Siegel & Kalinowski (1994) from Kalinowski & Witek (1985) and Miller & Hampton (1989a). t = thickness, l = intercepted length, p = density. Reproduced with permission from Cambridge University Press.

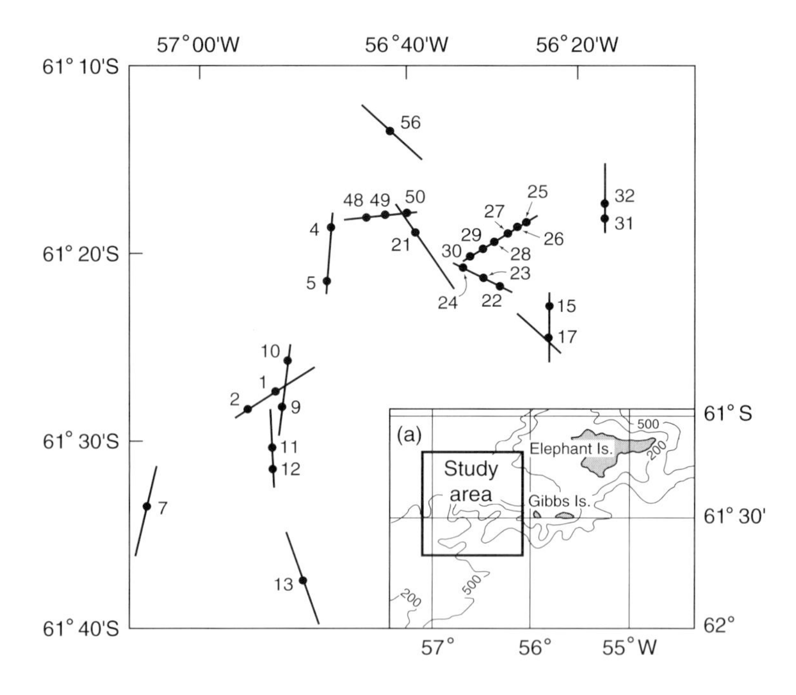

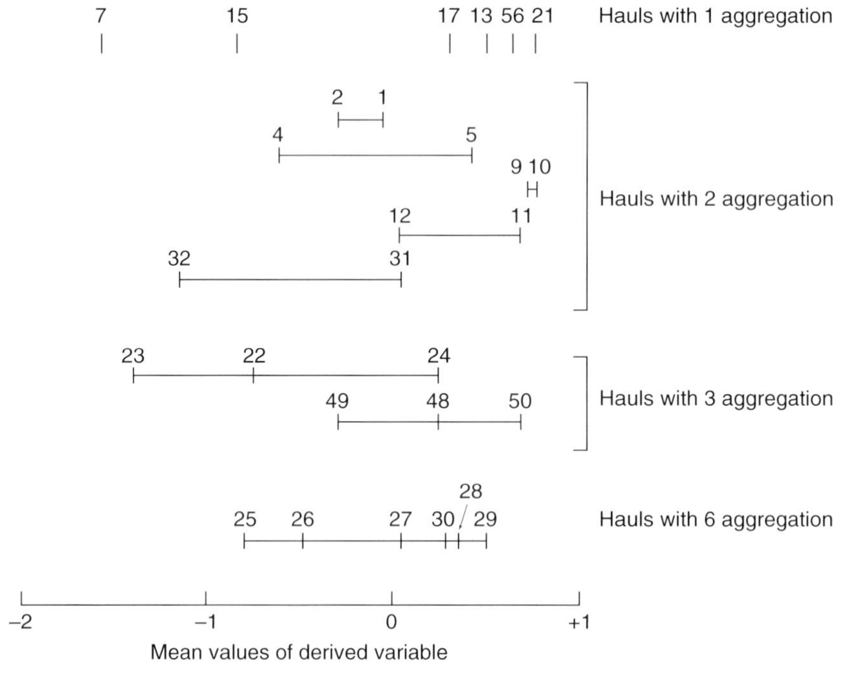

Fig. 4.2 Similarities and differences between 28 discrete swarms (shown as numbered dots) of *Euphausia superba* sampled during net hauls (shown as lines) near to Elephant Island, Antarctica. The degree of similarity between swarms is summarised using a discriminant function variate to combine variation due to interswarm differences in krill length, sex-ratio, maturity stage, alimentary tract fullness and exoskeleton hardness. Swarms that occur together in the same haul (connected with a line) are frequently more different from each other than from swarms that occur in different hauls. Modified with permission from Springer-Verlag, Watkins *et al.* (1986).

individuals (Pitcher *et al.*, 1982; Grunbaum, 1998). Therefore it is advantageous to swarm to find food but when food becomes limited and the swarm size is large the returns per individual are probably reduced and so it might be better for an individual to leave the swarm. However, if the swarm is threatened by a predator the best strategy may be for the swarm to remain as large as possible even if the food returns *per capita* decline (Ritz, 1994). Such trade-offs between individual and aggregation become even more complex in the case of some euphausiids where swarming as a predator response does not provide equal protection against all predators. Indeed in some cases predation, for instance by whales, can only take place because krill swarm (Brodie *et al.*, 1978; Murphy, 1995).

Information on the composition of krill swarms has also come from studies made by divers; in this case there have been observations that the krill are of a similar size but, more importantly has been the observation that in a swarm the individual krill were all swimming in the same direction, i.e. the swarm is polarised or shows parallel orientation (Hamner *et al.*, 1983; Hamner, 1984; Stretch *et al.*, 1988; Strand & Hamner, 1990; O'Brien, 1987; Fig. 4.3). These observations of the biological and behavioural properties of swarms of *E. superba* have lead to the suggestion at various times that the krill swarm should be considered as the basic unit of organisation in krill ecology (Watkins, 1986; Murphy *et al.*, 1988).

While the acoustic and direct observations seem to be to a large extent in

Fig. 4.3 Antarctic krill (*Euphausia superba*) in a swarm occurring on the northern shelf of South Georgia. Photograph taken with Camera Alive remotely controlled underwater camera by author.

agreement about the size and importance of krill swarms, the classification of other types of aggregation is more problematic. Thus although acoustically we may see features that are described as layers, irregular forms and superswarms (Kalinowski & Witek, 1985), the distinctions between such forms are often very subjective. Far less is known about the structural integrity of such features and it is much harder to relate such large aggregations with the understanding of small krill swarms controlled by interactions between individuals. Recent analysis of the biological properties and acoustic structure of some extensive krill layers now sheds some light on these discrepancies. Thus Watkins & Murray (1998) report on the characteristics of *Euphausia superba* in diffuse layers detected at night (see Fig. 4.4). These layers have considerable physical structure with variations in depth and thickness of the layer, indeed in some areas there appear to be minor gaps in the layer. The analysis of the type of krill within the layer using a multi-serial sampler that sampled more than 60 discrete samples shows that the variation in krill length, maturity and sex-ratio was as great as between discrete krill swarms that were detected within the same area during the day time. This indicates that the layer could well be composed of a series of krill swarms that had been brought so close together that they were touching but that there had been little mixing. In this case it therefore still seems valid to consider that these layers can be considered as a group of independent swarms that maintain their integrity. Little information is available on the internal biological structure of some of the other large-scale krill aggregations and it would appear that no layers and superswarms have been observed by divers. Perhaps the main reason for this is the relative rarity of such aggregations. Further information on the structure of these layers is required and should be a priority for small-scale and *in situ* observations to find out whether the heterogeneity within the layers observed here is a common phenomenon.

As suggested above, the attempts to classify different types of aggregations can lead to problems of interpretation. Indeed Miller & Hampton (1989a,b) point out that the size range of *Euphausia superba* swarms is very large (see also Miller *et al.*, 1993) and this may be due to inclusion of some swarms that perhaps should have been classified as layers. With the increasing availability of more computer intensive analysis techniques it is suggested that we should worry less about swarm classification and rather put more time and effort into descriptive techniques that can take account of large amounts of variation in spatial structure; examples that could prove fruitful for future research include the fractal dimensions analysis undertaken by Krause (1998) and also the increasing use of geostatistics (Murray, 1996).

Fig. 4.4 *(Opposite.)* The structure of a diffuse layer of *Euphausia superba* investigated using acoustics and a multi-sample LLHPR (see Chapter 2). The number of krill taken in each sample can be seen in relation to the depth of the net and the acoustic structure of the layer. The histograms show the variation in mean length of krill and percentage of male krill in each sample. Layer 1 is the layer shown at the top of the figure. Layer 2 was sampled on a separate night. The variability in length and percentage of male krill within a single layer is as great as the variability occurring between 28 discrete swarms sampled in the same area (see Fig. 4.2). Modified with permission from Springer-Verlag, Watkins & Murray (1998).

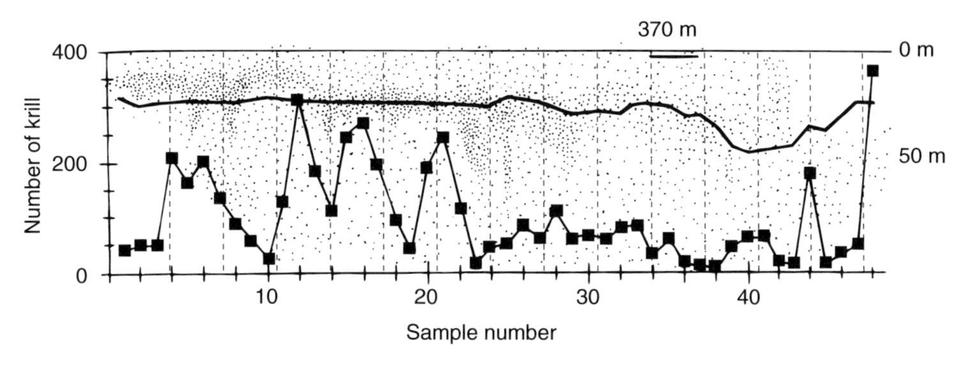

370 m
0 m
50 m
Number of krill
Sample number

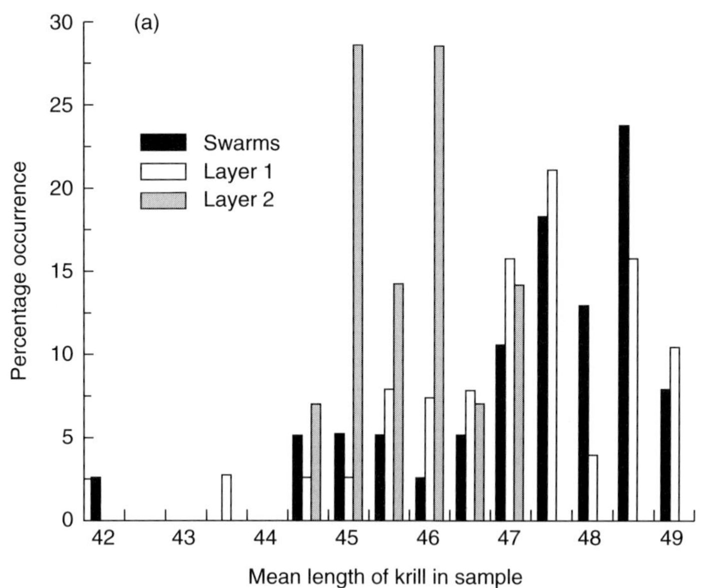

(a)
Swarms
Layer 1
Layer 2
Percentage occurrence
Mean length of krill in sample

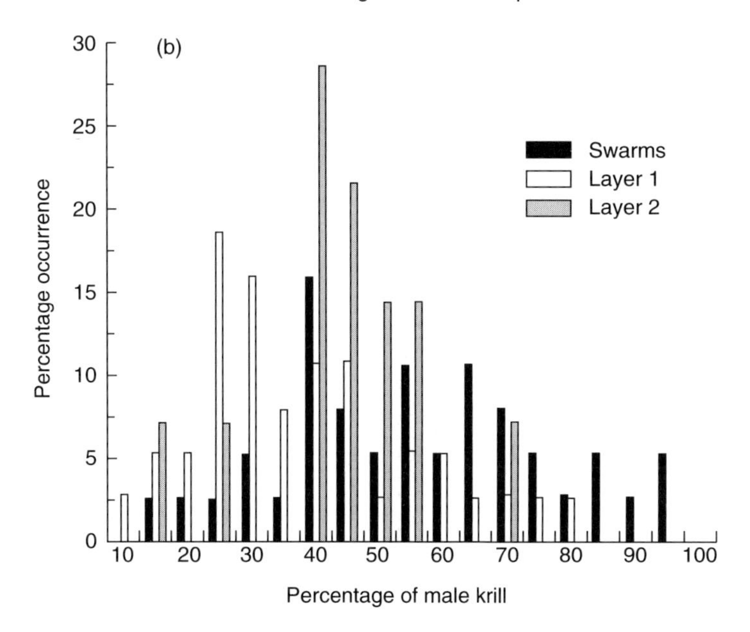

(b)
Swarms
Layer 1
Layer 2
Percentage occurrence
Percentage of male krill

The behavioural observations of parallel orientation between individuals and concerted predator avoidance behaviour (Hamner *et al.*, 1983; Hamner, 1984; Ritz, 1994) indicate that individuals in these small swarms act in unison. If this is so, then we might ask what is the maximum size of krill swarm that can exist and still be considered as a single unit? Ritz (1994) discusses the idea that there is an optimal group size for an aggregation but that the potential benefits to individuals wishing to join the group may push the size of the aggregation beyond the optimal size. If this is the case then can the krill control their group size (Levin, 1994)? Individual krill may be aware of the density of surrounding krill from information on nearest neighbour distance but beyond some critical size of swarm may have no control over the total group size (Levin, 1994). How does this critical size relate to the size of small swarms which are seen so frequently and to what extent do such basic swarm units then interact in larger aggregations? There seems to be a good case for bringing the fields of behavioural observations and remotely observed phenomenon together to increase our understanding of the way in which swarming works.

4.3 Aggregation density: differences and extremes

True densities in swarms

One of the key physical parameters used to describe krill aggregations is the density (usually expressed as number or mass of individual m^{-3}). In Table 4.1 a compilation of the estimates of density of krill swarms using acoustic and visual

Table 4.1 Estimates of krill density in swarms.

Species	Density	Collection method	References
Euphausia superba	20 000–60 000 n m^{-3}	scuba observations	Ragulin, 1969, Hamner *et al.*, 1983
Euphausia vallentini	70 000–80 000 n m^{-3}	scuba observations	Ragulin, 1969
Meganyctiphanes norvegica	9000–770 000 n m^{-3}	visual observations and bag sampler	Nicol, 1986
Nyctiphanes australis	3000–480 000 n m^{-3}	visual observations	O'Brien, 1988
Euphausia pacifica	10–72 000 n m^{-3}	visual observations	Hanamura *et al.*, 1984
Euphausia lucens	4000–5 000 000 n m^{-3}	visual observations	Nicol *et al.*, 1987
Euphausia superba	150–500 g m^{-3} (latest TS used)	acoustic survey	Lascara *et al.*, 1999
Euphausia superba	300–800 g m^{-3} (BIOMASS TS)	acoustic survey	Macaulay *et al.*, 1984
Euphausia superba	1534 g m^{-3} (BIOMASS TS)	acoustic survey	Higginbottom & Hosie, 1989

techniques has been produced; net haul estimates have been excluded from this table because of the effects of net avoidance (see Chapter 2). Visual estimates of *Euphausia superba* swarms may be up to 20 000 krill (Hamner *et al.*, 1983) or even as high as 60 000 krill m^{-3} while the highest published acoustic densities for Antarctic krill are in the region of 1500 krill m^{-3} (Higginbottom & Hosie, 1989). Densities from Ritz (1994), summarising visual and photographic density estimates for other species of euphausiid are even higher (Table 4.1), for instance up to 480 000 krill m^{-3} for *Nyctiphanes australis* and as high as 5 000 000 krill m^{-3} for *Euphausia lucens*.

While there is considerable variety in density estimates, however, it is generally acknowledged (Miller & Hampton, 1989a; Siegel & Kalinowski, 1994) that visual or photographically based estimates are considerably higher than acoustic estimates. From the data presented in Table 4.1 this difference may be as much as two orders of magnitude. Such differences are likely to be due to a combination of behavioural and methodological differences. Some of these differences may be explained by the considerable differences in the degree of dispersion that has been observed over a diurnal cycle (see for instance Everson, 1982). However some of these differences are undoubtably due to methodological effects. Accurate acoustic density estimates depend upon precise calibration and a realistic Target Strength (TS) relationship (see Chapter 2 for further details). The acoustic techniques have evolved considerably over the two decades in which observations on Antarctic krill have been collected and the present day acoustic data are generally exhaustively calibrated and validated (Chapter 2, Watkins & Murray, 1998; Foote *et al.*, 1987). However, the re-evaluation of TS in the early 1990s has had a major change on the calculation of krill biomass (Everson *et al.*, 1990; Foote *et al.*, 1990), thus all of the acoustic swarm density estimates quoted by Miller & Hampton (1989a, b) and many of those quoted by Siegel & Kalinowski (1994) use the TS derived during the BIOMASS Programme. This TS relationship underestimates the density by as much as an order of magnitude (Everson *et al.*, 1990; Trathan *et al.*, 1995).

Recent estimates of swarm density have been calculated using the most up-to-date TS relationships (Lascara *et al.*, 1999) but these density estimates are still low in comparison to the diver observed densities (Table 4.1). However, note that the highest published density estimate of the swarm observed acoustically by Higginbottom & Hosie (1989) was also calculated using the BIOMASS TS relationship; assuming that the more recent TS would increase the density fivefold results in a density estimate in the region of 9000 krill m^{-3} (also assuming that the weight of an individual krill is 1 g). Such a value is much closer to the visual density estimates. The accuracy of the diver density estimates is harder to determine; Ritz (1994) states that 'visual estimates and those made from plankton net catches are notoriously unreliable'. Visual and photographic density estimates are usually calculated using a packing model which requires the measurement of nearest neighbour distance (NND – the distance between adjacent krill). There is little information on precision or bias of estimates of NND (Watkins & Murray, 1998), although O'Brien (1989) indicates that some packing models may overestimate density by 20%. It is maintained therefore that there is still considerable work to be done to achieve a consensus on the range of density estimates that are found in krill swarms and aggregations, and studies using several independent methods of estimating krill swarm densities should be carried out.

The importance of 'huge' aggregations

The extensive acoustic data sets collected over the last 30 years (see for instance Miller & Hampton, 1989a, b; Miller *et al.*, 1993; Ross *et al.*, 1996; Lascara *et al.*, 1999) have sampled many thousands of krill aggregations and confirm, at least for *Euphausia superba,* that the majority of krill aggregations occur as small, discrete swarms (Miller & Hampton, 1989a, b; Watkins, 1986). The larger aggregations of *Euphausia superba* which occur as layers or irregular forms tend to be structures with a density substantially lower than the small swarms (Kalinowski & Witek, 1985; Watkins & Murray, 1998). However, now there does seem to be accumulating evidence that occasionally krill are found in aggregations that are both very extensive and dense. Such huge aggregations would appear to be relatively rare, for instance Higginbottom & Hosie (1989) carried out an acoustic survey of the Prydz Bay region in Antarctica which covered 1.3 million km and describe the occurrence of a large, dense aggregation with an estimated area of 5.7 km^2 (Table 4.2). This aggregation was two orders of magnitude more dense than any other aggregation observed in a total of six previous cruises carried out by the Australian Antarctic Division. Another dense and even more extensive superswarm of *Euphausia superba* off Elephant Island has been described in considerable detail (Mathisen &

Table 4.2 Characteristics of large aggregations of *Euphausia superba*.

Reference	Location	Horizontal area (km^2)	Vertical extent (m)	Average (max) density (g m^{-3})	Total biomass (tonnes)
Higginbottom & Hosie, 1989	Prydz Bay	5.7	60–80	94.7 (1534)*	57 000*
Macaulay *et al.*, 1984	Elephant Island	150	up to 250 m	300–500 (up to 800)*	> 1.2* 10^6*
Murray *et al.*, 1995	Bellingshausen Sea	2.8 (distance not area)	NA	70 (–)	37 000
Watkins, Murray & Brierley, unpublished	South Georgia	0.9 (distance not area)	50	230 (–)	145 000

* Note these calculations are based on the BIOMASS Target Strength relationship, use of the more recent CCAMLR TS relationship would increase the values by a factor of between 5 and 10.

Macaulay, 1983; Macaulay *et al.,* 1984) (Table 4.2). The occurrence of such large aggregations does not appear to be restricted only to *Euphausia superba*, thus Simard & Lavoie (1999) describe a large aggregation of *Meganyctiphanes norvegica* and *Thysanoessa rachi* at the head of the Gulf of St. Lawrence.

These aggregations contain substantial quantities of krill. Indeed the single dense swarm observed by Watkins, Murray & Brierley (BAS unpublished data) in 1997 around South Georgia accounted for more than 12% of the total biomass in an area of 80 000 km^2. Analyses of acoustic survey data in the shelf waters of the Antarctic Peninsula, showed that more than 80% of the krill biomass was associated with a small number of aggregations (less than 20% of the total) that were two- to four-fold larger and an order of magnitude more dense than average (Lascara *et al.*, 1999). The occurrence of such aggregations raises a number of important issues with regard to survey design and analysis, aggregation formation and implications for the functioning of the ecosystem.

In terms of ecosystem assessment, such aggregations create significant problems for the design of surveys to calculate biomass. The area occupied by such swarms is still small in relation to the spacing between transects frequently used by acoustic surveys, which may vary from about 10 km (Brierley *et al.*, 1997) to more than 100 km (Lascara *et al.*, 1999; Pauly *et al.*, 1996). Therefore the probability of encountering such aggregations would appear to be quite small. When encountered within a survey such large concentrations of biomass also cause significant problems for the assessment of variance of the biomass estimate and probably need to be treated separately from the usual transect data (Macaulay *et al.*, 1984; Murray, 1996).

How do such large aggregations arise? Because of their relative rarity we know little about the conditions under which such swarms form. Is this simply through the amalgamation of smaller swarms as was suggested for the low density layer described by Watkins & Murray (1998) or is it necessary for physical concentration of swarms into an area to occur as suggested by Witek *et al.* (1988). Interactions with the physical environment may be important in the formation of at least three of the huge aggregations described here. The Elephant Island superswarm was found on the shelf break (Macaulay *et al.*, 1984) while the Prydz Bay swarm was associated with a small seamount (Higginbottom & Hosie, 1989). The aggregation of euphausiids along the northern edge of the Laurentian Channel in the St. Lawrence Estuary may have resulted from the interaction between the negative phototaxis of the euphausiids and upwelling of the water circulation in the area (Simard *et al.*, 1986). Social interaction between krill is very important in the formation and maintenance of small swarms (Ritz, 1994) but both the role in these larger aggregations together with the possible role of predators on the formation of such large aggregations need to assessed. The dense swarm off South Georgia described above was seen in association with southern right whales and many fur seals (E.J. Murphy, pers. comm.). It may have been that the large, high density swarm was produced as a result of predator avoidance behaviour by the krill or herding behaviour by the whales (Hamner *et al.*, 1988).

In addition such aggregations are likely to have a great impact on predator prey interactions. Thus for instance in the Bransfield Strait, Antarctica, during a 20-day cruise over 60% of all the Cape Petrels and Antarctic Fulmars were seen at one krill aggregation and more than 75% of all Adelie Penguins were seen at another large krill aggregation (Heinemann *et al.*, 1989; Hunt, 1991). However, the duration and subsequent fate of such aggregations is not known. Therefore although such swarms may be very important when they occur, if such swarms only occur for a few days at a time and are relatively infrequent then their importance to predators over a breeding cycle may be far less important.

Given the importance to our understanding of krill biology, in particular the relative roles of physical and biological forces in the formation and maintenance of such aggregations, and also for assessing the status of the ecosystem it is vital that studies on these large aggregations should be undertaken whenever they are encountered as a matter of high priority.

4.4 Temporal and spatial structure

Diurnal time-scales

Much has already been written on the diurnal variation in the behaviour of krill aggregations (Mauchline & Fisher, 1969; Mauchline, 1980; Siegel & Kalinowski, 1994; Everson, 1982; Everson, 1983). More than 75% of all euphausiid species undertake some form of diurnal vertical migration (Mauchline, 1980) and this has been extensively documented for species such as *Meganyctiphanes norvegica* (see Onsrud & Kaartvedt, 1998; Buchholz *et al.*, 1995; Tarling *et al.*, 1998). In *Euphausia superba* krill commonly show a diurnal vertical migration towards the surface at night which may be linked to dispersal of the swarms at this time (see also Demer & Hewitt, 1995; Ross *et al.*, 1996; Godlewska, 1996). However, there can be considerable variation from this basic pattern. For instance krill swarms may remain as distinct swarms throughout the day and night (Watkins *et al.*, 1992) or swarms may be found at the surface during the daytime (Marr, 1962). It would appear that all krill that swarm also vertically migrate at least to some extent (Mauchline, 1980), and it is likely that both diurnal migration and swarming are linked to feeding or predator defence mechanisms (Ritz, 1994). It is therefore interesting to consider why these species exhibit both strategies; Ritz (1994) speculates that social (i.e. swarming) euphausiids might show different vertical responses to non-social ones and suggests that diurnal vertical migration in the facultative schoolers might be different when dispersed compared to while schooling all the time. However, it is not clear that this is occurring in the case of the seasonal differences in swarming and diurnal vertical migration presented by Ross *et al.* (1996) (see next section). It is likely that the complex and variable patterns of behaviour shown by swarms of *Euphausia superba* are an interplay between the various advantages and disadvantages resulting from aggregation and vertical migration, and that such balances change under differing environmental conditions and predation pressure. To date there has been no attempt to produce an integrated synthesis of diurnal vertical migration and swarming; such an undertaking, although daunting, is to be encouraged.

Godlewska (1996) has produced a very comprehensive synthesis of diurnal vertical migration in *Euphausia superba* using both data collected during Polish Antarctic cruises and also analysing much of the published literature. Vertical migration has been described by a model which combines the amplitudes of migration detected with a 12- and 24-hour periodicity. This analysis shows that the amplitude of vertical migration in *Euphausia superba* varies from only a few to 30 m. Such migrations are very small in comparison with distances of 400+ m described for tropical euphausiids (Andersen *et al.*, 1997). Food availability appears to be one of the most important factors affecting the extent of the migration amplitude; Godlewska (1996) found that under good feeding conditions the amplitude is large and duration is dominated by a 24-hour periodicity. As food becomes scarce, the amplitude decreases and the migration cycle has a greater component of the 12-hour periodicity (Godlewska, 1996). Once again, however, the complex interactions between the underlying causes and expression of diurnal migration are illustrated by the observation that the amplitude of vertical migration in euphausiid species at meso- and eutrophic stations in the north-east tropical Atlantic was significantly less than that observed in a low production oligotrophic station (Andersen *et al.*, 1997).

The role of light in initiating diurnal vertical migration has long been recognised (see Ringelberg, 1995). Godlewska (1996) confirms that in *Euphausia superba*, changes in light intensity rather than the actual light level appear to trigger the migration process. Other biotic and abiotic factors then serve to modify the extent of the migration. Such a process also appears to control the migrations occurring in other species. Thus Onsrud & Kaartvedt (1998) discuss the role of the physical environment, food and predators in modifying the diurnal vertical migration of *Meganyctiphanes norvegica*. Their results indicated that *M. norvegica* migrated irrespective of the level of phytoplankton or the presence of copepod prey at the daytime depths but that night-time migration into the surface layer was restricted when planktivorous fish were most abundant at the surface. Interestingly Onsrud & Kaartvedt (1998) also note that the migrating scattering layer appeared to follow a characteristic isolume during the ascent towards the surface which would appear to be unusual (cf. Ringelberg, 1995; Roe, 1983). The modifying effects of the physical environment on vertical migration may have profound implications for the distribution of euphausiids. In Gullsmarsfjorden (Sweden) *M. norvegica* did not migrate through a pycnocline at 50–60 m while *Thysanoessa raschii* did migrate up into the water above the sill (Bergstrom & Stromberg, 1997). Such differences in behaviour may have a marked effect on the horizontal distribution of species (see Kaartvedt, 1993). In addition to direct reactions to light level or change in light intensity, it is possible that migration is cued by an endogenous rhythm. The most recent evidence pointing to this comes from investigation of the effect of a lunar eclipse on the diurnal vertical migration in *M. norvegica* (Tarling *et al.*, 1999). In this study the secondary midnight sinking frequently observed in euphausiids was linked to the time of moon rise, but did not occur on the evening of a lunar eclipse when the moon effectively did not rise. Behaviour later in the month, when midnight sinking occurred at a relatively constant time and prior to moon rise, was attributed to an endogenous rhythm which was synchronised by the external cue.

Most of our information on diurnal migration and swarming is collected by observing a number of swarms for short periods of time; in the case of acoustic observations it is usually the time taken to steam over an individual aggregation, while in the case of diver observations the swarm is only likely to be observed for a matter of minutes. Consequently our knowledge of swarm dynamics is built up from a series of snapshots of many different swarms rather than an understanding of how individual swarms behave over longer periods of time. Many of the questions about how krill swarms form, maintain themselves and interact with the environment at small scales will probably only be resolved once extended observations on single swarms are carried out. The increased availability of non-invasive observation techniques (see Chapter 2 for more details) is likely to lead to substantial advances in this field. For instance, the use of ADCPs to measure the actual migration rates of individual scattering layers or swarms is becoming increasingly common (see for instance Tarling *et al.*, 1998; Tarling *et al.*, 1999).

Seasonal variation in aggregation and diurnal vertical migration

Most studies of aggregation and diurnal migration have been one-off studies and any information about seasonal changes has come from the comparison of cruises carried out to different protocols. Recently however, two major studies have been published where exactly the same methods and protocols have been used over a number of seasons (Ross *et al.*, 1996; Lascara *et al.*, 1999). Ross *et al.* (1996) carried out 12 cruises over the same acoustic survey grid west of the Antarctic Peninsula around Anvers Island and within the Bransfield Strait from February 1985 to December 1988. They observed aggregations of *Euphausia superba* throughout the year but there were seasonal differences in the number of aggregations, the total cross-sectional area of aggregations per kilometre and the size of the aggregations. During the spring and summer an increase in the total cross-sectional area per unit distance was attributed to the increase in the number of swarms. Maximum total cross-sectional area per unit distance occurred in the winter when the actual number of swarms seen was low but the size of the aggregations was very high (Fig. 4.5). The results of spring, summer, autumn and winter cruises using another survey grid west of the Antarctic Peninsula between 1991 and 1993 (Lascara *et al.*, 1999) revealed a cyclical change in aggregation dimensions and density similar to those observed by Ross *et al.* (1996). In the spring aggregations typically had cross-sectional areas between 700 and 2200 m^2 and mean biomass densities between 20 and 100 $g\ m^{-3}$; the

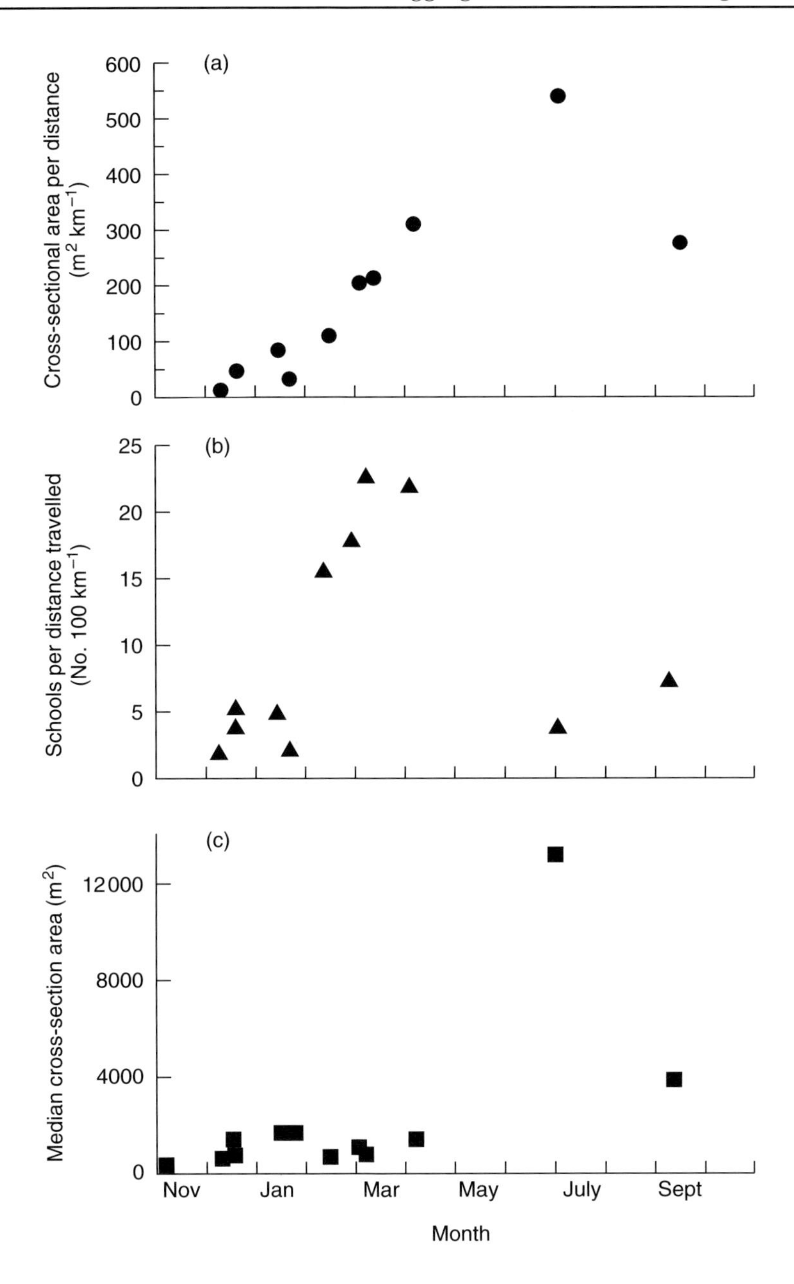

Fig. 4.5 Seasonal variation in Antarctic krill aggregation parameters collected on 12 Antarctic cruises to the west of the Antarctic Peninsula. Aggregation statistics for each cruise have been normalised to the total distance travelled. (a) vertical cross-sectional area (m^2) detected per km; (b) number of aggregations detected per km; (c) median cross-sectional area of all aggregations within a cruise. Figure from Ross *et al.* (1996).

smallest densest aggregations were found in the summer (area < 1000 m^2; mean biomass density 150–500 g m^{-3}), and then aggregations became larger and less dense through the autumn and by winter were characterised by aggregations with cross-sectional areas > 7000 m^2 and mean biomass densities < 10 g m^{-3}.

In these two studies there were seasonal differences in the depth distribution of swarms but the pattern observed was slightly different. Lascara *et al.* (1999) detected a seasonal shift in the depth distribution of the krill swarms. In the summer krill swarms were concentrated in the top 70 m of the water column while in the winter a substantial proportion of the swarms were found below 100 m. In the spring (and to a lesser extent during the autumn) swarms were still found in the top 70 m but the range of depths over which the swarms were found was greater. In contrast Ross *et al.* (1996) found that the overall depth distribution of the swarms did not change throughout the year; however, the pattern of diurnal vertical migration changed, with krill showing a marked diurnal migration only during the spring and autumn (Fig. 4.6). Seasonal patterns in diurnal vertical migration of other euphausiids have also been described. For instance in the Kattegat the timing of night-time dispersal and migration of *Meganyctiphanes norvegica* varied with day length, the period of dispersion being longer in the winter (Tarling *et al.*, 1998).

Spatial structure

For many observers krill swarms are likely to be seen only in two dimensions. The acoustic techniques in common use produce a vertical two-dimensional slice through the swarms (see for instance Everson, 1982; Everson, 1987). Similarly the observation of surface swarms produces a two-dimensional plan view of the shape (Nicol, 1986). Surface swarms are frequently very irregular in shape (Marr, 1962) and so any assumptions about the shape of a swarm in the unobserved third dimension may be very misleading. This is supported by the relatively limited number of observations of swarms detected by divers which reveal that frequently even the large and dense krill swarms are narrow in one dimension (Hamner, 1984). While knowledge of the shape of swarms may not be needed for biomass estimates using classical sampling methodology, the ability to build up a true three-dimensional picture of krill swarms would be immensely valuable to increase our understanding of small-scale spatial structure of krill swarms. Multi-beam sonar systems are now being used to make 3D observations on fish schools (Misund *et al.*, 1995; Gerlotto *et al.*, 1999) and may provide a suitable way of observing krill swarms in the near future.

4.5 Important processes in inaccessible regions

The frequent utilisation of acoustic technology has resulted in an avalanche of detailed information on aggregation characteristics over large areas of the ocean. However, there are a number of areas that are relatively inaccessible to the com-

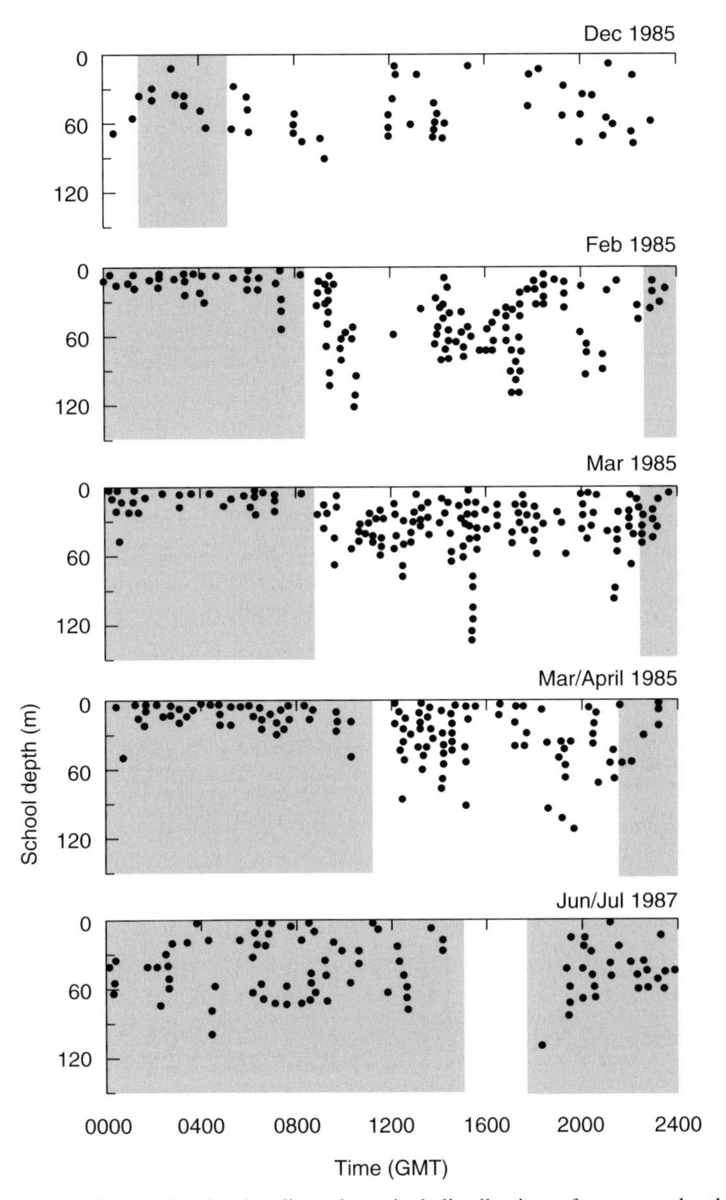

Fig. 4.6 Seasonal variation in the diurnal vertical distribution of average depths of swarms of *Euphausia superba* from 5 cruises west of the Antarctic Peninsula. The shaded area is from sunset to sunrise for 64°S. Figure from Ross *et al.* (1996).

monly used acoustic instruments. Two that are of considerable importance for understanding interactions within the ecosystem are the under-ice environment inhabited by *Euphausia superba* and the sea surface which may be inhabited by many euphausiid species during regular excursions to the surface at night or by euhausiids swarming at the surface during the day (Komaki, 1967; Nicol, 1984).

It is immediately obvious why the under-ice environment is difficult to sample; observations made by divers (O'Brien, 1987) and by ROVs (Bergstrom *et al.*, 1990) are time-consuming and generally limited in extent. Penetration by ships is costly, time-consuming and disruptive of the very environment to be studied. However, a significant proportion of the population of Antarctic krill spend part of their life cycle under the ice (Siegel & Loeb, 1995) and so there is a requirement to understand how krill swarms interact with the ice environment. The limited information available to date suggests that there may be quite different behavioural strategies. Thus krill may be dispersed and found in close association with the sea-ice (Marschall, 1988) while at other times krill are found in dense swarms well below the bottom of the ice (Daly & Macaulay, 1991). Once again it is likely that the actual behaviour patterns observed are due to the dynamic balance between predation risks and feeding requirements. The degree of interaction of krill with the ice environment is likely to have significant effects on both the survival and large-scale distribution of krill (Murphy *et al.*, 1998). It is therefore very important that further investigations on krill in relation to ice should be carried out. It may be less obvious why the surface layer should pose such a sampling problem, especially when we consider that much of the original information on swarming came from the observations of surface swarms (Hardy, 1936; Marr, 1962). In this case, however, the fact that transducers of acoustic systems are traditionally mounted in the bottom of the ship's hull (probably at least 5 m below the surface) means that often the only way to observe surface swarms is visually. This again is time-consuming and frequently only carried out in specialised studies. Daytime surface swarming has been described in a variety of species and it has been attributed to a number of causes related to feeding, reproduction and predator avoidance (Komaki, 1967). Such swarming not only affects regular biomass estimates made as part of ecosystem assessments, it also has a significant effect on the interactions between krill and predators.

Future research on krill swarming needs to address the processes occurring in both these relatively inaccessible regions. One possible area for development is the use of Autonomous Underwater Vehicles (AUVs) which may be used to collect information on swarms over larger spatial scales in a non-disruptive manner (Fernandes *et al.*, 2000).

References

Andersen, V., Sardou, J. & Gasser, B. (1997) Macroplankton and micronekton in the northeast tropical Atlantic: abundance, community composition and vertical distribution in relation to different trophic environments. *Deep-Sea Res. I* **44**, 193–222.

Bergstrom, B. & Stromberg, J.O. (1997) Behavioural differences in relation to pycnoclines during vertical migration of the euphausiids *Meganyctiphanes norvegica* (M. Sars) and *Thysanoessa raschii* (M. Sars). *J. Plankton Res.* **19**, 255–61.

Bergstrom, B.O., Hempel, G., Marschall, H.-P., North, A.W., Siegel, V. & Stromberg, J.-O.

(1990) Spring distribution, size composition and behaviour of krill *Euphausia superba* in the western Weddell Sea. *Polar Rec.* **26**, 85–89.

Brierley, A.S., Watkins, J.L. & Murray, A.W.A. (1997) Interannual variability in krill abundance at South Georgia. *Mar. Ecol. Prog. Ser.* **150**, 87–98.

Brodie, P.F., Sameoto, D.D. & Sheldon, R.W. (1978) Population densities of euphausiids off Nova Scotia as indicated by net samples, whale stomach contents, and sonar. *Limnol. Oceanogr.* **23**, 1264–67.

Buchholz, F., Buchholz, C., Reppin, J. & Fischer, J. (1995) Diel vertical migrations of *Meganyctiphanes norvegica* in the Kattegat – comparison of net catches and measurements with acoustic Doppler current profilers. *Helgolander Meeresuntersuchungen* **49**, 849–66.

Buchholz, F., Watkins, J.L., Priddle, J., Morris, D.J. & Ricketts, C. (1996) Moult in relation to some aspects of reproduction and growth in swarms of Antarctic krill, *Euphausia superba. Mar. Biol.* **127**, 201–208.

Daly, K.L. & Macaulay, M.C. (1991) Influence of physical and biological mesoscale dynamics on the seasonal distribution and behavior of *Euphausia superba* in the Antarctic marginal ice zone. *Mar. Ecol. Prog. Ser.* **79**, 37–66.

Demer, D.A. & Hewitt, R.P. (1995) Bias in acoustic biomass estimates of *Euphausia superba* due to diel vertical migration. *Deep-Sea Res. I* **42**, 455–75.

Everson, I. (1982) Diurnal variation in mean volume backscattering strength of an Antarctic krill (*Euphausia superba*) patch. *J. Plankton Res.* **4**, 155–61.

Everson, I. (1983) Variations in vertical distribution and density of krill swarms in the vicinity of South Georgia. *Mem. Natl Inst. Polar Res.*, Spec. Issue 27, 84–92.

Everson, I. (1987) Some aspects of the small scale distribution of *Euphausia crystallorophias. Polar Biol.* **8**, 9–15.

Everson, I., Watkins, J.L., Bone, D.G. & Foote, K.G. (1990) Implications of a new acoustic target strength for abundance estimates of Antarctic krill. *Nature* **345**, 338–40.

Fernandes, P.G., Brierley, A.S., Simmonds, E.J., Millard, N.W., McPhail, S.D., Armstrong, F., Stevenson, P. & Squires, M. (2000) Fish do not avoid survey vessels. *Nature* **404**, 35–36.

Foote, K.G., Everson, I., Watkins, J.L. & Bone, D.G. (1990) Target strengths of Antarctic krill (*Euphausia superba*) at 38 and 120 kHz. *J. Acoust. Soc. Am.* **87**, 16–24.

Foote, K.G., Knudsen, H.P., Vestnes, G., MacLennan, D.N. & Simmonds, E.J. (1987) Calibration of acoustic instruments for fish density estimation: A practical guide. *ICES Cooperative Research Report* **144**, 1–69.

Gerlotto, F., Soria, M. & Freon, P. (1999) From two dimensions to three: the use of multibeam sonar for a new approach in fisheries acoustics. *Can J. Fish. Aquat. Sci.* **56**, 6–12.

Godlewska, M. (1996) Vertical migrations of krill (*Euphausia superba* Dana). *Pol. Arch. Hydrobiol.* **43**, 9–63.

Grunbaum, D. (1998) Schooling as a strategy for taxis in a noisy environment. *Evol. Ecol.* **12**, 503–522.

Hamner, W.M. (1984) Aspects of schooling of *Euphausia superba. J. Crustacean Biol.* **4**, 67–74.

Hamner, W.M., Hamner, P.P., Strand, S.W. & Gilmer, R.W. (1983) Behavior of Antarctic krill, *Euphausia superba*: chemoreception, feeding, schooling, and molting. *Science* **220**, 433–35.

Hamner, W.M., Stone, G.S. & Obst, B.S. (1988) Behavior of southern right whales, *Eubalaena australis*, feeding on the Antarctic krill, *Euphausia superba. Fish. Bull.* **86**, 143–50.

Hanamura, Y., Endo, Y. & Taniguchi, A. (1984) Underwater observations on the surface swarm of a euphausiid, *Euphausia pacifica* in Sendai Bay, northeastern Japan. *La mer* **22**, 63–68.

Hardy, A.C. (1936) Observations on the uneven distribution of oceanic plankton. *Discovery Rep.* **9**, 511–38.

Harris, G.P., Griffiths, F.B. & Clementson, L.A. (1992) Climate and the fisheries off Tasmania – interactions of physics, food-chains and fish. *S. Afr. J. mar. Sci.* **12**, 585–97.

Haury, L.R., McGowan, J.A. & Wiebe, P.H. (1978) Patterns and processes in the time-space scales of plankton distributions. In *Spatial Patterns in Plankton Communities* (J. Steele, ed.), pp. 277–327. Plenum Press, New York.

Heinemann, D., Hunt, G. & Everson, I. (1989) Relationships between the distributions of marine avian predators and their prey, *Euphausia superba*, in Bransfield Strait and southern Drake Passage, Antarctica. *Mar. Ecol. Prog. Ser.* **58**, 3–16.

Higginbottom, I.R. & Hosie, G.W. (1989) Biomass and population structure of a large aggregation of krill near Prydz Bay, Antarctica. *Mar. Ecol. Prog. Ser.* **58**, 197–203.

Hunt, G.L. (1991) Marine ecology of seabirds in polar oceans. *Am. Zoologist* **31**, 131–42.

Kaartvedt, S. (1993) Drifting and resident plankton. *Bull. mar. Sci.* **53**, 154–59.

Kalinowski, J. & Witek, Z. (1985) BIOMASS Handbook, No. 27: *Scheme for classifying aggregations of Antarctic krill.* SCAR, Cambridge.

Kanda, K., Takagi, K. & Seki, Y. (1982) Movement of the larger swarms of Antarctic krill *Euphausia superba* off Enderby Land during 1976–1977 season. *J. Tokyo Univ. Fish.* **68**, 25–42.

Komaki, Y. (1967) On the surface swarming of Euphausiid crustaceans. *Pacific Sci.* **21**, 433–48.

Krause, D.C. (1998) Implications of a fractal distribution of plankton patchiness. In *IOC Workshop Report No. 142.* pp. 220–32. UNESCO, Paris.

Lascara, C.M., Hofmann, E.E., Ross, R.M. & Quetin, L.B. (1999) Seasonal variability in the distribution of Antarctic krill, *Euphausia superba*, west of the Antarctic Peninsula. *Deep-Sea Research Part I – Oceanographic Research Papers* **46**, 951–84.

Levin, S.A. (1994) Patchiness in marine and terrestrial systems: from individuals to populations. *Phil. Trans. R. Soc. Lond. Ser. B* **343**, 99–103.

Macaulay, M.C., English, T.S. & Mathisen, O.A. (1984) Acoustic characterisation of swarms of Antarctic krill (*Euphausia superba*) from Elephant Island and Bransfield Strait. *J. Crustacean Biol.* **4** (Spec. No. 1), 16–44.

Mangel, M. (1994) Spatial patterning in resource exploitation and conservation. *Phil. Trans. R. Soc. Lond. Ser. B* **343**, 93–98.

Marr, J.W.S. (1962) The natural history and geography of the Antarctic krill (*Euphausia superba* Dana). *Discovery Rep.* **32**, 33–464.

Marschall, H.-P. (1988) The overwintering strategy of Antarctic krill under the pack-ice of the Weddell Sea. *Polar Biol.* **9**, 129–35.

Mathisen, O.A. & Macaulay, M.C. (1983) The morphological features of a super swarm of krill, *Euphausia superba. Mem. Natl. Inst. Polar Res.*, Spec. Issue **27**, 153–64.

Mauchline, J. (1980) The biology of mysids and euphausiids. *Adv. mar. Biol.* **18**, 1–681.

Mauchline, J. & Fisher, L.R. (1969) The biology of euphausiids. *Adv. mar. Biol.* **7**, 1–454.

Miller, D.G.M. & Hampton, I. (1989a) BIOMASS Scientific Series, No. 9: *Biology and Ecology of the Antarctic krill* (Euphausia superba *Dana*): *A Review.* SCAR and SCOR, Scott Polar Research Institute, Cambridge, England.

Miller, D.G.M. & Hampton, I. (1989b) Krill aggregation characteristics: spatial distribution patterns from hydroacoustic observations. *Polar Biol.* **10**, 125–34.

Miller, D.G.M., Barange, M., Klindt, H., Murray, A.W.A., Hampton, I. & Siegel, V. (1993) Antarctic krill aggregation characteristics from acoustic observations in the South West Atlantic Ocean. *Mar. Biol.* **117**, 171–83.

Misund, O.A., Aglen, A. & Fronaes, E. (1995) Mapping the shape, size and density of fish schools by echo integration and a high resolution sonar. *ICES J. mar. Sci.* **52**, 11–20.

Murphy, E.J. (1995) Spatial structure of the Southern Ocean ecosystem: predator-prey linkages in Southern Ocean food webs. *J. Anim. Ecol.* **64**, 333–47.

Murphy, E.J., Morris, D.J., Watkins, J.L. & Priddle, J. (1988) Scales of interaction between Antarctic krill and the environment. In *Antarctic Ocean and Resources Variability* (D. Sahrhage, ed.), pp. 120–130. Springer-Verlag, Berlin.

Murphy, E.J., Watkins, J.L., Reid, K., Trathan, P.N., Everson, I., Croxall, J.P., Priddle, J., Brandon, M.A., Brierley, A.S. & Hofmann, E. (1998) Interannual variability of the South Georgia marine ecosystem: biological and physical sources of variation in the abundance of krill. *Fish. Oceanogr.* **7**, 381–90.

Murray, A.W.A. (1996) Comparison of geostatistical and random sample survey analyses of Antarctic krill acoustic data. *ICES J. mar. Sci.* **53**, 415–21.

Nicol, S. (1984) Population structure of daytime surface swarms of the euphausiid *Meganyctiphanes norvegica* in the Bay of Fundy. *Mar. Ecol. Prog. Ser.* **18**, 241–51.

Nicol, S. (1986) Shape, size and density of daytime surface swarms of the euphausiid *Meganyctiphanes norvegica* in the Bay of Fundy. *J. Plankton Res.* **8**, 29–39.

Nicol, S., James, A. & Pitcher, G. (1987) A first record of daytime surface swarming by *Euphausia lucens* in the southern Benguela region. *Mar. Biol.* **94**, 7–10.

O'Brien, D.P. (1987) Direct observations of the behaviour of *Euphausia superba* and *Euphausia crystallorophias* (Crustacea: Euphausiacea) under pack ice during the Antarctic spring of 1985. *J. Crustacean Biol.* **7**, 437–48.

O'Brien, D.P. (1988) Surface schooling behaviour of the coastal krill *Nyctiphanes australis* (Crustacea: Euphausiacea) off Tasmania, Australia. *Mar. Ecol. Prog. Ser.* **42**, 219–33.

O'Brien, D.P. (1989) Analysis of the internal arrangement of individuals within crustacean aggregations (Euphausiacea: Mysidacea). *J. Exp. Mar. Biol. Ecol.* **128**, 1–30.

Onsrud, M.S.R. & Kaartvedt, S. (1998) Diel vertical migration of the krill *Meganyctiphanes norvegica* in relation to physical environment, food and predators. *Mar. Ecol. Prog. Ser.* **171**, 209–219.

Pauly, T., Higginbottom, I., Nicol, S. & de la Mare, W.K. (1996) Results of a hydroacoustic survey of Antarctic krill populations in CCAMLR Division 58.4.1 carried out in January to April, 1996. CCAMLR, Hobart, WG-EMM-96/28.

Pitcher, T.J., Magurran, A.E. & Winfield, I.J. (1982) Fish in larger shoals find food faster. *Behav. Ecol. Sociobiol.* **10**, 149–51.

Ragulin, A.G. (1969) Underwater observations of krill. *Trudy VNIRO* **66**, 231–34.

Ricketts, C., Watkins, J.L., Morris, D.J., Buchholz, F. & Priddle, J. (1992) An assessment of the biological and acoustic characteristics of swarms of Antarctic krill. *Deep-Sea Res.* **39**, 359–71.

Ringelberg, J. (1995) Changes in light-intensity and diel vertical migration: a comparison of marine and freshwater environments. *J. Mar. Biol. Ass. UK* **75**, 15–25.

Ritz, D.A. (1994) Social aggregation in pelagic invertebrates. *Adv. mar. Biol.* **30**, 155–216.

Roe, H.S.J. (1983) Vertical distributions of euphausiids and fish in relation to light intensity in the Northeastern Atlantic. *Mar. Biol.* **77**, 287–98.

Ross, R.M., Quetin, L.B. & Lascara, C.M. (1996) Distribution of Antarctic krill and dominant zooplankton west of the Antarctic Peninsula. In *Foundations for Ecological Research West of the Antarctic Peninsula* (R.M. Ross, L.B. Quetin & E.E. Hofmann, eds), Antarctic Research Series, 70, pp. 199–217. American Geophysical Union, Washington.

Siegel, V. & Kalinowski, J. (1994) Krill demography and small-scale processes: a review. In *Southern Ocean ecology: the BIOMASS perspective* (S.Z. El-Sayed, ed.), pp. 145–63. Cambridge University Press, Cambridge.

Siegel, V. & Loeb, V. (1995) Recruitment of Antarctic krill *Euphausia superba* and possible causes for its variability. *Mar. Ecol. Prog. Ser.* **123**, 45–56.

Simard, Y. & Lavoie, D. (1999) The rich krill aggregation of the Saguenay–St. Lawrence Marine Park: hydroacoustic and geostatistical biomass estimates, structure, variability, and significance for whales. *Can. J. Fish. Aquat. Sci.* **56**, 1182–97.

Simard, Y., Ladurantaye, R. de & Therriault, J.-C. (1986) Aggregation of euphausiids along a coastal shelf in an upwelling environment. *Mar. Ecol. Prog. Ser.* **32**, 203–215.

Strand, S.W. & Hamner, W.M. (1990) Schooling behavior of Antarctic krill (*Euphausia superba*) in laboratory aquaria: reactions to chemical and visual stimuli. *Mar. Biol.* **106**, 355–59.

Stretch, J.J., Hamner, P.P., Hamner, W.M., Michel, W.C., Cook, J. & Sullivan, C.W. (1988) Foraging behavior of Antarctic krill *Euphausia superba* on sea ice microalgae. *Mar. Ecol. Prog. Ser.* **44**, 131–39.

Tarling, G.A., Buchholz, F. & Matthews, J.B.L. (1999) The effect of a lunar eclipse on the vertical migration behaviour of *Meganyctiphanes norvegica* (Crustacea: *Euphausiacea*) in the Ligurian Sea. *J. Plankton Res.* **21**, 1475–88.

Tarling, G.A., Matthews, J.B.L., Saborowski, R. & Buchholz, F. (1998) Vertical migratory behaviour of the euphausiid, *Meganyctiphanes norvegica*, and its dispersion in the Kattegat Channel. *Hydrobiologia* **376**, 331–41.

Trathan, P.N., Everson, I., Miller, D.G.M., Watkins, J.L. & Murphy, E.J. (1995) Krill biomass in the Atlantic. *Nature* **373**, 201–202.

Watkins, J.L. (1986) Variations in the size of Antarctic krill, *Euphausia superba* Dana, in small swarms. *Mar. Ecol. Prog. Ser.* **31**, 67–73.

Watkins, J.L. & Murray, A.W.A. (1998) Layers of Antarctic krill, *Euphausia superba*: are they just long krill swarms? *Mar. Biol.* **131**, 237–47.

Watkins, J.L., Buchholz, F., Priddle, J., Morris, D.J. & Ricketts, C. (1992) Variation in reproductive status of Antarctic krill swarms; evidence for a size-related sorting mechanism? *Mar. Ecol. Prog. Ser.* **82**, 163–74.

Watkins, J.L., Morris, D.J., Ricketts, C. & Priddle, J. (1986) Differences between swarms of Antarctic krill and some implications for sampling krill populations. *Mar. Biol.* **93**, 137–46.

Witek, Z., Kalinowski, J. & Grelowski, A. (1988) Formation of Antarctic krill concentrations in relation to hydrodynamic processes and social behavior. In *Antarctic Ocean and Resources Variability* (D. Sahrhage, ed.), pp. 237–44. Springer-Verlag, Berlin.

Chapter 5
Population Parameters

Volker Siegel and Stephen Nicol

5.1 Background on life history information

Introduction

Euphausiids, or krill, are important elements in many coastal and oceanic pelagic ecosystems and are targeted by commercial fisheries in a number of areas (Nicol & Endo, 1997). Krill populations fluctuate in response to environmental changes which as a result of fishing activities may have severe ecosystem implications. Understanding the population response to changes and managing the effect of fisheries requires information on the key population parameters of the species involved as well as their natural variability. Euphausiids are found in a range of habitats from the tropics to the poles but are most important in the ecosystems of polar and temperate waters which experience considerable seasonality (Mauchline & Fisher, 1969; Mauchline, 1980). Large seasonal environmental changes make it necessary for krill to develop strategies in growth, reproduction or overwintering to survive poor conditions and maintain stock stability. Of particular importance in attempting to clarify these strategies at both the individual and population level is the knowledge of basic body components, e.g. length, total wet mass, dry mass, and the formulation of length–weight relationships. This information may allow the development of indices of 'condition' and forms the backbone for production and energy budget determinations (Falk-Petersen & Hopkins, 1981).

Information on how krill species have adjusted to long winters with low temperatures, little food or light is critical to an understanding not only of the species survival but of the entire polar food webs. Other matters complicating the situation are seasonal and longer-term environmental changes in the system. If long-term climate variability is present, the question arises whether krill population dynamics will show enduring effects and how serious these effects are on the stock structure and stock abundance–biomass. Furthermore these changes in krill stocks are obviously critical for dependent higher trophic levels in the system with krill as the central component of the food web and at the basis of the trophic chain. These interactions are discussed at length in Chapter 7.

Management of emerging and established krill fisheries also requires information on a number of population parameters. The largest fishery, that for Antarctic krill, has the most sophisticated management regime and the generalised yield model used to set precautionary limits on the fishery requires primary information on the biomass, recruitment and growth rate of the krill population (Constable & de la Mare, 1996). This model is applicable to other fisheries and there have been indications that this approach may be adopted by other krill fisheries (Head, 1997) and this will require the collection of biological information on the species being harvested.

The present contribution summarises published, but widely scattered, basic information on key population parameters of post-larval krill: age, growth, mortality and recruitment, particularly for those krill species which are of commercial interest. The concentration is on those aspects of the life history which are critical to understanding processes associated with fisheries management or ecological processes. Information on other species is included where it helps to illustrate a point of interest. Much of the earlier material on these topics has been reviewed in two volumes (Mauchline & Fisher, 1969; Mauchline, 1980) and no attempt has been made to further summarise this literature.

Methodologies

Key population parameters of euphausiids have been investigated by studies at sea and in the laboratory. Both of these approaches have advantages and disadvantages and much of what we know of the better studied species of krill comes from a combination of both types of study. Field-based methods have used net-collected samples of krill, usually from surveys which have used an unbiased sampling approach, to estimate the population parameters of krill based on morphometrics and classification of sex and maturity stages. Most often, these have been infrequent or annual surveys although there have been some studies which have used monthly (or shorter) sampling intervals. Only a few studies have used very short sampling intervals to obtain estimates of changes in population parameters over a few days.

There are a number of methodological problems with net-based surveys and these are discussed by Watkins, Chapter 4. Nets are thought to be selective when dealing with euphausiids and are subject to avoidance (Everson & Bone, 1986). Fine-mesh nets may collect younger stages well but are thought to be evaded by the larger animals. There are, however, no regularly used techniques which can provide samples of krill for analysis or laboratory study other than nets. Samples from predators have been used to provide time series through a season, but there are large gaps in our knowledge as to what extent these samples are biased, e.g. by krill availability in a localised area or selectivity by the predator and its potential variability due to behavioural changes.

There are also uncertainties regarding the degree to which net surveys provide a representative sample of the population. Because krill are pelagic, often schooling organisms which can occur patchily in huge abundances, the ability of nets to adequately obtain a representative sample of the population has been questioned

(Miller & Hampton, 1989). High sampling density is the best guarantee that a synoptic picture of a krill population can be obtained; however, ensuring that the same population is sampled over time is more problematic. In some species of krill, there are thought to be resident populations in fjords which can be repeatedly sampled (Falk-Petersen & Hopkins, 1981). For many open ocean or shelf species, however, there is no certainty that the same population is being sampled from one survey to the next. Indeed, in the Antarctic, the prevailing theory is that Antarctic krill are subject to flux, being moved quite rapidly from one area to another by large-scale ocean currents (Murphy *et al.*, 1998). To assess changes in Antarctic krill populations, therefore, the assumption must be made that population processes inferred from small-scale surveys are affected by processes on a very much larger scale such that the sampling intervals are reflective of the residence time of krill in the larger-scale area.

Because of their small size, pelagic habit and high abundance, and their frequent moulting, krill are not amenable to mark-recapture studies so field-based experiments on the growth and maturity of individuals have not been attempted. Consequently, experimental investigations on krill have been restricted to laboratory studies and such studies into population parameters of krill have been confined to a small number of species. This is partly because of the difficulties that have been encountered in maintaining euphausiids in the laboratory for significant periods of time (Baker, 1963; Lasker & Theilacker, 1965; Komaki, 1966; Mackintosh, 1967; Clarke, 1976; Ikeda *et al.*, 1980; Ross, 1981; Miller *et al.*, 1983). Only *E. superba* (Ikeda, 1984; Ikeda, 1987a, b), and *E. pacifica* (Ross, 1982a; Ross, 1982b) have been reared from eggs to adults. *E. crystallorophias* (Ikeda, 1986), *Nematoscelis difficilis* (Gopalkrishnan, 1973) and *Euphausia lucens* (Pilar, 1984) have been reared through a significant portion of their life history. The paucity of information on living animals has meant that for most euphausiid species estimates of growth rates, development rates and longevity are based on analysis of preserved specimens.

Laboratory studies have a number of inherent problems which encourage caution when extrapolating their results to the wild population. Euphausiids are pelagic organisms so their collection and confinement in experimental apparatus imposes obvious stresses. Additionally, many species of krill, including all of those which have been experimentally studied, are aggregating or schooling organisms and their experimental manipulation in isolation may inflict unrealistic conditions on them.

Both field-based and laboratory-based studies have a number of caveats associated with them. A further problem that occurs when integrating the information that arises from both types of study lies in assessing which type of information is more useful in particular circumstances. These sorts of conflicts can be resolved in future by the use of more realistic laboratory-based experiments, the use of field-based experiments and the directed study of certain areas where the evidence from the two approaches are contradictory. An attempt has been made to highlight those areas which are a priority for future studies and where aspects of krill biology can be advanced by concerted study.

5.2 Age determination and population age structure

Introduction

Like all crustaceans krill possess an exoskeleton. The process of growth requires the shedding of this relatively inflexible structure so size increases are stepped rather than linear. Euphausiids are unusual amongst crustaceans because they moult regularly during their entire life even when they have reached maturity and have ceased growing and even if they are deprived of food (Nicol & Stolp, 1991). This regular periodic loss of their hard structures means that there are no known growth rings or incremental changes equivalent to those found in vertebrate bones, scales or otoliths. Consequently age determination of individuals or age structure of populations of all species of euphausiid has been somewhat fraught with difficulty.

Up until the late 1970s age determination of all species of euphausiids was determined from length–frequency analyses. The experimental finding that krill shrink, if starved (Ikeda & Dixon, 1982), raised suggestions that length-based ageing methods were possibly subject to more error than had been previously thought. All animals have the ability to shrink; however, Antarctic krill were shown to reduce their size in sub-optimal food conditions by moulting to a smaller size and to lose their sexual characteristics over time (Thomas & Ikeda, 1987). This finding introduced further uncertainties into the use of measures of length, combined with an assessment of the sexual maturity, as definitive indicators of age. These results precipitated a debate on ageing and mortality of krill that is yet to be resolved (Nicol, 1990). Further results of experimental studies on Antarctic krill showed evidence of shrinkage by populations of Antarctic krill in the laboratory (Nicol *et al.*, 1991) as well as by individuals. Groups of krill with polymodal size distributions tend to contract to a unimodal size distribution at the lower end of the original size range when kept in aquaria under low food conditions (Nicol *et al.*, 1991; Sun, 1997). This alteration of the shape of the size spectrum may come about through the differing food requirements of the smaller and larger animals – the small krill receiving a maintenance ration but the larger ones suffering from a sub-optimal diet and hence shrinking. This differential effect is that most likely to cause difficulties for length–frequency analyses if it occurs under natural conditions.

Although the issue of shrinkage has been discussed primarily with regard to Antarctic krill the questions raised through these experimental studies are equally applicable to other species. Shrinkage has been shown in the laboratory for five species of euphausiids: *E. pacifica* (Lasker, 1966), *E. superba* (Ikeda & Dixon, 1982), *M. norvegica* (Buchholz, 1985), *Thysanoessa inermis* (Dalpadado & Ikeda, 1989), *Nyctiphanes australis* (Hosie & Ritz, 1989). There is experimental evidence that all species of krill so far examined can shrink in the laboratory so the cautions expressed over the use of length–frequency analysis of Antarctic krill are probably applicable to other species.

Studies, using length–frequency analysis, have also revealed apparent reduction in mean or modal size which can be interpreted as winter shrinkage in particular size classes or in whole populations (Hollingshead & Corey, 1974; Bamstedt, 1976; Kulka & Corey, 1978; Falk-Petersen & Hopkins, 1981; Hopkins *et al.*, 1984; Falk-Petersen, 1985; Asthorsson, 1990; Dalpadado & Skjoldal, 1991; Bollens *et al.*, 1992; Dalpadado & Skjoldal, 1995; Dalpadado & Skjoldal, 1996). The observed decrease in mean sizes of these populations has also been attributed to overwinter mortality among the older (larger) animals or to immigration or emigration. However, reduction in mean length during winter has not been reported to the extent that the modal size structure of the length–frequency distribution was distorted sufficiently to disrupt the modal progression or distribution mixture analysis for these species (Falk-Petersen, 1985).

Ageing from length-frequency analysis

Despite reservations raised as a consequence of the results of experimental studies, most analyses of krill populations continue to rely on length–frequency analysis to provide measures of age and growth. Length–frequency analysis has two major assumptions: that the same population is being measured at each sampling and that the animals grow and mature in some predictable fashion, but both assumptions still have to be verified for pelagic organisms like euphausiids.

Only in a few isolated cases has there been direct evidence that the same local population of krill has been followed throughout a seasonal cycle. Several studies of the seasonal population biology of temperate euphausiids have been conducted in enclosed seas or fjords, in which the same population can be sampled repeatedly (Hollingshead & Corey, 1974; Kulka & Corey, 1978; Hopkins *et al.*, 1984; Falk-Petersen, 1985; Asthorsson, 1990; Bollens *et al.*, 1992).

In short-lived species the length–frequency distributions may ideally show a unimodal size distribution, representing a spawning event and the growth of the population over the season. Polymodal length–frequency distributions occur in warm or temperate water species, when discrete spawning pulses during the course of the year are separated by growth, e.g. *E. pacifica* off southern California (Brinton, 1976), *Nyctiphanes couchii* in the Bay of Biscay (Gros & Cochard, 1978), *T. longicaudata* (Lindley, 1978), *N. australis* off New Zealand (Jillett, 1971). These seasonal spawning events have a cohort structure and can often be followed over the entire life span of the species, which rarely extends over more than one year. In the case of continuous spawning over the year no modes would be obvious from the size distribution and cohorts are no longer detectable. Examples for these non-structured length–frequency distributions are tropical species such as *Nyctiphanes simplex* and *Euphausia eximia* off the west coast of Baja California (Gomez, 1995).

Cold water species from Antarctic or subarctic regions also show polymodal length–frequency distributions. Most of these species have a single relatively short breeding season every year, e.g. *E. superba* (Marr, 1962; Mauchline & Fisher, 1969); *E. triacantha* (Baker, 1959); *M. norvegica* (Einarsson, 1945). Each spawning event then represents a cohort or age group and for younger size classes leads to a clear mode in their length–frequency distribution. Therefore, in such cases it can be concluded that the polymodal patterns indicate a multi-year life cycle of the species. If there are no size dependent net selection factors (methodological biases) or prey size selection factors by predators (natural mortality influences) one can generally assume that the length–frequency distribution of a cohort represents a normal distribution.

The maximum lifespan of krill has been estimated by simple modal progression analysis, which follows the sequential progression of the peaks in length–frequency samples. This works for species with few cohorts and a relatively fast growth rate, for example: *T. inermis*, *T. raschii* and *E. pacifica* (Einarsson, 1945; Bollens *et al.*, 1992). If there are too many cohorts produced in one year for a short-lived species, or if too many cohorts of a long-living species survive for several years and the growth rate of the older animals slows down substantially, then the modes become less and less distinguishable, because of overlapping normal distributions.

Since size distribution mixtures of species with a multi-year life cycle consist of a number of normal distributions, statistical methods have been developed to facilitate the separation of mixture components into normal distributions. Graphical analysis on probability paper (Harding, 1949) gives the mean length of the size groups (Matthews, 1973). For older age classes this method is highly subjective and does not allow an estimation of the reliability of the analyses.

An interactive computer program called MIX reduced the subjectivity; the results include estimates of the proportions of the single components, the mean length, and the standard deviation of the mean as well as the standard errors of all estimates (Macdonald & Pitcher, 1979). Furthermore, a goodness of fit between the observed and the expected distribution mixture is given. A similar approach was described by (Bhattacharya, 1967) and applied to *T. inermis* populations (Dalpadado & Skjoldal, 1995). Even if the maximum number of age groups for long-lived species like *E. superba* cannot be fully resolved, due to overlapping oldest size groups, these methods can at least give the minimum number of age groups in the population.

For quantitative analyses of distribution mixtures (e.g. recruitment estimates) length–frequency data may be inadequate, because, in euphausiids they are not normally distributed and length–frequency data represent standardised, but pooled, averaged data. This could strongly bias the quantitative analysis of the different age classes. For this reason the statistical properties of the distribution of density values have been analysed and the MLMIX method was established to create the length–density distribution (de la Mare, 1994). He also modified the Macdonald and Pitcher least-square approach of fitting the data to the maximum likelihood method. As a result the quantitative proportions as well as the mean length of the distribution mixture components can be calculated including a statistical test for the goodness of fit to the data. This technique has yet to be applied to populations of other species of euphausiids.

Lifespan and size at age

Despite the issues raised above, methods using analyses of length–frequency data have found wide application for the determination of age and growth in krill. Statistical methods applied to distribution mixtures were first successfully used for Antarctic krill by Aseev (1984) (graphical method on probability paper) and Siegel (1986, 1987) applying the Macdonald and Pitcher approach. These authors found a much longer lifespan for this species than the previously suggested 2-year life cycle (Rudd, 1932; Marr, 1962; Mauchline & Fisher, 1969) (Table 5.1). Analyses from other studies and areas in principal confirmed the results of a multi-year life cycle (Hosie *et al.*, 1988; Pakhomov 1995a; Wang *et al.*, 1995). *E. superba* populations around the continent show very similar results on length-at-age data with two deviations (Table 5.2). Firstly, krill at high latitudes like the inner Weddell Sea region seem to be smaller than those from the more seasonal pack-ice zone. Secondly, in these high latitude areas, the oldest krill are age 4+, whereas they reach at least age 5+ in the Atlantic and in parts of the Indian Ocean sector (Siegel, 1987; Hosie *et al.*, 1988). This difference was first noted by Marr (1962) and juvenile Weddell Sea krill indeed differ in mean length from more northerly waters. However, older age classes in the Weddell Sea are very similar in size to Antarctic Peninsula and Indian Ocean krill. Second, the largest age classes in the Indian Ocean sector are absent (Hosie *et al.*, 1988; Pakhomov 1995a), either indicating a statistical problem of fitting the distribution mixture to the observed data, or indicating a missing age class. This may be because these very old age classes strongly overlap in size and occur only in low abundance. However, the high variability in mean length of the 4+ group (Hosie *et al.*, 1988), and the very similar mean length values for all younger classes may support the conclusion that, at least in some years, age class 5+ animals have been attributed erroneously to 4+, even though older animals did exist in the population. The detailed work of Lu & Wang (1996) on the separation of distribution mixtures clearly demonstrates that a high increase in the significance level of the fit is possible when assuming the existence of a 5+ age group instead of a maximum age of 4+. It is also evident that the mean length-at-age for the younger age classes is very stable and the results are very robust whether the procedure includes or excludes a 5+ age group. In this case only the mean length of the 4+ group is obviously affected.

Whether there are age classes beyond 5+ for the Antarctic krill, those cannot be resolved with the present methods because there is too much overlap in the distribution of size groups. However, the age group 5+ is already below 1% of the stock size, so that even older age groups would not contribute substantially to the population or strongly influence population dynamic parameters.

Other euphausiid species age structure

The sympatric species *T. inermis* and *T. raschii* have a wide boreal to low arctic distribution, in the Atlantic from temperate waters off Nova Scotia and the North

Table 5.1 Longevity and age at maturation for different euphausiid species in different geographical regions; F = females, M = males, R = recruits.

Species	Geogr. region	Max age	Age at maturity	Age at 1st spawning	Reference
Thysanoessa raschii	N North Sea, S Norway	1+	1+	1+	Lindley, 1980
Thysanoessa raschii	boreal	2+	1+	1+	Einarsson, 1945
Thysanoessa raschii	subarctic	2+	2+	2+	Einarsson, 1945
Thysanoessa raschii	arctic	2+	2+	no spawning	Einarsson, 1945
Thysanoessa raschii	Scotland	2+			Mauchline, 1966
Thysanoessa raschii	S Barents Sea	2+	1+	2+	Drobysheva, 1987
Thysanoessa raschii	N Norway	2 yr 3 mo		few 1 + F, few 1 + M	Falk-Petersen & Hopkins, 1981
Thysanoessa raschii	E Greenland	3+	2+	2+	Lindley, 1980
Thysanoessa inermis	N North Sea, S Norway	1+	1+	1+	Lindley, 1980
Thysanoessa inermis	N North Sea	1+			Jones *et al.*, 1967
Thysanoessa inermis	W Norway	1+			Jörgensen & Matthews, 1975
Thysanoessa inermis	boreal	2+	1+	1+	Einarsson, 1945
Thysanoessa inermis	subarctic	2+	2+	2+	Einarsson, 1945
Thysanoessa inermis	arctic	2+	2+	no spawning	Einarsson, 1945
Thysanoessa inermis	Scotland	2+			Mauchline, 1966
Thysanoessa inermis	S Barents Sea	2+	1+	2+	Drobysheva, 1987
Thysanoessa inermis	N Norway	2+			Berkes, 1976
Thysanoessa inermis	Canada, Nova Scotia	2+			Kulka & Corey, 1978
Thysanoessa inermis	N Norway	2 yr 3 mo		few 1 + F, most 1 + M	Falk-Petersen & Hopkins, 1981
Thysanoessa inermis	Barents Sea	3+	2+	2+	Dalpadado & Skjoldal, 1991, 1995, 1996
Thysanoessa inermis	Greenland	3+	3+	3+	Lindley, 1980
Thysanoessa inermis	E Greenland	3+			Wiborg, 1968
Thysanoessa inermis	N Pacific, NE Aleutians	3+	3+?	3+?	Nemoto, 1957
Thysanoessa inermis	N Pacific, SE Aleutians	2+	2+	2+	Nemoto, 1957
Meganyctiphanes norvegica	Mediterranean	1+	1+	1+	Lindley, 1982
Meganyctiphanes norvegica	Scotland	2+	1+	1+	Mauchline, 1960, Mauchline & Fischer, 1969
Meganyctiphanes norvegica	W Norway	2+			Sager & Sammler, 1984

Meganyctiphanes norvegica	Skagerrak	2+	1+	1+	Einarsson, 1945
Meganyctiphanes norvegica	Skagerrak	2+	1	1+	Boysen & Buchholz, 1984
Meganyctiphanes norvegica	W Norway	2+			Lindley, 1980
Meganyctiphanes norvegica	N Norway	2 yr 6 mo		no spawning	Falk-Petersen & Hopkins, 1981
Euphausia pacifica	Oregon	1 yr	1 yr	1 yr	Smiles & Pearcy, 1971
Euphausia pacifica	Oregon, Washington	1+			Brodeur & Pearcy, 1992
Euphausia pacifica	S California	6–8 mo spring R	4 mo	4 mo	Brinton, 1976
Euphausia pacifica	S California	10–13 mo fall R	7 mo	7 mo	Brinton, 1976
Euphausia pacifica	NE Japan	15 mo			Iguchi *et al.*, 1993
Euphausia pacifica	Canada BC	19–22 mo			Nicol & Endo, 1997
Euphausia pacifica	NW Pacific	2+	1+		Brinton, 1976
Euphausia pacifica	NW Pacific, Kamchatka	2+	1+	1+	Ponomareva, 1963
Euphausia pacifica	NW Pacific	2+	1+		Nemoto, 1957
Euphausia pacifica	N Pacific, Aleutians	2+			Iguchi & Ikeda, 1995
Euphausia pacifica	N Japan	2 + F, 1 + M	1+	1+	Endo, 1981, cf. Nicol & Endo, 1997
Euphausia pacifica	SW Japan	21 mo			Iguchi *et al.*, 1993
Euphausia superba	Antarctic, S. Georgia	4+			Ivanov, 1970
Euphausia superba	Antarctic Peninsula	5+	2 + F, 3 + M	2* F, 3 + M	Siegel, 1986, 1987, Siegel & Loeb, 1994
Euphausia superba	Antarctic Peninsula	5+			Kawaguchi *et al.*, 1997
Euphausia superba	Antarctic, Indian Ocean	5+			Aseev, 1984
Euphausia superba	Antarctic, Indian Ocean	4+	3 + F, 3 + M	3 + f, 3 + M	Hosie *et al.*, 1988
Euphausia superba	Antarctic, Indian Ocean	4+	3+	3+	Pakhomov, 1995a
Euphausia superba	Antarctic, Indian Ocean	5+			Wang *et al.*, 1995
Euphausia superba	Antarctic, Indian Ocean	5+			Lu & Wang, 1996
Euphausia superba	Antarctic, Indian Ocean	6+	0 = larv; 1 + = juv		Ettershank, 1983, 1985
Euphausia superba	Antarctic, Scotia Sea	6–7 yr			Rosenberg *et al.*, 1986

Table 5.2 Mean length-at-age data (TL in mm) for different euphausiid species in different regions.

Species	Geogr. region	Season
Thysanoessa raschii	N Atlantic boreal	spawning season
Thysanoessa inermis	Barents Sea	June
Thysanoessa inermis	Barents Sea	Oct
Thysanoessa inermis	Barents Sea	Jan
Thysanoessa inermis	Barents Sea	May
Thysanoessa inermis	Barents Sea	Oct
Thysanoessa inermis	Barents Sea	Jan
Thysanoessa inermis	Barents Sea	end of growth season
Meganyctiphanes norvegica	Skagerrak	winter, range of size classes
Meganyctiphanes norvegica	Scotland	winter, range of size classes
Meganyctiphanes norvegica	Norway	winter, range of size classes
Meganyctiphanes norvegica	N Atlantic	winter (mean TL)
Meganyctiphanes norvegica	N Atlantic	winter (means for diff. years)
Meganyctiphanes norvegica	Kattegat	winter (mean TL)
Euphausia pacifica	S California	Feb. cohort after 7 months
Euphausia pacifica	S California	June cohort after 10 months
Euphausia pacifica	S California	1st spawning
Euphausia pacifica	S California	2nd spawning
Euphausia pacifica	S California	3rd spawning
Euphausia pacifica	Oregon	Sep. cohort after 13 months
Euphausia pacifica	S Japan	0 group in winter, 1+ at 21 mo
Euphausia pacifica	NW Pacific	spring
Euphausia pacifica	NW Pacific	
Euphausia pacifica	NE Japan	0 group in winter, 1+ summer
Euphausia superba	Antarctic Peninsula	Feb–March
Euphausia superba	Antarctic Peninsula	March–April
Euphausia superba	Antarctic Peninsula	May/June
Euphausia superba	Antarctic Peninsula	October, spring
Euphausia superba	Antarctic, Weddell Sea	Jan–March, summer
Euphausia superba	Antarctic, Weddell Sea	Oct–Nov. winter/spring
Euphausia superba	Antarctic, Indian Ocean	Dec–March
Euphausia superba	Antarctic, Indian Ocean	Dec–Feb
Euphausia superba	Antarctic, Indian Ocean	Jan–Feb
Euphausia superba	Antarctic, Indian Ocean	March–April
Euphausia superba	Antarctic, Indian Ocean	

Sea to almost high arctic waters off Greenland and the northern Barents Sea. In the Pacific their distribution ranges from north of 40°N in the western part and north of 50°N in the eastern part extending into the Bering and Beaufort Seas. Due to their wide latitudinal range these species show an interesting succession of population age structures (Table 5.1).

At the limit of their southern distribution *T. inermis* and *T. raschii* attain a maximum age of one year, dying after the first spawning season. In boreal waters around Scotland, Iceland, western Norway and Nova Scotia maximum age of the species increases to two years. Age-at-maturity and first spawning still occurs at the

Age 0	1+	2+	3+	4+	5+	Reference
	13	21				Einarsson, 1945
	13.4	20.8				Dalpadado & Skjoldal, 1991
12.1	19.8	22.5				Dalpadado & Skjoldal, 1991
11.7	20.3	24.5				Dalpadado & Skjoldal, 1991
	13.1	22.1	24.2			Dalpadado & Skjoldal, 1991
9.5	12.8	17.1	22.1			Dalpadado & Skjoldal, 1991
9.9	12.6	16.6	21.5			Dalpadado & Skjoldal, 1991
10.5	13.9	18.5	22.7			Dalpadado & Skjoldal, 1995
13–25	27–35					Poulsen, 1926
24–26	36–37	42–48				Mauchline, 1960
14–27	29–34	43–47				Wiborg, 1971
18	25	32				Einarsson, 1945
	20–23	31–36				Lindley, 1982
20	31	36				Boysen & Buchholz, 1984
19						Brinton, 1976
19						Brinton, 1976
11.5						Brinton, 1976
16						Brinton, 1976
20						Brinton, 1976
	22					Smiles & Pearcy, 1971
10	20					Iguchi *et al.*, 1993
	13–14	19				Ponomareva, 1963
	17–18	22				Nemoto, 1957
10–12	16–18					Taki & Ogishima, 1997
3–8.5	29.0–31.8	38.0–41.3	45.5–47.6	51.0–52.3	55.5–57.1	Siegel, 1986
8.0	30.0	39.4	44.5	50.0		Siegel, 1987
6.7	28.7	36.0	43.6	50.3	54.2	Siegel, 1987
12.7	30.5	40.0	46.0			Siegel, 1987
1.0–2.0	20.8–25.2	39.2	44.5	50.1		Siegel, 1987
	26.4	39.3	44.8			Siegel, 1987
	29.6–32.2	37.4–38.3	43.4–44.8	47.8–50.7	52.4–55.4	Aseev, 1984
1.2–7.3	19.0–30.8	34.2–40.9	41.2–48.8	51.0–54.9		Hosie *et al.*, 1988
	22.3–28.4	39.0–41.8	44.2–46.4	49.2–51.5	54.8	Wang *et al.*, 1995
	25.7–30.6	32.8–40.7	40.3–48.3	49.2–55.0		Pakhomov, 1995a
	28.4	41.8	46.4	50.3	54.8	Lu & Wang, 1996

age of 1+ and part of the population survives for another year and spawns for a second time. Off northern Norway these species further demonstrate their developmental plasticity. Maximum age is still 2+, and most males become mature as 1-year-olds; however, only very few females spawn at the age of 1+ and most spawn once at the age of 2+ (Falk-Petersen & Hopkins, 1981). This trend continues in the Barents Sea and for the Pacific in the southeastern Aleutian region, where individual krill reach a maximum age of 3+ (Nemoto, 1957; Lindley, 1980). Here they mature and spawn at 2+, however, the number of second time spawners at the age of 3+ must be very low, because this oldest age class is rarely reported (see Table

5.1.). Greenland and the northeastern Aleutians are probably the most environmentally extreme locations, because both species require three years to reach maturity and only spawn for a single season. Further north in high arctic waters they still mature, but spawning does not occur and the populations must be replaced by advection from the south.

Well-documented mean length-at-age data for *T. inermis* and *T. raschii* are only available from Barents Sea studies (Table 5.2). Although Einarsson (Einarsson, 1945) did not identify the 3+ age group, his mean length-at-age data for *Thysanoessa* krill in the North Atlantic are similar to those from the Barents Sea (Dalpadado & Skjoldal 1991; Dalpadado & Skjoldal 1995). The maximum age of these species changes with latitude, but not the length-at-age. In spite of this there are records indicating that *Thysanoessa* grow to a larger maximum size close to their northern distribution limits (Einarsson, 1945).

M. norvegica has a wide distribution range in the Atlantic and its southern range extends into the Mediterranean Sea, where it has a maximum life expectancy of one year and a single spawning season during its life cycle. Off western Norway *M. norvegica* lives to the age of 2+ and, as in the Skagerrak and off Scotland, spawns for the first time at 1+ and performs a possible repeat spawning as a 2-year-old. Unlike the *Thysanoessa* species, the maximum age of *M. norvegica* does not increase off northern Norway, and in the southern Barents Sea, it is still documented as 2 years and 6 months, when it does not appear to spawn there (Falk-Petersen & Hopkins, 1981; Drobysheva, 1987). Although the mean length-at-age for most areas seems to be consistent, older age classes seem to be greater in mean length in Scottish and Norwegian waters (Table 5.2). Mean length-at-age data from the North Atlantic for different years show only minor inter-annual differences (Lindley, 1982).

E. pacifica also shows large variation in maximum age. It experiences the shortest life span of 6 to 8 months along the coast of California which is at its start of the southern limit of its distribution. After 4 months the spring recruits reach maturity and spawn for the first time with the potential of repeat spawning every 2 months. In the same area autumn cohorts survive to the age of 10 to 13 months. However, maximum size of the cohorts is very similar (Brinton, 1976). In Japanese waters maximum age increases to 15 to 21 months (Iguchi *et al.*, 1993). Off the coast of Oregon and Washington longevity is similar to that in Japanese waters and mean length-at-age for 0+ and 1+ krill is comparable on both sides of the Pacific (Smiles & Pearcy, 1971).

At higher latitudes (Aleutians, Kamchatka) *E. pacifica* matures and spawns for the first time at 1+ and part of the population survives the following winter and spawns again at 2+ . Although maximum size of the species seems to be similar for most of the distribution range, the mean length-at-age in high latitude regions is obviously smaller. The mean length of 1+ krill off Oregon and Japan is 20 to 22 mm, whereas in northern regions it is 13 to 18 mm and reaches 19 to 22 mm mean length at 2+.

In principal, maximum age and age at sexual maturity increase with latitude. For

E. pacifica and *Thysanoessa* spp, however, an intermediate step can be observed in mid latitudes. The maximum age of *Thysanoessa* does not change, but most males become mature at 1+, whereas only very few females spawn at the age of 1+ (see above). For *E. pacifica* both sexes reach maturity and spawn as 1+, but only females survive for another season and spawn a second time as 2+ (Endo, 1981; Nicol & Endo, 1997) indicating a flexible response to latitudinal changes. Further, *E. superba* shows a marked difference in sexual maturation; in this case females mature and spawn first at the age of 2+, while male maturation is retarded by one year to age class 3+ (Siegel & Loeb, 1994).

Other techniques for estimating krill age and longevity

Age pigments

The prospect of Antarctic krill shrinking during winter prompted investigations into the use of fluorescent age pigment accumulation as an indicator of age (Ettershank, 1983). Fluorescent age pigments, including lipofuscin, are substances which accumulate with time in the cells of all organisms so far examined (Hammer & Braum, 1988). They are thought to be the products of cellular lipid peroxidation processes; accumulating at a rate dependent on the metabolic rate of the animals. They are fluorescent, emitting light at 420–470 nm when excited by wavelengths of 340–370 nm. Although initial studies using fluorescent age-pigments as an indicator of age held promise (Ettershank, 1983), subsequent investigations demonstrated methodological problems which raised doubts about the reliability of the technique used (Nicol, 1987; Mullin & Brooks, 1988). Studies examining fluorescence of soluble age-pigment in other organisms have yielded equivocal results (Hirche & Anger, 1987; Hammer & Braum, 1988; Hill & Radkte, 1988; Vernet *et al.*, 1988) but the utility of this technique for ageing *E. superba* is yet to be proven. Laboratory studies with *E. superba* indicated that some soluble fluorescent age-pigments accumulated with age in a population that shrank significantly in length over the period of a year in captivity (Nicol *et al.*, 1991). Studies with captive populations of other species of crustaceans have shown significant accumulation of age-pigments when examined using image analysis and fluorescence microscopy (Sheehy *et al.*, 1994; Wahle *et al.*, 1996) and this technique might prove more reliable than the solvent extraction method if applied to euphausiids. It seems unlikely that the rather time-consuming methodologies for quantifying levels of fluorescent age-pigments will supplant the more convenient and rapid length-frequency analysis. More likely, they may become a useful way of ground truthing the results of more conventional techniques.

Multiple morphometrics

Studies on *E. superba* and *T. macrura* have attempted to use multiple morphometrics to obtain more information on krill population age structure than can be obtained by conventional length–frequency analysis (Farber-Lorda, 1990; Farber-Lorda, 1991; Farber-Lorda, 1994). Although this approach yielded some promising preliminary results in separating out groups of adults of both sexes, it has not been adopted in other studies.

Eyeballs

A novel approach to ageing in *E. superba* has involved utilising some measure of the size of the eyeball relative to the body size (Sun *et al.*, 1995; Sun & Wang, 1995a, b, 1996; Sun, 1997). The premise of this approach is that the compound eye is a complex structure and would be conserved if the animal is shrinking. Consequently, the eye of an adult krill which had shrunk would be relatively larger when compared to the eye of a juvenile of the same size which had been growing. The initial results of this approach showed that laboratory-shrunk Antarctic krill had relatively larger eyes (from crystalline cone counts) than might be predicted from their overall body length. Further, the number of crystalline cones in the eyes of krill obtained from a sample caught in spring (the period following the greatest supposed food scarcity and hence the greatest likelihood of shrinkage) was higher than might be predicted for their size and resembled those of shrunken laboratory krill (Sun *et al.*, 1995). Subsequently, eyeball diameter has been shown to be a good substitute for crystalline cone counts and eyeball diameter appears to remain constant when krill shrink in the laboratory (Sun & Wang, 1996; Sun, 1997). Although this technique has shown promise there are no recent reports of attempts to apply it to field samples from other areas. One important practical outcome of this method is that it appears possible to use the crystalline cone number or eyeball diameter to distinguish older krill from new recruits. This is sufficient information to estimate the key parameters of recruitment and mortality (de la Mare 1994a, b).

Maintenance

The potential age of Antarctic krill has been demonstrated by laboratory studies which have indicated that they can live in captivity for over 6 years giving a total longevity of at least 7–8 years (Ikeda, 1987). There have been no reported studies of maintaining other species of krill for extended periods and for most species, the estimated life-span is derived from size frequency analysis.

Growth

Growth rates

Krill growth rates have been estimated in a number of ways. The most frequently used method is the analysis of sequential length–frequency plots from samples obtained throughout a seasonal cycle. Experimental studies have provided useful comparative measures of laboratory growth rates for larvae, juveniles and adults. Finally, a combination of field and laboratory measurements – the instantaneous growth rate (IGR) methodology (Nicol *et al.*, 1993; Quetin *et al.*, 1994) has produced some recent results for Antarctic krill and this technique may be applicable to other species. Daily growth rates are often difficult to compare not only because of different methods of measurement, but also because growth is a function of size, age, maturation, feeding condition and environmental condition. Furthermore, growth in many species has a strong seasonal component. Most laboratory experiments have utilised constant environmental conditions which may lead to unrealistically high or low growth rates.

Field estimates of summer growth rates in *E. superba* populations have been calculated from time series of length–frequency data, (Miller & Hampton, 1989). Only two studies have been reported which track changes in the mean length of krill populations over short periods of time. Kanda *et al.* (1982) observed a change in mean size of a population of *E. superba* from 41.13 to 41.98 mm in 12 days corresponding to a daily change of 0.063 mm. Over a 30-day moult cycle this would represent a size change at moult of ~4.6% which is consistent with laboratory measurements. Clarke & Morris (1983) measured the modal length change of a population of *E. superba* sampled over a 6-day period and observed a 2 mm increase corresponding to a 0.33 mm day^{-1} increase. The krill in their study were ~30 mm long so for a 30-day moult cycle the percentage increase at moult would be 33%, considerably higher than other estimates. Other studies on Antarctic *E. superba* have used data obtained from much more widely spaced sampling intervals which may introduce a degree of error since it is not known whether the same population is being sampled over time. McClatchie (1988) predicted growth rates of between 0.041 to 0.125 mm day^{-1} for adult and juvenile *E. superba* in Admiralty Bay during summer (~4.1–12.5% increase in length per moult, again assuming a 30-day moult cycle) which are similar figures to those derived from the IGR measurements (Quetin & Ross, 1991). Growth rates of *E. superba* have been calculated from modal peaks in the length frequencies from the South West Indian Ocean sector (Farber-Lorda, 1994). The estimated growth rate for each of the identified year classes was: 1+: 0.12–0.133, 2+: 0.077–0.08, 3+: 0.048–0.057, 4+: 0.03–0.034, 5+: 0.019–0.022 – in mm day^{-1} – equivalent to up to 18% per moult.

The growth rate of other species of krill has been estimated mostly from length–frequency analyses. The data for *M. norvegica* indicate that growth is similar in the Skagerrak and Kattegat and in North Atlantic waters (Poulsen, 1926; Einarsson, 1945; Lindley, 1982; Boysen & Buchholz, 1984). Although growth is only part of the energy budget, daily growth rates in g C day^{-1} showed that most of the carbon goes into body growth, the carbon net growth efficiency is highest in the *E. pacifica* calyptopis stage (60 to 74%), but decreases to 10% in the adults. About 3 to 4% goes into moults and 10 to 18% into reproduction during the adult stage, which is on average more than used for growth at this age (Ross 1982a, b). This may explain how *T. inermis* and *T. raschii* can reach a larger maximum size but at the same time postpone maturation and spawning by one or even two years in subarctic regions (Einarsson, 1945; Lindley, 1980; Dalpadado & Skjoldal, 1991, 1995). Even for these species the high arctic waters are so unfavourable that they no longer reproduce, leading to the conclusion that there are no true arctic euphausiid species (Einarsson, 1945; Dalpadado & Skjoldal, 1996; Weigmann-Haass, 1997).

Field-derived estimates of growth rates of *E. superba* indicate that daily growth rates decrease with age. Larvae during the first year of life grow at 0.13 to 0.16 mm d^{-1}, juveniles of age 1+ at 0.12 to 0.148 mm d^{-1}, while older animals grow only at 0.025 to 0.07 mm d^{-1} (see Table 5.3). Growth parameter estimates for krill sampled during the 'Discovery' investigations were determined by Rosenberg *et al.* (1986) using the von-Bertalanffy growth function:

$$L_t = L_\infty\left(1 - e^{-K(t-t_0)}\right)$$

Where L_t is the length at age t, L_∞ is the asymptotic length, K is the coefficient of catabolism and t_0 is the theoretical origin of the equation.

The k values for the yearly as well as the monthly data were highly consistent. However, the estimated yearly k values of the juvenile krill were approximately 25% of those for the high growth season indicating that maximum growth of *E. superba* (up to 0.179 mm day^{-1}) occurs over little more than three months (Rosenberg *et al.*, 1986). This conclusion is similar to that derived from analyses of growth data using a seasonal VBGF which concluded that krill growth in the Antarctic Peninsula area is probably confined to the period from late October to the end of January/early February (Siegel, 1982). These data fit the theoretical growth curves established by Mauchline (1980), if the annual growth period is limited to 90 days.

In the Bering Sea growth rate for postlarval *T. raschii* and *T. inermis* is about 2% dry body weight per day, but is higher in warmer years, often > 5% (Table 5.3) when they grow larger (Smith, 1991). In the North Sea *T. inermis* grew at a rate of 1.5% dry body weight per day during March to June (Lindley & Williams, 1980). Growth rates were highest (almost four times as high as the average) early in the season during and shortly after the spring bloom. In northern Norway in spring and summer *T. inermis* and *T. raschii* grew at an average rate of 1.2% and 1.7% dry body weight per day, respectively, (Hopkins *et al.*, 1984). Lipid synthesis in *T. inermis* is about 0.2% body weight per day, which obviously is a high proportion of the total daily growth rate (Smith, 1991).

E. pacifica is short-lived in Californian waters, about 8 months for winter and early spring cohorts, but growth can exceed 3 to 3.5 mm/month. Late spring and summer cohorts have a life expectancy of 12 months and growth rates are similar to the winter cohorts except for the slowing down in growth to 1.5 mm/month from October at the start of autumn (Brinton, 1976). Longer living animals (15 to 21 months) in Japanese waters show a similar rapid growth during spring, March to May, (Nishikawa *et al.*, 1995) but zero growth from November to March (Taki &

Table 5.3 Growth rates of euphausiid species; IGR = instantaneous growth rates IGR* are measured in mm (or % of TL) per intermoult period

Species	Geogr. region	0 group	Growth rate in mg WW/day 1+ group	older groups	life-time	References
Meganyctiphanes norvegica	Iceland	0.238	0.31		0.274	Mauchline & Fischer, 1969
Meganyctiphanes norvegica	Scotland	0.378	0.476		0.427	Mauchline & Fischer, 1969
Meganyctiphanes norvegica	Spain	0.383			0.383	Mauchline & Fischer, 1969
Thysanoessa raschii	Scotland	0.126	0.184		0.155	Mauchline & Fischer, 1969
Thysanoessa inermis	N Iceland/Greenland	0.040	0.127	0.230	0.132	Mauchline & Fischer, 1969
Thysanoessa inermis	S Iceland	0.093	0.110		0.101	Mauchline & Fischer, 1969
			Growth rate in % WW/day			
Thysanoessa raschii, 1st + 2nd winter	N Norway	−0.09	−0.21			Hopkins *et al.*, 1984
Thysanoessa raschii, summer	N Norway		1.53	0.68		Hopkins *et al.*, 1984
Thysanoessa inermis, 1st + 2nd winter	N Norway	0.11	−0.03			Hopkins *et al.*, 1984
Thysanoessa inermis, summer	N Norway		1.02	0.36		Hopkins *et al.*, 1984
			Growth rate in ug DW/day			
Thysanoessa inermis, adult	North Sea, mean		107.7			Lindley & Williams, 1980
Thysanoessa inermis, adult	North Sea, before bloom		44.1			Lindley & Williams, 1980
Thysanoessa inermis, adult	North Sea, during bloom		152.5			Lindley & Williams, 1980
Thysanoessa inermis, larvae	North Sea, mean	1.5				Lindley & Williams, 1980
Euphausia superba, larvae	Antarctic Peninsula	8.3				Huntley & Brinton, 1991
			Growth rate in ug C/day			
Euphausia pacifica, calyptopis	Washington, 8–12°C	0.29–1.43				Ross, 1982
Euphausia pacifica, furcilia	Washington, 8–12°C	1.37–5.41				Ross, 1982
Euphausia pacifica, juvenile	Washington, 8–12°C	4.08–11.95				Ross, 1982
Euphausia pacifica, adult	Washington, 8–12°C	7.79–33.83				Ross, 1982
Euphausia superba, C1 to F5 larvae	Antarctic Peninsula	2.93–13.61				Huntley & Brinton, 1991

Continued

Table 5.3 *Continued*

Species	Geogr. region	Growth rate in mg WW/day				References
		0 group	1+ group	older groups	life-time	
		Growth rate in % DW/day				
Thysanoessa raschii, Nauplius 1	Bering Sea	6.5–13.5				Smith, 1991
Thysanoessa raschii, Calyptopis 1	Bering Sea	9.0–11.5				Smith, 1991
Thysanoessa raschii, subadult M	Bering Sea	1.1–4.7				Smith, 1991
Thysanoessa raschii, subadult F	Bering Sea	1.3–9.4				Smith, 1991
Thysanoessa raschii, adult M	Bering Sea	1.8–4.0				Smith, 1991
Thysanoessa raschii, adult F	Bering Sea	0.6–2.4				Smith, 1991
Thysanoessa raschii, calyptopis	Bering Sea	14.0				Vidal & Smith, 1986
Thysanoessa raschii, subadult	Bering Sea		3.0			Vidal & Smith, 1986
Thysanoessa inermis, subadult F	Bering Sea		1.0–6.0			Smith, 1991
Thysanoessa inermis, adult M	Bering Sea		1.3			Smith, 1991
Thysanoessa raschii, 1st + 2nd winter	N Norway	−0.02	−0.29			Hopkins *et al.*, 1984
Thysanoessa raschii, summer	N Norway		1.74	0.98		Hopkins *et al.*, 1984
Thysanoessa inermis, 1st + 2nd winter	N Norway	−0.03	−0.05			Hopkins *et al.*, 1984
Thysanoessa inermis, summer	N Norway		1.19	0.73		Hopkins *et al.*, 1984
		Growth rate in mm TL/day				
Euphausia pacifica	Oregon	0.053–0.096				Smiles & Pearcy, 1971
Euphausia pacifica, February cohort	S California	0.085				Brinton, 1976
Euphausia pacifica, July cohort	S California	0.073				Brinton, 1976
Euphausia pacifica, September cohort	S California	0.062				Brinton, 1976
Euphausia pacifica, December cohort	S California	0.090				Brinton, 1976
Euphausia pacifica	Washington	0.095			0.065	Bollens *et al.*, 1992
Euphausia pacifica, spring	Canada BC	0.05–0.08				Fulton & LeBrasseur, 1984
Euphausia pacifica	Canada BC	0.094			0.038	Heath, 1977 cf. Bollens *et al.*, 1992
Euphausia pacifica, < 10 mm, summer	Washington	0.08–0.09				Cooney, 1971 cf. Bollens *et al.*, 1992
Euphausia pacifica, > 10 mm, summer	Washington	0.05–0.08				Cooney, 1971 cf. Bollens *et al.*, 1992

Euphausia pacifica	Canada BC	0.095	0.075			Hulsizer, 1971 cf. Bollens *et al.*, 1992
Euphausia pacifica, summer	Canada BC	0.066	0.047			Hulsizer, 1971 cf. Bollens *et al.*, 1992
Euphausia pacifica, winter	Canada BC	0.001–0.004				Hulsizer, 1971 cf. Bollens *et al.*, 1992
Euphausia pacifica, larvae, summer	N Japan	0.133				Nishikawa *et al.*, 1995
Euphausia pacifica, winter	N Japan	0.000				Taki & Ogishima, 1997
Euphausia pacifica, summer	N Japan		0.056			Taki & Ogishima, 1997
Euphausia pacifica, larvae, summer	S Japan	0.100				Iguchi *et al.*, 1993
Euphausia superba	Antarctic, experimental				0.0345	Murano *et al.*, 1979
Euphausia superba, 22–44 mm	Antarctic, experimental			0.070		Ikeda *et al.*, 1985
Euphausia superba	Antarctic, experimental				0.024–0.044	Poleck & Denys, 1982
Euphausia superba, juvenile	Antarctic, experimental		0.047			Ikeda & Thomas, 1987a, b
Euphausia superba, larvae, summer	Antarctic, experimental	0.06–0.11				Ikeda, 1984
Euphausia superba, larvae, winter	Antarctic, experimental	0.02				Elias, 1990, cf. Quetin *et al.*, 1994
Euphausia superba, 32 mm	Antarctic, experimental			0.083–0.156		Buchholz *et al.*, 1989
Euphausia superba, 40 mm	Antarctic Peninsula, IGR*			0.056		Quetin & Ross, 1991
Euphausia superba, 40 mm	Antarctic Peninsula, IGR*			−0.04 to −0.812		Quetin & Ross, 1991
Euphausia superba, larvae, winter	Antarctic Peninsula, IGR*	0.017				Ross & Quetin, 1991
Euphausia superba, juveniles/adults respe	Antarctic, Indian Oc. IGR*		2.4 to 9.5%	0.3/7.3%		Nicol *et al.*, 1993
Euphausia superba, growth season	Antarctic, theoretical				0.130	Mauchline, 1980
Euphausia superba, mean	Antarctic, theoretical				0.032	Mauchline, 1980
Euphausia superba, summer	Antarctic Peninsula	0.130		0.025		McClatchie, 1988
Euphausia superba, winter	Antarctic Peninsula			0.010–0.048		McClatchie, 1988
Euphausia superba, larvae, summer	Antarctic Peninsula	0.163				Huntley & Brinton, 1991
Euphausia superba, larvae, winter	Antarctic, Atlantic	0.070				Daly, 1990
Euphausia superba, max. growth season	Antarctic, Atlantic		0.148			Rosenberg *et al.*, 1986
Euphausia superba	Antarctic Peninsula		0.120	0.070	0.033	Siegel, 1986
Euphausia superba	Antarctic, Indian Ocean			0.058		Kanda *et al.*, 1982
Euphausia superba	Antarctic, Indian Ocean		0.120–0.133	0.019–0.080		Pakhomov, 1995a
Euphausia superba, 24–36 mm	Antarctic, Atlantic		0.133			Clarke & Morris, 1983
Euphausia superba	Antarctic, Indian Ocean		0.109	0.039–0.078		Hosie *et al.*, 1988
Euphausia superba, larvae, winter	Antarctic, Indian Ocean	0.047				Hosie & Stolp, 1989

Ogishima, 1997). In the higher latitudes of the NW Pacific *E. pacifica* shows a slower growth rate and longer life cycle compared to lower latitudes (California and Oregon) (Nemoto, 1957; Ponomareva, 1963). Growth rates are almost twice as high in the south and the number of generations per year also decreases from California to the north. In Japanese waters *E. pacifica* reaches the same maximum size (Iguchi *et al.*, 1993). Rapid growth of *E. pacifica* off Oregon and California may be related to the high productivity of the upwelling events and the lack of large seasonal temperature fluctuations (Smiles & Pearcy, 1971). Under experimental conditions increasing temperatures did indeed shorten intermoult periods in *E. pacifica*, but growth increments did not change (Iguchi & Ikeda, 1995). Optimum temperature for growth in this species is thought to be about 10–13°C.

Growth in euphausiids probably is not temperature dependent, but related to food availability (Clarke & Peck, 1991). Daily growth rates can be quite high at the height of the production season which indicates that a substantial (but not necessarily perfect) degree of temperature compensation in growth rate occurs. If growth in polar species is affected by seasonal resource limitation rather than temperature, some predictions can be made:

(1) Growth will be seasonal in herbivores and less so in carnivores under the same environmental conditions.
(2) Maximum growth rates will be comparable between polar species and similar species from warmer waters (Clarke & Peck, 1991).

Good examples for the first prediction are the euphausiids *E. superba*, *T. inermis* and *T. raschii* which show high growth rates modulated by a strong seasonal pattern (e.g. Rosenberg *et al.*, 1986; Siegel, 1986, 1987; Dalpadado & Skjoldal, 1996; Falk-Petersen *et al.*, 1981; Falk-Petersen & Hopkins, 1981). The species *M. norvegica* and *Euphausia triacantha* are known to feed throughout the winter as omnivores or even pure carnivores and although growth slows down at the end of the winter these species grow continuously in length with a much less pronounced seasonal growth pattern (Boysen & Buchholz, 1984, Falk-Petersen, 1985; Baker, 1959; Siegel, 1987). Interestingly, the data for *M. norvegica* also indicate that growth is similar for different geographical regions like the Skagerrak and Kattegat and North Atlantic (Poulsen, 1926; Einarsson, 1945; Lindley, 1982; Boysen & Buchholz, 1984).

Similar maximum growth rates have been reported from a range of species from polar and more temperate waters. *E. superba* daily growth rates estimated from field samples during the summer growth period (Table 5.3), can be as high as those theoretically predicted by Mauchline (1980) for an annual growth period of 90 days. From these results it is apparent that the growth potential of the Antarctic krill *E. superba* is comparable to other species like *Thysanoessa* spp or *M. norvegica* in more temperate waters. It is not the potential growth rate in *E. superba* that is slower during the summer growth period, but the shorter annual growth period in polar latitudes that is responsible for lesser annual increase in size. Those results support the view that krill growth performance is more or less independent of temperature within the range to which the species is adapted, but it is the food resource supply that drives the growth performance. Similarly, the rapid growth of *T. inermis* shows that the species must have a highly efficient biosynthesis despite the low summer-water temperature. Fatty acid biosynthesis from cold-water Norway was triple that of specimens from temperate Scotland. From this Falk-Petersen (1985) drew the conclusion that the constraints leading to an overall reduced annual growth in high latitudes are not physico-chemical, but ecological.

Seasonality and winter stagnation

Parameter C of the seasonal von-Bertalanffy growth function (seasonal VBGF in the form of $L_t = L_\infty\ (1 - e^{-[k(t-t_o) + CU/(2\mu)\ \sin 2\mu\ (t-t_S)]})$ with t_s as the starting point of the sinus oscillation) describes the amplitude of growth oscillation which normally ranges between 0 and 1; low values express a low seasonality in the growth amplitude, values near 1 indicate a stagnation of growth in winter, larger values mean a high amplitude and possible degrowth in winter. Seasonal VBGF curves in some cases show that growth of *T. inermis* ceases during winter (Dalpadado & Skjoldal, 1991, 1996) or between March and October for *E. superba* (Siegel, 1986) (Table 5.4). The only seasonal VBGF curve for *M. norvegica* shows a C value that suggests a lack of any seasonal stagnation in growth (Sager & Sammler, 1984).

M. norvegica is known to feed throughout the winter as a carnivore and although growth slows down a little at the end of the winter, growth in length is continuous (Boysen & Buchholz, 1984). Wet weight of individuals was relatively stable, while dry weight decreased at the end of winter (Falk-Petersen, 1985). In the Mediterranean, there is a rapid increase in body size during the first year followed by a tenfold reduction in growth rate in the second (Labat & Cuzin-Roudy, 1996). This pattern was attributed to a concentration on somatic growth during the first year and gonadal production during the second. The sizes of the two-year classes in the Mediterranean are somewhat smaller than those found in the North Atlantic and the North Sea despite the growth rates being similar. The patterns of growth for this species are likely to vary considerably in this species because of its wide geographic range extending over regimes with different productivity.

Laboratory estimates

The growth rates of *E. superba* derived from laboratory studies have been summarised by Buchholz (1991) and a range of positive growth increments up to 21% per moult have been reported. Most of these laboratory growth studies have relied on measurements of length taken from consecutively moulted exoskeletons which gives a coarse-scale picture of growth over a whole moult cycle. One laboratory study has used a version of the 'instantaneous growth rate technique', measuring the length of the krill and their newly produced exuviae (Buchholz *et al.*, 1989) and an

Table 5.4 Parameters of von Bertalanffy growth function. L inf = asymptotic length, k = growth constant, ts = starting point of oscillation, to = origin of growth curve, C = amplitude of growth oscillation, sVBGF = seasonalised version

Species	Geogr region	VBGF parameters t = 1 year, 1* = per day						References
		L inf	k	ts	to	C		
Thysanoessa inermis	Barents Sea	45.0	0.16	0.88	−0.41	2.80	sVBGF	Dalpadado & Skjoldal, 1996
Thysanoessa inermis	Barents Sea	30.3	0.53	0.20	−0.29	−1.55	sVBGF	Dalpadado & Skjoldal, 1991
Thysanoessa longicaudata	Barents Sea	20.11	0.56	1.09	−0.27	1.47	sVBGF	Dalpadado & Skjoldal, 1996
Meganyctiphanes norvegica	W Norway	40.5–41.2	0.751–0.776		−0.049/−0.057	0.47–0.51	sVBGF	recalc. from Sager & Sammler, 1984
Euphausia pacifica	British Columbia	30.19	0.0085				t*	Tanasichuk, 1998
Euphausia superba	Antarctic, Scotia Sea	60.0	0.43–0.47					Rosenberg *et al.*, 1986
Euphausia superba	Antarctic Peninsula	63.8–64.2	0.384–0.401		−0.004/−0.106			Siegel, 1986
Euphausia superba	Antarctic Peninsula	61.0	0.4728	−0.027	0.141	0.959	sVBGF	Siegel, 1986
Euphausia superba	Antarctic Peninsula	61.3–63.3	0.354–0.478		0.06–0.155			Siegel, 1987
Euphausia superba	Antarctic Weddell Sea	56.6	0.532		0.072			Siegel, 1987
Euphausia superba	Antarctic, N Indian Ocean	61.3–65.1	0.417–0.440		−0.044/−0.101			Pakhomov, 1995a
Euphausia superba	Antarctic, S Indian Ocean	59.1–60.1	0.466–0.467		−0.002/−0.044			Pakhomov, 1995a

average increase in length of 3.8% per moult was observed under conditions of high food availability.

Laboratory growth rates of *E. pacifica* have been reported for individuals from populations in widely separated parts of its range. Iguchi & Ikeda (1995) report growth rates of 1.1–2.0% per moult with moulting rates of between 2 and 25 days. Long-term food deprivation results in a relatively constant shrinkage rate of 0.033 mm per day (nearly −3.3% per moult cycle) in juvenile *E. superba* (Ikeda *et al.*, 1985). This figure is similar to the maximum rate of shrinkage demonstrated by Antarctic krill in IGR experiments that have been allowed to proceed beyond five days (Nicol *et al.*, 1993). Thus under experimental conditions krill can apparently achieve their maximal shrinkage rate within a matter of days of food limitation and, in the laboratory, krill have survived periods of starvation-induced shrinkage for 211 days (Ikeda & Dixon, 1982).

Instantaneous growth rate (IGR)

A methodology for the measurement of IGR was outlined by Quetin & Ross (1991) and their results demonstrated both shrinkage and growth in freshly caught krill. This technique was further modified by Nicol *et al.* (1993) to examine the effects of capture on the measurements of growth rate and to increase the sample size (Nicol *et al.*, 1999). This technique is beginning to provide information on the growth rate of a wide range of sizes of krill which can be related to environmental conditions and to the observed changes in size of the whole population (Ross *et al.*, 2000).

The instantaneous growth rate technique relies on collecting live krill in good condition. On board, the krill are kept in individual containers and are checked regularly for moults during the five days following capture. During this period the growth rate of krill appears to be stable and is thought to reflect the natural growth rate; after five days the growth rate starts to decline and is thought to reflect the effect of confinement in the laboratory (Nicol *et al.*, 1993). The growth rate is estimated from the difference in length of the uropods on the moults and the whole post-moult krill. Since Antarctic krill moult every 20–30 days at 0°C (Buchholz, 1991) only 1 in 20 to 1 in 30 krill will moult on any given day of the experiment. This methodology therefore requires large numbers of animals of which only a small fraction will be used for actual measurements of initial growth rates. The key to obtaining greater numbers of growth rate measurements is to increase the number of krill that can be maintained individually at one time (Nicol *et al.*, 2000). Using small (250 ml), perforated jars to hold individual krill in a large (1000 l) flow-through tank permits a sample size of over 1000 krill per experiment, a much larger sample size than has been the case in earlier studies.

The IGR technique has shown positive growth of adult *E. superba* during the late summer and autumn with increases in body length at moulting of between 1.75 and 4.4% (Quetin & Ross, 1991) and, in winter, shrinkage of between −0.16 and −2.03% of body length per moult. The growth rates of Antarctic krill in the Indian

Ocean sector of the Southern Ocean were measured in four summers (Nicol *et al.*, 1993) using the IGR technique. Mean growth rates ranged from 0.35 to 7.34% per moult in adults and 2.42 to 9.05% in juveniles. The mean growth rates of adult and juvenile krill differed between areas and between the different years of the investigation. When food was restricted under experimental conditions, individual krill began to shrink immediately and mean population growth rates decreased gradually, becoming negative after as little as seven days. Populations of krill which exhibited higher initial growth rates began to shrink later than those which had initially been growing more slowly. IGR measurements on Antarctic krill in the southeastern Indian Ocean sector using the modified technique with large sample sizes (up to 1000 animals moulting per experiment) provided results which again indicated that juvenile krill exhibit the fastest growth rates (a mean of 6% per moult) (Nicol *et al.*, 2000). Males were growing slightly slower than juveniles (mean 4.3%), early stage females were growing at a higher rate (mean 2.4%) whereas gravid and spent females were growing very slowly (mean 0.9%). All stages were growing at a rate that was similar to the rate predicted from demographic measurements of the mean change in size of each of the maturity and sex groups over the sampling period.

The IGR technique appears to offer a method which provides data on growth rates that can be compared to field-derived methods using length frequency analysis from the same populations. This will allow some of the conflicting results on krill growth that have emerged from field and laboratory studies to be examined. Although the technique has been developed for Antarctic krill it should be applicable to other species which are robust enough to withstand capture and experimental confinement. For smaller temperate or tropical species, the more rapid moulting rates may actually enhance the sensitivity of this technique by providing far greater sample sizes.

Shrinkage

The intense seasonality of Antarctic regions has led to investigations into how Antarctic krill populations survive the winter. Due to the low food availability and the low lipid content in *E. superba* Ikeda & Dixon (1982) assumed that feeding ceases during winter and their experimental results indicated that if this happens they will shrink. A reduction in metabolic rate could also be the mechanism that permits krill to overwinter without feeding (Quetin & Ross, 1991). Body shrinkage would then account for only 4% of the energy required. Both scenarios are based on the assumption that krill do not feed during the winter but this may not always be the case. Feeding on ice-algae has now been reported several times (Daly & Macaulay, 1988; Marschall, 1988; Stretch *et al.*, 1988; Lancraft *et al.*, 1991) and krill may also feed on zooplankton or protozoa (Siegel, 1988; Lancraft *et al.*, 1991; Hopkins & Torres, 1989; Siegel, 1989; Lancraft *et al.*, 1991). Winter feeding by *E. superba* is also indicated by observed C:N ratios and ammonium excretion rates (Huntley *et al.*, 1994). Shrinkage, starvation and even reduced metabolism of Antarctic krill may be unusual phenomena in winter under natural conditions.

Winter is the period when food is most likely to be limiting and hence this is the period when overall shrinkage of krill populations would be likely to occur. Daly (1990) reported significant growth of larval *E. superba* during winter in the marginal ice zone but there are few unequivocal data sets which demonstrate either growth or shrinkage of adult or larval Antarctic krill over winter. Krill captured under sea-ice over winter showed a decrease in mean size and a restriction in the size range of krill present (Kawaguchi *et al.*, 1986). They also found little evidence of feeding in the water column, because there was only low chlorophyll a concentrations throughout the winter and the oxygen consumption figures corresponded to those of starving, shrinking krill (Ikeda & Dixon, 1982). *E. superba* in Admiralty Bay (Antarctic Peninsula region) may show winter shrinkage (Ettershank, 1983) but this change may have been a result of either water movements into and out of the bay bringing krill of different origin and life history or of differential size-related mortality (McClatchie *et al.*, 1991).

Reduction in the mean size of *E. superba* populations at the end of the winter has not been observed in the few studies that have been sampled in this time period, nor has there been a reported reduced range of sizes in each adult age-group. If shrinkage does occur over winter and it occurs in a way that preserves the order of the lengths between age groups across the population, then the method of decomposition of length–frequency distributions into separate components of length for each age class remains valid (de la Mare, 1994a, b).

Using the IGR technique individual Antarctic krill have been observed shrinking during winter (Quetin & Ross, 1991) although it is not certain whether shrinkage is a widespread or common enough phenomenon to significantly affect the mean size of the population. The major question is how frequently (if ever) krill encounter food shortage sufficient to engender shrinkage and do they do so for sufficient periods to affect the size structure of the population. Only *in situ* studies in winter and spring will be able to answer these questions.

A reduction in mean length of *T. inermis*, mainly of age class 2+ during winter, has been reported (Dalpadado & Skjoldal, 1996) but the mean size of this age group never became as small as the mean summer length of age group 1+ (which also shrank in winter), so a critical overlap in the length distribution between these two cohorts never occurred and the age–size groups remained identifiable.

Euphausiids are obviously capable of shrinkage but when and where this occurs is a matter for further research. How such shrinkage affects the analysis of population dynamics of krill depends on whether the shrinkage occurs in a systematic fashion, preserving the year classes or whether it tends to merge year classes. In analyses of euphausiid populations it should be recognised that they are capable of shrinking and that this should be taken into account as a potential source of uncertainty in life-history models.

5.4 Length–weight relationships

A number of studies have reported relationships between the various body-length measurements and weight measures for euphausiids (Table 5.5). Measurements of length–weight relationships are affected by preservation; *M. norvegica* lost about 15% of the fresh weight when stored in formalin or even 21% when preserved in alcohol (Kulka & Corey, 1982). These losses appear to be species-specific. When *T. inermis* was transferred from formalin into alcohol, specimens lost another 30% of their wet weight; more than one would expect from figures obtained for *M. norvegica*. It is clear that practical problems also exist in the accuracy of weight measurements of fresh specimens onboard vessels.

Length–weight relationships in male and female *M. norvegica* in the Kattegat show seasonal differences (Boysen & Buchholz, 1984). Although males are heaviest in spring and autumn, females are heavier than males in spring, but after spawning in late summer (April to September) are lighter. However, they attain the same weight as males in autumn. In western Norwegian fjords monthly differences in length–weight relationships of *M. norvegica* are observed, but differences between males and females are not apparent, although males are, on average, heavier in late summer (Bamstedt, 1976). *M. norvegica* appears to be heavier in summer and autumn than in winter and spring. Significant seasonal differences in length–weight relationships were also reported from *E. pacifica* in Japanese waters (Iguchi *et al.*, 1993).

In populations in Northern Norway *T. inermis* shows monthly differences in length–weight relationships and length–lipid relationships and no differences in length–protein, while *T. raschii* shows significant differences in all three regressions (Falk-Petersen, 1985). The decrease in dry weight is first seen in December and lasts to March and it indicates that a negative energy balance lasts from mid-winter until spring. Interestingly, *M. norvegica* showed significant differences in monthly length to wet-weight relationships; however, absolute values for dry weight and protein contents did not significantly change with season. The differences in the various relationships of these three species may reflect different overwintering strategies in the subarctic environment. *T. inermis* builds up lipid reserves as an energy resource for the winter, *T. raschii* partly utilises lipids stores but additionally feeds on detritus whereas *M. norvegica* is carnivorous (Falk-Petersen, 1985).

Length–weight relationships for the Antarctic krill *E. superba*. have been summarised previously (Morris *et al.*, 1988; Siegel, 1992b). We have condensed some of this large amount of information into a few length–weight relationships to facilitate their practical use (Table 5.5). The first general observation on these relationships is that length–weight relationships do not differ between males and females during spring (November), when animals are in the pre-spawning stage (IIIA), but still are of low weight. Generally in December there is a significant increase in weight with a simultaneous progression of krill maturity to gravid stages; males and females still do not show significant differences in length–weight relationships. After spawning (January to March) the weight of males and females drops immediately with females losing significantly more weight than males. From this low level both sexes

Table 5.5 Length–weight relationships for different euphausiids, weight in mg and length in mm; a = regression constant, b = regression coefficient; WW = wet weight, DW = dry weight, TL = total length, CL = carapace length, M = male, F = female

Species	Geogr. region	Length/Weight relationship parameters					References
		a	log a	b	WW/DW	Length	
Thysanoessa inermis	N Norway		−2.580	3.330	WW	TL	Dalpadado & Sjkoldal, 1991
Meganyctiphanes norvegica	Canada, Nova Scotia		−0.136	2.730	WW	CL	recalc. from Kulka & Corey, 1982
Thysanoessa inermis	Canada, Nova Scotia		−0.342	2.882	WW	CL	recalc. from Kulka & Corey, 1982
Meganyctiphanes norvegica	Canada, Nova Scotia		−1.005	2.717	DW	CL	recalc. from Kulka & Corey, 1982
Thysanoessa inermis	Canada, Nova Scotia		−1.083	2.884	DW	CL	recalc. from Kulka & Corey, 1982
Maganyctiphanes norvegica	Kattegat, February	0.00144		3.029	DW	TL male	Boysen & Buchholz, 1984
Meganyctiphanes norvegica	Kattegat, February	0.00036		3.397	DW	TL female	Boysen & Buchholz, 1984
Meganyctiphanes norvegica	Kattegat, May	0.00023		3.533	DW	TL male	Boysen & Buchholz, 1984
Meganyctiphanes norvegica	Kattegat, May	0.00019		3.629	DW	TL female	Boysen & Buchholz, 1984
Meganyctiphanes norvegica	Kattegat, August	0.00145		3.028	DW	TL male	Boysen & Buchholz, 1984
Meganyctiphanes norvegica	Kattegat, August	0.00219		2.848	DW	TL female	Boysen & Buchholz, 1984
Meganyctiphanes norvegica	Kattegat, November	0.00014		3.729	DW	TL male	Boysen & Buchholz, 1984
Meganyctiphanes norvegica	Kattegat, November	0.00009		3.876	DW	TL female	Boysen & Buchholz, 1984
Meganyctiphanes norvegica	Canada, St. Lawrence	0.00029		3.500	DW	TL	Sameoto, 1976
Thysanoessa raschii	Canada, St. Lawrence	0.000717		3.170	DW	TL	Sameoto, 1976
Thysanoessa inermis	Canada, St. Lawrence	0.000712		3.380	DW	TL	Sameoto, 1976
Thysanoessa inermis	N Norway, May		1.202	2.203	WW	CL	Falk-Petersen, 1985
Thysanoessa inermis	N Norway, May		0.134	2.338	DW	CL	Falk-Petersen, 1985
Thysanoessa inermis	N Norway, August		0.682	2.543	WW	CL	Falk-Petersen, 1985
Thysanoessa inermis	N Norway, August		0.084	2.885	DW	CL	Falk-Petersen, 1985
Thysanoessa inermis	N Norway, November		0.746	2.532	WW	CL	Falk-Petersen, 1985
Thysanoessa inermis	N Norway, November		0.094	3.007	DW	CL	Falk-Petersen, 1985
Thysanoessa inermis	N Norway, February		0.657	2.607	WW	CL	Falk-Petersen, 1985
Thysanoessa inermis	N Norway, February		0.065	3.130	DW	CL	Falk-Petersen, 1985
Thysanoessa raschii	N Norway, May		0.521	2.667	WW	CL	Falk-Petersen, 1985
Thysanoessa raschii	N Norway, May		0.039	3.235	DW	CL	Falk-Petersen, 1985
Thysanoessa raschii	N Norway, August		0.634	2.654	WW	CL	Falk-Petersen, 1985
Thysanoessa raschii	N Norway, August		0.142	2.666	DW	CL	Falk-Petersen, 1985

Continued

Table 5.5 *Continued*

Species	Geogr. region	Length/Weight relationship parameters					References
		a	log a	b	WW/DW	Length	
Thysanoessa raschii	N Norway, November		0.409	2.917	WW	CL	Falk-Petersen, 1985
Thysanoessa raschii	N Norway, November		0.056	3.265	DW	CL	Falk-Petersen, 1985
Thysanoessa raschii	N Norway, February		0.449	2.858	WW	CL	Falk-Petersen, 1985
Thysanoessa raschii	N Norway, February		0.054	3.208	DW	CL	Falk-Petersen, 1985
Meganyctiphanes norvegica	N Norway, May		0.151	3.310	WW	CL	Falk-Petersen, 1985
Meganyctiphanes norvegica	N Norway, July		0.698	2.714	WW	CL	Falk-Petersen, 1985
Meganyctiphanes norvegica	N Norway, November		0.744	2.711	WW	CL	Falk-Petersen, 1985
Meganyctiphanes norvegica	N Norway, February		0.779	2.693	WW	CL	Falk-Petersen, 1985
Meganyctiphanes norvegica	N Norway, all year		0.039	3.028	DW	CL	Falk-Petersen, 1985
Meganyctiphanes norvegica	W Norway, February	0.097		3.004	DW	CL	Bamstedt, 1976
Meganyctiphanes norvegica	W Norway, May	0.092		2.954	DW	CL	Bamstedt, 1976
Meganyctiphanes norvegica	W Norway, August	0.361		2.480	DW	CL	Bamstedt, 1976
Meganyctiphanes norvegica	W Norway, November	0.074		3.218	DW	CL	Bamstedt, 1976
Thysanoessa inermis	N Atlantic	0.0013		2.916	DW	TL	Lindley, 1978
Thysanoessa raschii	N Atlantic	0.0013		2.916	DW	TL	Lindley, 1978
Euphausia pacifica	Japan, summer		−2.580	3.310	WW	TL	Iguchi *et al.*, 1993
Euphausia pacifica	Japan, winter		−2.270	3.020	WW	TL	Iguchi *et al.*, 1993
Euphausia pacifica	Japan	0.00099		3.156	DW	TL	Iguchi & Ikeda, 1995
Thysanoessa inermis	N Atlantic		−1.170	2.770	WW	CL	Matthews & Hestad, 1977 (cf. Mauchline, 1980)
Meganyctiphanes norvegica	Nova Scotia		−3.540	3.500	DW	TL	Sameoto, 1976b
Meganyctiphanes norvegica	W Norway, November		−0.99	2.820	DW	TL	Barattelid & Matthews, 1978
Meganyctiphanes norvegica	W Norway, December		−0.85	2.910	DW	TL	Barattelid & Matthews, 1978
Meganyctiphanes norvegica	W Norway, June		−0.42	2.350	DW	TL	Barattelid & Matthews, 1978
Thysanoessa raschii	Bering Sea, March		−1.192	3.056	DW	CL	Vidal & Smith, 1986
Thysanoessa raschii	Bering Sea, April		−1.210	3.257	DW	CL	Vidal & Smith, 1986
Thysanoessa raschii	Bering Sea, May		−0.914	2.874	DW	CL	Vidal & Smith, 1986
Euphausia superba	Antarctic Peninsula, winter		−3.250	3.270	DW	TL	Huntley *et al.*, 1994
Euphausia superba	Antarctic Peninsula, January		−2.909	2.866	DW	TL	Huntley *et al.*, 1994
Euphausia superba, M	Ant. Pen. Nov. Pre-spawning	0.00076		3.071	DW	TL	Siegel, 1992
Euphausia superba, F	Ant. Pen. Nov. pre-spawning	0.00105		2.965	DW	TL	Siegel, 1992

Euphausia superba, M	Ant. Pen., gravid	0.00019	3.435	DW	TL	Siegel, 1986
Euphausia superba, F	Ant. Pen., gravid	0.00025	3.357	DW	TL	Siegel, 1986
Euphausia superba, M	Ant. Pen., February, spent	0.00009	3.694	DW	TL	Siegel, 1986
Euphausia superba, F	Ant. Pen., February, spent	0.00031	3.306	DW	TL	Siegel, 1986
Euphausia superba, M	Ant. Pen., adult	0.00401	2.790	DW	TL	Morris *et al.*, 1988
Euphausia superba, F	Ant. Pen., gravid	0.00199	3.043	DW	TL	Morris *et al.*, 1988
Euphausia superba, M	Ant. Pen.	0.00238	2.927	DW	TL	Morris *et al.*, 1988
Euphausia superba, F	Ant. Pen., non-gravid	0.00139	3.073	DW	TL	Morris *et al.*, 1988
Euphausia superba, M	Ant. Pen.	0.00613	3.077	WW	TL	Morris *et al.*, 1988
Euphausia superba, F	Ant. Pen., gravid	0.00975	2.981	WW	TL	Morris *et al.*, 1988
Euphausia superba, M	Ant. Pen., adult	0.01729	2.815	WW	TL	Morris *et al.*, 1988
Euphausia superba, F	Ant. Pen., non-gravid	0.01088	2.908	WW	TL	Morris *et al.*, 1988
Euphausia superba, M	Ant. Pen. Nov. Pre-spawning	0.00315	3.207	WW	TL	Siegel, 1989
Euphausia superba, F	Ant. Pen. Nov. Pre-spawning	0.0043	3.102	WW	TL	Siegel, 1989
Euphausia superba, M	Ant. Pen., gravid	0.00083	3.561	WW	TL	Siegel, 1986
Euphausia superba, F	Ant. Pen., gravid	0.00115	3.457	WW	TL	Siegel, 1986
Euphausia superba, M	Ant. Pen., February, spent	0.00111	3.507	WW	TL	Siegel, 1986
Euphausia superba, F	Ant. Pen., February, spent	0.00211	3.302	WW	TL	Siegel, 1986
Euphausia superba, M	Ant. Pen., winter, resting stage	0.00328	3.176	WW	TL	Siegel, 1989
Euphausia superba, F	Ant. Pen., winter, resting stage	0.00441	3.084	WW	TL	Siegel, 1989

further steadily decline in weight to a minimal and similar winter level (Siegel, 1989). In summary, this indicates that differences in male and female length–weight relationships are observed during a very brief period immediately after spawning (with males heavier than females) and a significant difference between the spawning season (high) and the rest of the year (low).

Morris *et al.* (1988) detailed variability of length–weight relationships in *E. superba* and they found that the moult cycle has little effect upon length–weight relationships. In their opinion a simple division of krill into males and females without considering maturity stages does not improve the precision of the length–weight relationships. However, separating first gravid females from males or non-gravid females or secondly adult males from other males would improve the precision of weight predictions.

Weight at length has been examined as a 'condition index' for *E. superba* in the South-west Indian Ocean (Farber-Lorda, 1994) using a complex allometric morphometric relationship. He found no significant differences as a function of sex or developmental stage in either total length and weight or carapace length. The changes in the relationship between the carapace length and dry weight have also been used as a 'condition factor' to examine overwintering in *E. superba* (Kawaguchi *et al.*, 1986). They showed systematic decreases in both carapace length and dry weight over winter and an increase in both of these measurements in spring. A more simple approach has indicated regional differences in the weight of juvenile Antarctic krill of the same length in the South-east Indian Ocean (Nicol *et al.*, 2000) but whether this technique is generally applicable as a condition index is uncertain.

Based on the above observations, it can be recommended that for greatest utility:

(1) During the summer spawning period separate length–weight relationships should be applied for gravid females, non-gravid females, adult male and other male krill.
(2) During the post-spawning season separate length–weight relationships should be applied for the same categories as under (1) and additionally for spent females.
(3) During winter and spring general length–weight relationships can be used which should be on a monthly basis in the pre-spawning period, although statistically the differences are not significantly different before the occurrence of gravid stages.
(4) Where data are available only on sex and size of the animals, during and shortly after the spawning period, separate length–weight relationships should be applied for male and female krill. A greater improvement in precision is gained by considering maturity stages separately.

5.5 Natural mortality

There are very few quantitative estimates of natural mortality in the published literature on euphausiid species (Table 5.6). For non-Antarctic species changes in abundance of single modes over a period of time are presented as a reduction in relative frequencies. Mortality of *E. pacifica* larvae is high in Californian waters. Brinton (1976) estimates that only 16% of the larvae survive per month. After the short larval phase the mortality rate is lower, about 67% of the juveniles survive to adults per month, but mortality increases again after spawning and adult survival is only about 60% per month for the adults until the next spawning event.

Mortality rates have been reported for *E. pacifica* (Jarre-Teichmann, 1996) with M = 3.0 y-1 off California, M = 8.7 y-1 for Oregon and 0.6 to 1.9 y-1 for British Columbia. These differences are difficult to explain, although the maximum life span is different in these areas. As expected the lowest mortality rate is seen for the animals living for 19–22 months off British Columbia. However, stocks off Oregon and California reach the same maximum age of about one year, but their life cycle is different. One cohort is reproducing about three times during its life off California, but seems to reproduce only once off Oregon (Table 5.6), for which a much lower mortality rate would be expected. One possible explanation could be that the mortality rate estimate from Oregon was derived from a period of strong El Niño, which may have increased the mortality of the species beyond normal levels.

Mortality of *Thysanoessa* species in the Barents Sea is lowest in winter, when the feeding pressure of the main predator stock (cod and capelin) are minimal. Age group 0+ has the highest summer growth rate, but also the highest mortality rate (98%). Age group 1+ shows a mortality rate of approximately 50%, which still leaves enough spawners (2+) in the next season to sustain the population. After spawning mortality is thought to increase rapidly so that it was suspected that hardly any 2+ animals survive (Drobysheva, 1987). However, other studies have found a mortality rate of 55% y-1 for 2+ aged animals of *T. inermis* (Lindley, 1980), and the presence of the 3+ age group has been reported, although in very low numbers in the Barents Sea (Dalpadado & Skjoldal, 1991; Dalpadado & Skjoldal, 1995; Dalpadado & Skjoldal, 1996). In Scottish waters *T. raschii* shows a slightly higher mortality rate (75%) for the 1-year-olds than in the Barents Sea (Mauchline & Fisher, 1969).

Estimates of mortality rates for *E. superba* have been carried out, using different methods. The results vary between 0.38 and 5.5 for the postlarval population (Table 5.6). Some of these seriously underestimated krill longevity. Since maximum age is a function of natural mortality, this underestimation of longevity led to the highest values. The Beverton and Holt (1959) approach takes into account the size composition of the catches under the assumption that the population is fully represented in the samples (Pakhomov, 1995b) but this is not always the case. If age classes are under-represented by recruitment failure or by net selectivity for larger animals, or if sampling is not covering the geographical range of all size groups (Siegel, 1988), then mortality rates can be seriously biased.

Mortality is inversely related to the longevity of the individuals of the population and hence with the growth constant parameter 'k' of the VBGF. High growth rates are usually associated with high mortality rates, and low growth rates with low mortality rates, so that the ratio, M/k, is approximately constant for a given species.

Table 5.6 Natural mortality rates (M) of euphausiids. As indicated under References in some cases M had to be calculated from data mentioned in these papers; y = year, mo = month.

Species	Geogr. region	Age group or develop. stage	Mortality M	Per y or mo	Method	References
Euphausia pacifica	California	larvae 0+	1.83	mo		calc. from Brinton, 1976
Euphausia pacifica	California	juvenile, 0+	0.40	mo		calc. from Brinton, 1976
Euphausia pacifica	California	adult, 0+	0.51	mo		calc. from Brinton, 1976
Euphausia pacifica	California	lifetime	3.00	y		cf. Jarre-Teichmann, 1996
Euphausia pacifica	Oregon	lifetime	8.70	y		cf. Jarre-Teichmann, 1996
Euphausia pacifica	Brit. Columbia	lifetime	0.60–1.90	y		cf. Jarre-Teichmann, 1996
Thysanoessa spp	Barents Sea	0+	3.90	y		calc. from Drobysheva, 1987
Thysanoessa spp	Barents Sea	1+	0.69	y		calc. from Drobysheva, 1987
Thysanoessa inermis	N Atlantic	1+	0.80	y		Lindley, 1980
Thysanoessa raschii	NW Atlantic	1+	0.29–1.2	y		Lindley, 1980
Thysanoessa raschii	Scotland	1+	1.39	y		Mauchline & Fisher, 1969
Meganyctiphanes norvegica	Scotland	1+	0.69	y		Mauchline & Fisher, 1969
Meganyctiphanes norvegica	Scotland	2+	2.30–3.00	y		Mauchline & Fisher, 1969
Euphausia superba	Antarctic, Indian Ocean		5.50	y	Edmondson's	Kawakami & Doi, 1979
Euphausia superba	Antarctic, Scotia Sea	<2 year olds	2.31	y	size group progression	Brinton & Townsend, 1984
Euphausia superba	Antarctic, Scotia Sea	2 and 3 year olds	0.51	y	size group progression	Brinton & Townsend, 1984
Euphausia superba	Antarctic Peninsula	postlarvae	0.78–1.17	y	linearised catch curve	Siegel, 1986
Euphausia superba	Antarctic Peninsula	postlarvae	0.88–0.96	y	linearised catch curve	Siegel, 1992a
Euphausia superba	Antarctic Peninsula	postlarvae	0.66–0.92	y	Alagajara, max age: M	Siegel, 1992b
Euphausia superba	Antarctic Peninsula	postlarvae	0.94–0.99	y	Pauly, VBGF: M	Siegel, 1986
Euphausia superba	Antarctic, Scotia Sea	postlarvae	0.81–1.35	y	VBGF K: M	Priddle *et al.*, 1988
Euphausia superba	Antarctic, Scotia Sea	postlarvae	0.5	y	1-cumulative length frequency	Basson & Beddington, 1989
Euphausia superba	Antarctic, Scotia Sea	postlarvae	0.45–0.65	y	length-depend. predation curve	Basson & Beddington, 1989
Euphausia superba	Antarctic, Scotia Sea	postlarvae	0.75–1.13	y		Maklygin, 1992 in CCAMLR Rep
Euphausia superba	Antarctic Peninsula	postlarvae	0.38–1.22	y	Pauly, VBGF: M; max age: M	Basson, 1994
Euphausia superba	Antarctic, Indian Ocean	postlarvae	0.52–0.57	y	Richter & Efanov	Pakhomov, 1995a
Euphausia superba	Antarctic, Indian Ocean	postlarvae	0.72–0.87	y	Alverson & Carney	Pakhomov, 1995a
Euphausia superba	Antarctic, Indian Ocean	postlarvae	0.76–2.92	y	Beverton & Holt	Pakhomov, 1995a
Euphausia superba	Antarctic, Indian Ocean	year 1	1.09	y	Zikov & Slepokurov	Pakhomov, 1995a
Euphausia superba	Antarctic, Indian Ocean	year 2	0.65	y	Zikov & Slepokurov	Pakhomov, 1995a
Euphausia superba	Antarctic, Indian Ocean	year 3	0.57	y	Zikov & Slepokurov	Pakhomov, 1995a
Euphausia superba	Antarctic, Indian Ocean	year 4	0.77	y	Zikov & Slepokurov	Pakhomov, 1995a
Euphausia superba	Antarctic, Indian Ocean	year 5	1.54	y	Zikov & Slepokurov	Pakhomov, 1995a

For a large range of fish species, the ratio of M/k lies between 1.5 and 2.5 . Preliminary estimates for Euphausiids range from M/k = 1.5 to 2.5 based on the regression equation of Pauly (1980) and between 0.83 and 1.67 for *E. superba* obtained from the maximum age calculations (Basson, 1994). The range of M values for *E. superba* from both approaches is 0.38 to 1.22.

Reviews of the methodology used in earlier studies concluded that the realistic values of natural mortality for the postlarval life cycle may range between M = 0.66 and 1.35 (Siegel, 1992a; Pakhomov, 1995b). It is also likely that M varies between geographical regions and interannually, and it is known to vary during the lifetime of a species as already indicated above for *T. inermis*. Possible differences in mortality can be attributed to differing predator demands, as well as short-term variability or long-term trends of environmental parameters. There may also be differences in natural mortality of different developmental stages or age groups. Larval mortality, however, is not of direct concern for management; the basic requirement is the estimate of M for those age classes typically affected by the fishery.

Using the Zikov & Slepokurov (1982) approach Pakhomov (1995b) estimated that the natural mortality rate for *E. superba* during the first year is relatively high (M = 1.11 to 1.12), it decreases thereafter during the maturation period (M = 0.52 to 0.65) and increases slowly for older age groups, reaching a maximum during the last year of life (M = 1.29 to 2.41). These results strongly confirm the dependence of natural mortality on age. Similar studies in different regions and different years are certainly needed, to evaluate possibilities of interannual or geographical variation in mortality rates of certain age groups and consequently year classes, leading finally also to variability in krill recruitment processes.

5.6 Recruitment variation

Recruitment of euphausiids has only recently been estimated following the introduction of a technique for *E. superba* (de la Mare, 1994a,b). For some other species there are only descriptive comments on the success of single or few cohorts. Recruitment of *E. pacifica* in Californian waters occurs after about 30 days, when the larvae enter the juvenile phase. At least four generations are generally reported each year. Due to the short life-span and the few cohorts, the maximum stock size of this species is reached immediately after successful recruitment of a single cohort. In general there is no spawning stock-recruitment relationship. In most years highest recruitment is observed for the spring and summer cohorts, while autumn and winter cohorts show low recruitment in these waters (Brinton, 1976). On the other hand the July to December recruits live longer (10–13 months) than the shorter-living February to June recruits (6–8 months).

While up to four or more *E. pacifica* cohorts recruit off California every year, there is only one large pulse of larvae recruiting in spring off the coast of Washington and a distinct but less abundant cohort in late summer (Bollens *et al.*, 1992). The reduced level of inferred spawning in summer is discussed by these authors as a result of the much reduced phytoplankton level at this time of the year, although *E. pacifica* is known to show omnivory. There is apparently no recruitment during winter as is the case off California. Low temperatures cannot be responsible, because reproduction of this species occurs in the northwestern Pacific at even lower temperatures. Another possibility could be that survival of overwintering adult *E. pacifica* is low due to predation in spring and summer and only few 1+ adults remain to spawn in late summer. At the same time only few 0-aged animals are able to mature fast enough to spawn during their first year of life (Ross *et al.*, 1982).

In protected fjords of British Columbia recruitment occurs after around 47 days, which is later than off California (Tanasichuk, 1998). In contrast to Washington waters there is evidence for 4 to 6 cohorts recruiting every year with generally two less successful spawning events in winter with no changes in spawning patterns and frequencies between pre-post-ENSO and ENSO years. Maximum *E. pacifica* abundance in El Niño years, however, was related to exceptional recruitment success (Tanasichuk, 1998). Interestingly this author also reports an inverse stock-recruitment relationship and suggests that high parental stock densities can be responsible for high cannibalism and as a consequence reduce the larval stock substantially. Tanasichuk (1998) believes that good recruitment is not directly triggered by warm water, but is caused by strong upwelling in the area, preceding the warm water events This idea would be in conformity with Smiles & Pearcy (1971), who argue that off Oregon and California both spawning and recruitment success depend strongly on upwelling conditions and related primary production. The upwelling period would be the critical time for larvae to get enough phytoplankton for a better growth rate. Small changes in growth rates of krill larvae may cause high fluctuations in recruitment (Daly 1990). A combination of predation, food limitation and possibly cannibalism for eggs and larvae may restrict the period of successful recruitment for *E. pacifica*.

In northern Japanese waters recruitment takes place after two months at 8 mm length (Nishikawa *et al.*, 1995). Japan, like California, is at the southern distribution limit of this species, but at its southwestern distribution limit *E. pacifica* produces only one generation every year (Iguchi *et al.*, 1993). The conditions may be less favourable in these waters than in the productive upwelling areas along the north American coast. Small changes in the influence of the cold and productive Oyashio current, versus the warm less productive Kuroshio current, negatively affect the *E. pacifica* stock (Nishikawa *et al.*, 1995; Taki & Ogishima, 1997; Yamamura *et al.*, 1998). However, the analysis of quantitative aspects of *E. pacifica* recruitment in Japanese waters has still to be performed.

In the Barents Sea a long-term 12–14 year periodicity of overall krill density was reported to be possibly coupled with a high-frequency 2–3 year variation in krill recruitment (Boytsov & Drobysheva, 1987). Recruitment was defined as the transition from the larval into the juvenile age group 0+ and changes in recruitment were explained by influences in water temperature on the youngest age group, so that the effects can be seen with a time lag of one year (Drobysheva, 1987). This would allow predictions about possible recruitment success of *T. inermis* stocks.

Recruitment of *E. superba* was first defined as the proportion of recruits known as the R_1 rate which is the ratio of numbers in age class 1 to the numbers in age class 1 and all age classes above (de la Mare, 1994a). This procedure can also be applied for age class 2. Since krill distribution is extremely patchy, neither stock density estimates nor density for each length class of standardised length–frequency distributions show a Gaussian distribution. De la Mare (1994b) analysed the underlying statistics of these distributions and developed a new approach to create density-at-length (instead of length–frequency) data from scientific surveys as a basis for recruitment estimates. However, some of the survey results on R_1 and R_2 given by de la Mare (1994b) for the Atlantic sector are not representative, because these calculations did not consider the incomplete survey coverage of the stock and its underlying geographical non-random distribution pattern of different age groups (Siegel *et al.*, 1997).

One disadvantage of using the proportion of recruits is the possible misinterpretation of the values; a high proportionate value has different meanings when stock density is either very high or very low. By standardising the number of recruits to a given density (number of recruits/1000 m^3), this new index was defined as absolute recruitment of one-year-olds (RI_1, (Siegel *et al.*, 1998)).

Another disadvantage of the proportional index is that it does not necessarily provide an indicator of reproductive success. This is because it does not refer to the reproductive stock of the previous year, but to the year of the survey when recruitment takes place, and the assumption of a closed krill population considered to be in equilibrium is far from being realistic. In reality large variations in stock and spawning stock size are quite apparent from krill net surveys, although these changes in abundance usually occur gradually over several years and not from one year to the next (Siegel *et al.*, 1997, 1998). However, some conclusions can still be drawn from the presently defined recruitment indices. The longest time series exists for the Antarctic Peninsula region where proportional recruitment peaks were observed for the year classes 1980–81, 1981–82, 1987–88, 1994–95. It is also evident that most values from scientific surveys since the mid-1980s show generally poor recruitment results which might explain part of the low stock size in the 1990s (Siegel *et al.*, 1997, 1998). No stock-recruitment relationship could be found for Antarctic krill (Siegel & Loeb, 1995) and the very strong 1994–95 year class demonstrated that even at a low spawning stock level strong cohorts can be produced.

Estimates of recruitment by the proportional method require unbiased sampling regimes. Hence they are not strictly applicable to data derived from fisheries operations. Given the low density of sampling for most areas, however, fisheries-derived indices can provide information that can augment scientific studies. Results for the R_2 values derived from Japanese Antarctic krill fishery data from the South Shetland Island region show significant correlation with the R_1 and R_2 recruitment indices derived from scientific surveys (Kawaguchi *et al.*, 1997). In contrast R_1 values derived from the fishery were not significantly correlated with the scientific data, probably due to net selectivity of the commercial trawl and/or incomplete coverage of the areas where 1+ age-class krill tend to occur.

Immediately after the first recruitment estimates had been made, it became obvious that the results differed substantially between years and regions. Various parameters had been tested to find possible links between krill recruitment and the biological and physical environment. The most promising variable to explain the high variability in krill recruitment was sea-ice (Kawaguchi & Sarake, 1994; Siegel & Loeb, 1995), especially the spatial extent and duration of winter sea-ice cover. Since sea-ice seems to affect different krill population dynamics at a number of scales, the following conceptual model was developed by Siegel & Loeb (1995) derived from significant correlations of various parameters.

Extended sea-ice cover and a large spatial extent during winter is favourable for an early onset of the spawning season, probably due to the ice-algae resource available to krill during late winter to early spring. Protection from predation may be another reason for a higher survival rate. As a consequence early reproduction leads to a successful spawning event in summer. Larvae that hatch early in summer show a high rate of survival during the feeding and growth season and are well conditioned before their first winter. At the same time, salps which are thought to compete for phytoplankton are scarce following winters of extensive ice cover. A severe winter also favours the survival of the krill larvae under the ice, so that at the end of the larval phase when entering the juvenile stage as one-year-olds (recruits) we observe a high recruitment rate for that cohort.

In case of poor winter sea-ice conditions, salps as opportunistic short-living animals develop rapidly in early spring and compete with krill for phytoplankton (Loeb *et al.*, 1997). With efficient competitors in the system and no additional sea-ice algae resource, krill may not be able to fulfil its energy requirement for growth and an early spawning. Delayed or prolonged spawning leads to further competition for newly hatched larvae with salps, and larvae show slower growth and therefore according to Daly (1990) lower survival rates. An unsuccessful spawning event will consequently be followed by poor recruitment in the next spring even if the following winter sea-ice conditions are favourable for larval survival.

Geographical differences in krill recruitment are quite obvious from Table 5.7. As an example year classes 1980–81 and 1985–86 were highly successful in the Atlantic sector, while recruitment in the Indian Ocean was a failure during the same years. On the other hand year class 1988–89 experienced high recruitment in the Indian Ocean, but it was low in the Atlantic. This is no surprise, because individual regions of the Southern Ocean show significant interannual variability in ice coverage with no coherence between the Antarctic Peninsula region and the Indian Ocean sector (Stammerjohn & Smith, 1996). Due to these differences in the periodicity of ice conditions for different sub-regions of the Southern Ocean, krill recruitment indices have to be estimated and interpreted separately.

Populations of euphausiids are likely to suffer the impacts of long-term changes

Table 5.7 Recruitment of Antarctic krill, *Euphausia superba*; year gives the time when the year-class was born.

Year	Proportional recruitment (R)						Absolute R
	R1	R2	R1	R2	R2		RI1
	de la Mare, 1994b		Siegel *et al.*, 1997, 1998		Kawaguchi *et al.*, 1997		N/1000 m^3
	Indian Ocean		Antarctic Peninsula		Livingston I.	Elephant I.	Antarctic Pen.
1975/76				0.144			
1976/77			0.048				16.71
1977/78							
1978/79		0.096		0.069	0.336	0.451	
1979/80	0.167	0.561	0.559		0.000	0.086	90.21
1980/81	0.001		0.757		0.462		245.93
1981/82		0.557	0.470	0.663	0.585	0.427	130.05
1982/83	0.016	0.431	0.030	0.0001	0.156	0.142	4.59
1983/84	0.528		0.0001	0.214	0.147	0.072	0.01
1984/85		0.231	0.175		0.191	0.165	
1985/86	0.025			0.633	0.316	0.334	
1986/87			0.156	0.291	0.194	0.360	4.31
1987/88			0.651	0.275	0.000	0.000	51.98
1988/89		0.556	0.057	0.063	0.000	0.000	0.88
1989/90	0.314		0.099	0.345		0.083	0.64
1990/91		0.02	0.375	0.587	0.574	0.129	7.45
1991/92	0.064		0.000	0.012	0.719	0.169	0.00
1992/93			0.068	0.029	0.000	0.000	2.24
1993/94			0.046	0.125	0.058	0.000	3.37
1994/95			0.622	0.837			74.62
1995/96			0.198	0.384			42.07
1996/97			0.120				3.06
1997/98							
1998/99							

caused by global warming or ozone depletion. Increasing air temperature will affect sea-surface temperature and ice conditions in the Southern Ocean. Consequently this will affect krill recruitment and krill stock size in the long term. Increased UV-B is another variable that might damage near surface krill concentrations, especially larvae, and increase mortality rates and again affect recruitment rates and overall krill biomass (Jarman *et al.*, 1999; Newman *et al.*, 1999). Current low recruitment rates over many years have to be regarded with some concern. The krill stocks and the population dynamic parameters have to be monitored carefully to gain a better understanding of the development of these stocks as key elements of the ecosystems. The key question is have we been observing high interannual variability over the past decade or is this the starting point of a downward trend?

5.7 Conclusions

A large amount of effort has been devoted to the study of krill, particularly Antarctic krill, over the last 20 years. This has resulted in a number of developments in methodology and in knowledge of certain key areas of their life history parameters. Examples of these developments are the techniques for maintaining krill in laboratories, the statistical methods developed for recruitment analysis and the improvement in the tools available for length–frequency analysis. Despite these advances there are still a number of basic aspects of the life history of these organisms that require detailed study if we are to advance our understanding of krill population dynamics. Specifically, improved, rapid techniques for ageing krill would be extremely useful, validation of existing growth rate methodologies is necessary and experimental studies into population parameters of species other than *E. superba* are desirable. Additionally, further studies into the recruitment and mortality of a range of species of krill are required. Generally, there is still a major problem in reconciling the results of field studies with those from the laboratory and this should be a topic for future research. Many of the outstanding problems in krill research could be addressed through concerted short-term (< 1 month) studies on identified populations (or swarms) of krill which can be sampled repeatedly to provide a time series of demographic measurements from an identified population. The resulting information might then be compared to experimental results on individuals obtained from the same population.

References

Aseev, Y.P. (1984) Size structure of krill populations and life span in the Indian Ocean Sector of the Antarctic. *Hydrobiology* **6**, 89–94.

Asthorsson, O.S. (1990) Ecology of the euphausiids *Thysanoessa raschii*, *T. inermis* and *Meganyctiphanes norvegica* in Isafjord-Deep, northwest-Iceland. *Mar. Biol.* **107**, 147–57.

Baker, A. de C. (1959) The distribution and life history of *Euphausia triacantha* Holt and Tattersall. *Discovery Rep.* **29**, 309–340.

Baker, A. de C. (1963) The problem of keeping planktonic animals alive in the laboratory. *J. Mar. Biol. Ass. UK* **43**, 291–94.

Bamstedt, U. (1976) Studies on the deep-water pelagic community of Korsfjorden, Western Norway; changes in the size and biochemical composition of *Meganyctiphanes norvegica* (Euphausiacea) in relation to its life cycle. *Sarsia* **61**, 15–30.

Basson, M. (1994) Towards a distribution of M/K for krill (*Euphausia superba*) required for the stochastic krill yield model. CCAMLR-WG-Krill. 94/11, 16.

Basson, M. & Beddington, J. R. (1989) In *University Research in Antarctica. Proceedings of British Antarctic Survey Antarctic Special Topic Award Scheme Symposium.* (R. B. Heywood, ed.), pp. 51–55. British Antarctic Survey, Cambridge.

Berkes, F. (1976) Ecology of Euphausiids in the Gulf of St. Lawrence. *Journal of the Fisheries Research Board of Canada* **33**, 1894–1905.

Beverton, R.J.H. & Holt, S.J. (1959) A review of the lifespans and mortality rates of fish in nature, and their relation to growth and other physiological factors. *CIBA Foundation Colloquium on Aging* **5**, 142–77.

Bhattacharya, C. (1967) A simple method of resolution of a distribution into Gaussian components. *Biometrics* **23**, 115–35.

Bollens, S.M., Frost B.W. & Lin, T.S. (1992) Recruitment, growth, and diel vertical migration of *Euphausia pacifica* in a temperate fjord. *Mar. Biol.* **114**, 219–28.

Boysen, E. & Buchholz, F. (1984) *Meganyctiphanes norvegica* in the Kattegat – studies of the annual development of a pelagic population. *Mar. Biol.* **79**, 195–207.

Boytsov, V.D. & Drobysheva, S.S. (1987) Effect of hydrometeorological factors on the regularity of the long-term variations in euphausiid (Crustacea, Euphausiacea) abundance in the southern Barents Sea. (Anonymous). In *Proceedings of the third Soviet-Norwegian Symposium, Murmansk, 26–28 May 1986. The effect of oceanographic conditions on distribution and population dynamics of commercial fish stocks in the Barents Sea.* Institute of Marine Research, Bergen, pp. 91–100.

Brattelid T.E. & Mathews J.B.L. (1978) Studies on the deep-water pelagic community of Korsfjorden, western Norway. The dry weight and calorie content of Euchaeta norvegica (Copepoda), *Boreomysis arctica* (Mysidacea), and *Meganyctiphanes norvegica* (Euphausiacea). *Sarsia* **63**, 203–11.

Brinton, E. (1976) Population biology of *Euphausia pacifica* off southern California. *Fish. Bull.* **74**, 733–62.

Brinton, E. & Townsend, A.W. (1984) Regional relationships between development and growth in larvae of Antarctic krill, Euphausia superba from field samples. *J Crust. Biol.* **4** (Spec No 1), 224–46.

Brodeur, R. D. & Pearcy, W. G. (1992) Effects of environmental variability on trophic interactions and food web structure in a pelagic upwelling ecosystem. *Marine Ecology Progress Series* **84**, 101–19.

Buchholz, F. (1985) Moult and growth in euphausiids. (W.R. Siegfried, P. Condy & R.M. Laws, eds), *Antarctic nutrient cycles and food webs.* In Proceedings of the 4th Symposium on Antarctic Biology, Springer, Berlin. pp. 339–45.

Buchholz, F. (1991) Moult cycle and growth of Antarctic krill *Euphausia superba* in the laboratory. *Mar. Ecol. Prog. Ser.* **69**, 217–29.

Buchholz, F., Morris D.J. & Watkins, J.L. (1989) Analysis of field moult data: prediction of intermoult period and assessment of seasonal growth in Antarctic krill *Euphausia superba* Dana. *Antarctic Sci.* **1**, 310–316.

Clarke, A.C. (1976) Some observations on Krill (*Euphausia superba* Dana) maintained alive in the laboratory. *Brit. Ant. Surv. Bull.* **43**, 111–18.

Clarke, A. & Morris, D.J. (1983) Towards an energy budget for krill: the physiology and biochemistry of *Euphausia superba* Dana. *Polar Biol.* **2**, 69–86.

Clarke, A. & Peck, L.S. (1991) The physiology of polar marine zooplankton. *Polar Res.* **10**, 355–69.

Constable, A.J. & de la Mare, W.K. (1996) A generalised model for evaluating yield and the long-term status of fish stocks under conditions of uncertainty. *CCAMLR Sci.* **3**, 31–54.

Dalpadado, P. & Ikeda, T. (1989) Some observations on moulting, growth and maturation of krill (*Thysanoessa inermis*) from the Barents Sea. *J. Plankton Res.* **11**, 133–39.

Dalpadado, P. & Skjoldal, H.R. (1991) Distribution and life history of krill from the Barents Sea. *Polar Res.* **10**, 443–60.

Dalpadado, P. & Skjoldal, H.R. (1995) Distribution and life cycle of krill north of 73°N in the Barents Sea, 1984–1992. *Fisken og Havet* **16**, 1–50.

Dalpadado, P. & Skjoldal, H.R. (1996) Abundance, maturity and growth of the krill species *Thysanoessa inermis* and *T. longicaudata* in the Barents Sea. *Mar. Ecol. Progr. Ser.* **144**, 175–83.

Daly, K.L. (1990) Overwintering development, growth and feeding of larval *Euphausia superba* in the Antarctic marginal ice edge zone. *Limnol. Oceanogr.* **35**, 1564–76.

Daly, K.L. & Macaulay, M.C. (1988) Abundance and distribution of krill in the ice edge zone of the Weddell Sea, austral spring 1983. *Deep-Sea Res.* **35**, 21–41.

de la Mare, W.K. (1994a) Estimating krill recruitment and its variability. *CCAMLR Sci.* **1**, 55–69.

de la Mare, W.K. (1994b) Modelling krill recruitment. *CCAMLR Sci.* **1**, 49–54.

Drobysheva, S.S. (1987) Populational characteristics of abundant Barents Sea Euphausiacea. *ICES C.M.* **12**, 1–15.

Einarsson, H. (1945) Euphausiacea. 1. North Atlantic Species. *Dana Rep.* **27**, 1–185.

Endo, Y. (1981) Ecological studies on the Euphausiids occurring in the Sanriku Waters with special reference to their life history and aggregated distribution., *Tohoku University, Sendai*, 166 pp.

Ettershank, G. (1983) Age structure and cyclical annual change in the Antarctic krill, *Euphausia superba* Dana. *Polar Biol.* **2**, 189–93.

Ettershank, G. (1985) Population age structure in males and juveniles of the Antarctic krill, *Euphausia superba Dana. Polar Biology* **4**, 199–201.

Everson, I. & Bone, D.G. (1986) Effectiveness of the RMT8 system for sampling krill (*Euphausia superba*) swarms. *Polar Biol.* **6**, 83–90.

Falk-Petersen, S. (1985) Growth of the Euphausiids *Thysanoessa inermis*, *Thysanoessa raschii*, and *Meganyctiphanes norvegica* in a subarctic fjord, North Norway. *Can. J. Fish. Aquat. Sci.* **42**, 14–22.

Falk-Petersen, S., Gatten, R.R., Sargent, J.R. & Hopkins, C.C.E. (1981) Ecological investigations on the zooplankton of Northern Norway: seasonal changes in the lipid class composition of *Meganyctiphanes norvegica*, *Thysanoessa inermis* and *Thysanoessa raschii*. *J. Exp. Mar.Biol. Ecol.* **54**, 209–224.

Falk-Petersen, S. & Hopkins, C.C.E. (1981) Ecological investigations of the zooplankton community of Balsfjorden, Northern Norway: population dynamics of the euphausiids *Thysanoessa inermis*, *Thysanoessa raschii* and *Meganyctiphanes norvegica* in 1976 and 1977. *J Plank Res* **3**, 177–92.

Farber-Lorda, J. (1990) Somatic length relationships and ontogenetic morphometric differentiation of *Euphausia superba* and *Thysanoessa macrura* of the southwest Indian Ocean during summer (February 1981). *Deep-Sea Res.* **37**, 1135–43.

Farber-Lorda, J. (1991) Multivariate approach to the morphological and biochemical differentiation of Antarctic krill (*Euphausia superba* and *Thysanoessa macrura*). *Deep-Sea Res.* **38**, 771–79.

Farber-Lorda, J. (1994) Length–weight relationships and coefficient of condition of *Euphausia superba* and *Thysanoessa macrura* (Crustacea: Euphausiacea) in southwest Indian Ocean during summer. *Mar. Biol.* **118**, 645–50.

Fulton, J. & Le Brasseur, R. (1984) Euphausiids of the continental shelf and slope of the Pacific coast of Canada. *La Mer* **22**, 182–90.

Gomez, J.G. (1995) Distribution patterns, abundance and population dynamics of euphausiids *Nyctiphanes simplex* and *Euphausia eximia* off the west coast of Baja California, Mexico. *Mar. Ecol. Prog. Ser.* **119**, 63–76.

Gopalkrishnan, K. (1973) Development and growth studies of the euphausiid *Nematoscelis difficilis* (Crustacea) based on rearing. *Bull. Scripps Inst. Oceanogr.* **20**, 1–87.

Gros, P. & Cochard, J. (1978) Biologie de *Nyctiphanes couchii* (Crustacea, Euphausiacea) dans le secteur nord du golfe de Gascogne. *Ann. Inst. Oceanogr.* **54**, 25–46.

Hammer, C. & Braum, E. (1988) Quantification of age pigments (lipofuscin). *Comp. Biochem. Physiol.* **90**, 7–17.

Harding, J.P. (1949) The use of probability paper for the graphical analysis of polymodal frequency distributions. *J. Mar. Biol. Ass. UK* **28**, 141–53.

Head, E., ed. (1997) *Proceedings of the workshop on ecosystem considerations for krill and other forage fisheries*. Canadian Stock Assessment Proceedings Series, Fisheries and Oceans Science, Canada, Halifax.

Hill, K.T. & Radkte, R.L. (1988) Gerontological studies of the damselfish, *Dascyllus abisella*. *Bull. Mar. Sci.* **42**, 424–34.

Hirche, H.-J. & Anger, K. (1987) The accumulation of age pigments during larval development of the spider crab, *Hyas arenaeus* (Decapoda, Majidae). *Comp. Biochem. Phys.* **88B**, 777–82.

Hollingshead, K.W. & Corey, S. (1974) Aspects of the life history of *Meganyctiphanes norvegica* (M. Sars) Crustacea (Euphausiacea) in Passamaquoddy Bay. *Can. J. Zool.* **52**, 495–505.

Hopkins, C.C.E., Tande, K.S. Gronvik S. & Sargent, J.R. (1984) Biological investigations of the zooplankton community of Balsfjorden, northern Norway: an analysis of growth and overwintering tactics in relation to niche and environment in *Metridia longa* (Lubbock), *Calanus finmarchicus* (Gunnerus), *Thysanoessa inermis* (Kroyer) and T. raschii (M. Sars). *J. exp. Mar. Biol. Ecol.* **82**, 77–99.

Hopkins, T.L. & Torres, J.J. (1989) Midwater food web in the vicinity of a marginal ice zone in the western Weddell Sea. *Deep-Sea Res.* **36**, 565–69.

Hosie, G.W. & Ritz, D.A. (1989). Body shrinkage in the sub-tropical euphausiid *Nyctiphanes australis* G. O. Sars. *J. Plank. Res.* **11**, 595–98.

Hosie, G.W. & Stolp, M. (1989) Krill and zooplankton in the western Prydz Bay region, September–November 1985. *Polar Biology* **2**, 34–45.

Hosie, G.W., Ikeda, T. & Stolp, M. (1988) Distribution, abundance and population structure of the Antarctic krill (*Euphausia superba* Dana) in the Prydz Bay region, Antarctica. *Polar Biol.* **8**, 213–24.

Huntley, M. & Brinton, E. (1991) Mesoscale variation in growth and early development of *Euphausia superba* Dana in the western Bransfield Strait region. *Deep-Sea Research* **38**, 1213–40.

Huntley, M.E., Nordhausen, W. & Lopez, M.D.G. (1994) Elemental composition, metabolic activity and growth of Antarctic krill *Euphausia superba*, during winter. *Mar. Ecol. Prog. Ser.* **107**, 23–40.

Iguchi, N. & Ikeda, T. (1995) Growth, metabolism and growth efficiency of a euphausiid crustacean *Euphausia pacifica* in the southern Japan Sea, as influenced by temperature. *J. Plank. Res.* **17**, 1757–69.

Iguchi, N., Ikeda, T. & Imamura, A. (1993) Growth and life cycle of a euphausiid crustacean (*Euphausia pacifica* Hansen) in Toyama Bay, southern Japan Sea. *Bull. Japan Sea Nat. Fish. Res. Inst.* **43**, 69–81.

Ikeda, T. (1984) Development of the larvae of the Antarctic krill (*Euphausia superba* Dana) observed in the laboratory. *J. Exp. Mar. Biol. Ecol.* **75**, 107–117.

Ikeda, T. (1986) Preliminary observations on the development of the larvae of *Euphausia crystallorophias* Holt and Tattersall in the laboratory. *Memoirs of the National Institute of Polar Research*, Special Issue **40**, 183–86.

Ikeda, T.P.G.T. (1987a) Longevity of the Antarctic krill (*Euphausia superba* Dana) based on a laboratory experiment. *Proc. Natl. Inst. Polar Res. Symp. Polar Biol.* **1**, 56–62.

Ikeda, T. (1987b) Mature Antarctic krill (*Euphausia superba* Dana) grown from eggs in the laboratory. *J. Plank. Res.* **9**, 565–69.

Ikeda, T. & Dixon, P. (1982) Body shrinkage as a possible over-wintering mechanism of the Antarctic krill, *Euphausia superba* Dana. *J. Exp. Mar. Biol. Ecol.* **62**, 143–51.

Ikeda, T. & Thomas, P.G. (1987a) Longevity of the Antarctic krill (*Euphausia superba* Dana) based on a laboratory experiment. *Proceedings of the NIPR Symposium on Polar Biology* **1**, 56–62.

Ikeda, T. & Thomas, P.G. (1987b). Moulting interval and growth of juvenile Antarctic krill fed different concentrations of the diatom *Phaenodactylum tricornutum* in the laboratory. *Polar Biology* **7**, 339–343.

Ikeda, T., Dixon, P. & Kirkwood, J.M. (1985) Laboratory observations of moulting, growth and maturation in Antarctic krill (*Euphausia superba* Dana). *Polar Biol.* **4**, 1–8.

Ikeda, T., Mitchell, A.W., Carlton, J.H. & Dixon, P. (1980) Transport of living Antarctic zooplankton to a tropical laboratory: a feasibility study. *Aust. J. Mar. Freshw. Res.* **31**, 271–74.

Ivanov, B.G. (1970) On the biology of the Antarctic krill (*Euphausia superba Dana*). *Marine Biology* **7**, 340–51.

Jarman, S., Elliott, N., Nicol, S., McMinn, A. & Newman, S. (1999) The base composition of the krill genome and its susceptibility to damage by UV-B. *Antarctic Sci.* **11**, 23–26.

Jarre-Teichmann, A. (1996) Initial estimates on krill. *Fish. Centre Res. Rep.* **4**, 20.

Jillett, J.D. (1971) Zooplankton and hydrography of Hauraki Gulf, New Zealand. *Mem. New Zealand Oceanogr. Inst.* **53**, 1–103.

Jones, L.T., Forsyth, D.C.T. & Cooper, G.A. (1967) The occurrence of the two-spined form of *Thysanoessa inermis* (Crustacea: Eupahsusiacea) in the North Sea. *Bull. Mar. Ecol.* **6**, 181–84

Jorgensen, C. & Matthews, J.B.L. (1975) Ecological studies on the deep water pelagic community of Korsfjorden, western Norway. Population dynamics of six species of euphausiids in 1968 and 1969. *Sarsia* **59**, 67–84.

Kanda, K., Takagi, K. & Seki, Y. (1982) Movement of the larger swarms of Antarctic krill *Euphausia superba* off Enderby Land during 1976–77 season. *J. Tokyo Univ. Fish.* **68**, 24–42.

Kawaguchi, S. & Sarake, M. (1994) Relationship between recruitment of the Antarctic krill and the degree of ice cover near the South Shetland Islands. *Fish. Sci.* **60**, 123–24.

Kawaguchi, K., Ishikawa, S. & Matsuda, O. (1986) The over wintering of Antarctic krill

Euphausia superba Dana under the coastal fast ice off the Ongul Islands in Lutzow-Holm Bay, Antarctica. *Mem. Natl. Inst. Polar Res.*, Special Issue **44**, 67–85.

Kawaguchi, S., Ichii, T. & Naganobu, M. (1997) Catch per unit effort and proportional recruitment indices from Japanese krill fishery data in Subarea 48.1. *CCAMLR Sci.* **4**, 47–63.

Kawakami, T. & Doi, T. (1979) Natural mortality of krill and density in swarms. In Comprehensive report on the population of krill, *Euphausia superba* in the Antarctic. (ed Doi T.) pp 19–21. Tokai regional Fisheries research Laboratory.

Komaki, Y. (1966) Technical notes on keeping euphausiids live in the laboratory, with a review of experimental studies on euphausiids. *Inf. Bull. Planktol. Japan* **13**, 95–105.

Kulka, D.W. & Corey, S. (1978) The life history of *Thysanoessa inermis* (Kroyer) in the Bay of Fundy. *Can. J. Zool.* **56**, 492–506.

Kulka, D.W. & Corey, S. (1982) Length and weight relationships of euphausiids and caloric values of *Meganyctiphanes norvegica* (M. Sars) in the Bay of Fundy. *J. Crust. Biol.* **2**, 239–47.

Labat, J. P. & Cuzin-Roudy, J. (1996) Population dynamics of the krill *Meganyctiphanes norvegica* (M. Sars, 1857) (Crustacea: Euphausiacea) in the Ligurian Sea (NW Mediterranean Sea). Size structure, growth and mortality modelling. *J. Plank. Res.* **18**, 2295–312.

Lancraft, T.M., Hopkins, T.L., Torres, J.J. & Donnelly, J. (1991) Oceanic micronektonic/macrozooplanktonic community structure and feeding in ice covered Antarctic waters during the winter (AMERIEZ 1988). *Polar Biol.* **11**, 157–67.

Lasker, R. (1966) Feeding, growth, respiration and carbon utilisation of a euphausiid crustacean. *J. Fish. Res. Board Can.* **23**, 1291–317.

Lasker, R.T. & Theilacker, G.H. (1965) Maintenance of euphausiid shrimps in the laboratory. *Limnol. Oceanogr.* **10**, 287–88.

Lindley, J.A. (1978) Population dynamics and production of euphausiids. I. *Thysanoessa longicaudata* in the North Atlantic Ocean. *Mar. Biol.* **46**, 121–30.

Lindley, J.A. (1980) Population dynamics and production of euphausiids. II. *Thysanoessa inermis* and *T. raschii* in the North Sea and American coastal waters. *Mar. Biol.* **59**, 225–33.

Lindley, J.A. (1980) Population dynamics and production of euphausiids. II. *Thysanoessa inermis and T. raschi* in the North Sea and American coastal waters. *Marine Biology* **59**, 225–33.

Lindley, J.A. (1982) Population dynamics and production of euphausiids. III. *Meganyctiphanes norvegica* and *Nyctiphanes couchii* in the North Atlantic and the North Sea. *Mar. Biol.* **66**, 37–46.

Lindley, J.A. & Williams, R. (1980) Plankton of the Fladen Ground during FLEX '76. II. Population dynamics and production of *Thysanoessa inermis* (Crustacea: Euphausiacea). *Mar. Biol.* **57**, 79–86.

Loeb, V., Siegel, V., Holm-Hansen, O.R., Hewitt, R., Fraser, W., Trivelpiece, W. & Trivelpiece, S. (1997) Effects of sea-ice extent and salp or krill dominance on the Antarctic food web. *Nature* **387**, 897–900.

Lu, B. & Wang, R. (1996) Further study on distribution mixture analysis and its application to age-groups analysis of Antarctic krill. *Oceanologica et Limnologica Sinica* **27**, 179–86.

Macdonald, P.D.M. & Pitcher, T.J. (1979) Age-groups from size-frequency data: a versatile and efficient method of analyzing distribution mixtures. *J. Fish. Res. Board Can.* **36**, 987–1001.

Mackintosh, N.A. (1967) Maintenance of living *Euphausia superba* and frequency of moults. *Norsk Hvalfangsttid* **56**, 97–102.

Maklygin, L.G. & Latogursky, V.I. (1992) Possible approaches to the evaluation of the Antarctic krill mortality. *SC-CAMLR*. **WG-KRILL–92/8.**

Marr, J.W.S. (1962) The natural history and geography of the Antarctic krill (*Euphausia superba* Dana). *Discovery Rep.* **32**, 33–464.

Marschall, H.-P. (1988) The overwintering strategy of Antarctic krill under the pack-ice of the Weddell Sea. *Polar Biol.* **9**, 129–35.

Matthews, J.B.L. (1973) Ecological studies on the deep-water pelagic community of Korsfjorden, western Norway. Population dynamics of *Meganyctiphanes norvegica* (Crustacea, Euphausiacea) in (1968) and 1969. *Sarsia* **61**, 15–30.

Mathews, J.B.L. & Hestad, L. (1977) Ecological studies on the deep-water pelagic communityof Korsfjorden, western Norway. Length/weight relationships for some macriplankton organisms. *Sarsia* **63**, 57–63.

Mauchline, J. (1960) The biology of the euphausiid crustacean *Meganyctiphanes norvegica* (M Sars). *Proc R Soc Edinb B (Biol)*. **67**, 141–79.

Mauchline, J. (1966) The biology of *Thysanoessa raschii* (M Sars) with a comparison of its diet with that of *Meganyctiphanes norvegica* (M Sars). *In Some contemporary studies in Marine Science* (ed Barnes, H.). pp. 493–510. London: Allen and Unwin.

Mauchline, J. (1980) The biology of mysids and euphausiids. *Adv. Mar. Biol.* **18**, 1–677.

Mauchline, J. & Fisher, L.R. (1969) The biology of euphausiids. *Adv. Mar. Biol.* **7**, 1–454.

McClatchie, S. (1988) Food limited growth of *Euphausia superba* in Admiralty Bay, South Shetland Islands, Antarctica. *Cont. Shelf Res.* **8**, 329–45.

McClatchie, S., Rakusa-Suszczewski, S. & Filcek, K. (1991) Seasonal growth and mortality of *Euphausia superba* in Admiralty Bay, South Shetland Islands, Antarctica. *ICES J. Mar. Res.* **48**, 335–42.

Miller, D.G. & Hampton, I. (1989) Biology and ecology of the Antarctic krill (*Euphausia superba* Dana): a review. *BIOMASS Scientific Series* **9**, 1–66.

Miller, D.G.M., Horstman, D.A., Wehmeyer, H.H. & Kuster, S. (1983) Maintenance of living krill *Euphausia superba* in simple aquaria. *S. Afr. J. Mar. Sci.* **1**, 65–69.

Morris, D.J., Watkins, J.L., Ricketts, C., Buchholz, F. & Priddle, J. (1988) An assessment of the merits of length and weight measurements of Antarctic krill *Euphausia superba*. *Br. Antarctic Survey Bull.* **79**, 27–50.

Mullin, M.M. & Brooks, E.R. (1988) Extractable lipofuscin in larval marine fish. *Fish. Bull. US* **86**, 407–415.

Murano, M., Segawa, S. & Kato, M. (1979) Moult and growth of the Antarctic krill in the laboratory. *Trans. Tokyo Univ. Fish* **3**, 99–106.

Murphy, E.J., Watkins, J.L., Reid, K., Trathan, P.N., Everson, I., Croxall, J.P., Priddle, J., Brandon, M.A., Brierley, A.S. & Hofmann, E. (1998) Interannual variability of the South Georgia marine ecosystem: biological and physical sources of variation in the abundance of krill. *Fish. Oceanogr.* **7**, 381–90.

Nemoto, T. (1957) Foods of baleen whales in the North Pacific. *Sci. Repts. Whales Res. Inst. Tokyo* **12**, 33–89.

Newman, S., Nicol, S., Ritz, D. & Marchant, H. (1999) Susceptibility of Antarctic krill (*Euphausia superba* Dana) to ultraviolet radiation. *Polar Biol.* **22**, 50–55.

Nicol, S. (1987) Some limitations on the use of the lipofuscin ageing technique. *Mar. Biol.* **93**, 609–614.

Nicol, S. (1990) The age-old problem of krill longevity. *BioSci.* **40**, 833–36.

Nicol, S. & Endo, Y. (1997) *Krill Fisheries of the World*, 100 pp. FAO, Rome.

Nicol, S. & Stolp, M. (1991) Molting, feeding and fluoride concentration of the Antarctic krill, *Euphausia superba* Dana. *J. Crust. Biol.* **11**, 10–16.

Nicol, S., Kitchener, J., King, R., Hosie, G.W. & de la Mare, W.K. (2000) Population structure and condition of Antarctic krill (*Euphausia superba*) off East Antarctica (80–150°E) during the Austral summer of 1995/6. *Deep-Sea Res.* In press.

Nicol, S., Stolp, M. & Hosie, G.W. (1991) Accumulation of fluorescent age pigments in a laboratory population of Antarctic krill (*Euphausia superba* Dana). *J. Exp. Mar. Biol. Ecol.* **146**, 153–61.

Nicol, S., Stolp, M., Cochran, T., Geijsel, P. & Marshall, J. (1993) Growth and shrinkage of Antarctic krill *Euphausia superba* from the Indian Ocean sector of the Southern Ocean during summer. *Mar. Ecol. Prog. Ser.* **89**, 175–81.

Nishikawa, J., Tsuda, A., Ishigaki, T. & Terazaki, M. (1995) Distribution of euphausiids in the Kuroshio front and warm water tongue with special reference to the surface aggregation of *Euphausia pacifica*. *J. Plank. Res.* **17**, 611–29.

Pakhomov, E.A. (1995a) Demographic studies of Antarctic krill (*Euphausia superba*) in the Cooperation and Cosmonaut Seas (Indian sector of the Southern Ocean). *Mar. Ecol. Prog. Ser.* **119**, 45–61.

Pakhomov, E.A. (1995b) Natural age-dependent mortality rates of Antarctic krill *Euphausia superba* Dana in the Indian sector of the Southern Ocean. *Polar Biol.* **15**, 69–71.

Pauly, D. (1980) On the interrelationships between natural mortality, growth parameters, and mean environmental temperature in 175 fish stocks. *J. Cons. Int. Explor. Mer.* **39**, 175–92.

Pilar, S.C. (1984) Laboratory studies on the larval growth and development of *Euphausia lucens* (Euphausiacea). *S. Afr. J. Mar. Sci.* **2**, 43–48.

Poleck, T.P & Denys, C.J. (1982) Effect of temperature on the molting, growth and maturation of the Antarctic krill *Euphausia superba* (Crustacea: Euphausiacea) under laboratory conditions. *Marine Biology* **70**, 255–65.

Ponomareva, L.A. (1963) The euphausiids of the North Pacific, their distribution and ecology. *Dokl.Akad. Nauk. SSSR*, 1–142.

Poulsen, E.M. (1926) Om den store lyskrebs (*Meganyctiphanes norvegica*) betydning som fiskfode i Skagerrak. *Dansk Fisk Tid.* **24**, 286–89.

Priddle, J., Croxall, J.P., Everson, I. *et al.* (1988) Large-scale fluctuations in distribution and abundance of krill – A discussion of possible causes. In *Antarctic Ocean and Resources Variability* (ed Sahrage, D.) Berlin: Springer-Verlang.

Quetin, L. & Ross, R.M. (1991) Behavioural and physiological characteristics of the Antarctic krill, *Euphausia superba*. *Am. Zoologist* **31**, 49–63.

Quetin, L.B., Ross, R.M. & Clarke, A. (1994) Krill energetics: seasonal and environmental aspects of the physiology of *Euphausia superba*. In *Southern Ocean Ecology: the BIOMASS perspective* (S.Z. El-Sayed, ed.), pp. 165–84. Cambridge University Press, Cambridge.

Rosenberg, A.A., Beddington, J.R. & Basson, M. (1986) Growth and longevity of krill during the first decade of pelagic whaling. *Nature* **324**, 152–53.

Ross, R.M. (1981) Laboratory culture and development of *Euphausia superba*. *Limnol. Oceanogr* **26**, 235–46.

Ross, R.M. (1982) Energetics of *Euphausia pacifica*. I. Effects of body carbon and nitrogen and temperature on measured and predicted production. *Mar. Biol.* **68**, 1–13.

Ross, R. M. & Quetin, L. B. (1991) Ecological physiology of larval euphausiids, *Euphausia superba* (Euphausiacea). *Memoirs of the Queensland Museum* **31**, 321–33.

Ross, R.M. *et al.* (1982) Energetics of *Euphausia pacifica*. II. Complete carbon and nitrogen budgets at 8° and 12°C throughout the life span. *Mar. Biol.* **68**, 15–23.

Ross, R.M., Daly, K.L. & English, T.S. (1982) Reproductive cycle and fecundity of *Euphausia pacifica* in Puget Sound, Washington. *Limnol. Oceanogr.* **27**, 304–14.

Ross, R.M., Quetin, L.B., Vernet, M., Baker, K.S. & Smith, R.C. (2000) Growth limitation in young *Euphausia superba* under field conditions. *Limn. Oceanogr.* in press.

Rudd, J.T. (1932) On the biology of the southern Euphausiidae. *Hvalradets Skrifter* **2**, 1–105.

Sager, G. & Sammler, R. (1984) Seasonal length growth of the Norwegian krill (*Meganyctiphanes norvegica*) after data from Wiborg (1966–1969). *Zoologische Jahrbucher/Abteilung for Anatomie und Ontogenie* **112**, 79–84.

Sameoto, D.D. (1976) Respiration rates, energy budgets and moulting frequencies of three species of euphausiids found in the Gulf of St. Lawrence. *Journal of the Fisheries Research Board of Canada* **33**, 2568–76.

Sheehy, M.J.R., Greenwood, J.G. & Fielder, D.R. (1994) More accurate chronological age determination of crustaceans from field situations using the physiological age marker, lipofuscin. *Mar. Biol.* **121**, 237–45.

Siegel, V. (1982) Investigations on krill (*Euphausia superba*) in the Southern Weddell Sea. *Meeresforschung* **29**, 244–52.

Siegel, V. (1986) Structure and composition of the Antarctic krill stock in the Bransfield Strait (Antarctic Peninsula) during the Second International BIOMASS Experiment (SIBEX). *Archiv fur Fischereiwiss* **37**, 51–72.

Siegel, V. (1987) Age and growth of Antarctic Euphausiacea (Crustacea) under natural conditions. *Mar. Biol.* **96**, 483–95.

Siegel, V. (1988) A concept of seasonal variation of krill (*Euphausia superba*) distribution and abundance west of the Antarctic Peninsula. *Antarctic Ocean and Resources Variability* (D. Sahrhage, ed.), pp. 219–230. Springer-Verlag, Berlin.

Siegel, V. (1989) Winter and spring distribution and status of the krill stock in the Antarctic Peninsula waters. *Archiv fur Fischereiwiss* **39**, 45–72.

Siegel, V. (1992a) Assessment of the krill (*Euphausia superba*) spawning stock off the Antarctic Peninsula. *Archiv fur Fischereiwiss* **41**, 101–130.

Siegel, V. (1992b) Review of length–weight relationships for Antarctic krill, *Selected Scientific Papers, 1992*, SC-CAMLR-SSP/9, pp. 145–55. CCAMLR, Hobart, Australia.

Siegel, V. & Loeb, V. (1994) Length and age at maturity of Antarctic krill. *Antarctic Sci.* **6**, 479–82.

Siegel, V. & Loeb, V. (1995) Recruitment of Antarctic krill *Euphausia superba* and possible causes for its variability. *Mar. Ecol. Prog. Ser.* **123**, 45–56.

Siegel, V., de la Mare, W.K. & Loeb, V. (1997) Long-term monitoring of krill recruitment and abundance indices in the Elephant Island area (Antarctic Peninsula). *CCAMLR Sci.* **4**, 19–35.

Siegel, V., Loeb, V. & Groger, J. (1998) Krill (*Euphausia superba*) density, proportional and absolute recruitment and biomass in the Elephant Island region (Antarctic Peninsula) during the period 1977–1997. *Polar Biol.* **19**: 393–98.

Smiles, M.C. & Pearcy, W.G. (1971) Size structure and growth rate of *Euphausia pacifica* off the Oregon Coast. *Fish. Bull.* **69**, 79–86.

Smith, S.L. (1991) Growth, development and distribution of the euphausiids *Thysanoessa raschii* (M. Sars) and *Thysanoessa inermis* (Kroyer) in the southeastern Bering Sea. *Polar Res.* **10**, 461–78.

Stammerjohn, S.E. & Smith, R.C. (1996) Spatial and temporal variability of Western Antarctic Peninsula sea ice coverage. (R.M. Ross, E.E. Hofmann, L.B. Quetin, eds), *Ecological Research West of the Antarctic Peninsula*, **70**, 81–104. American Geophysical Union, Washington D.C.

Stretch, J.J., Hamner, P.P., Hamner, W.M., Michel, W.C., Cook, J. & Sullivan, C.W. (1988) Foraging behaviour of Antarctic krill *Euphausia superba* on sea ice microalgae. *Mar. Ecol. Prog. Ser.* **44**, 131–39.

Sun, S. (1997) Using Antarctic krill as an indicator of environmental interannual change. *Korean J. Polar Res.* **8**, 97–103.

Sun, S. & Wang, R. (1995a) Ageing the Antarctic krill. *Antarctic Res.* **7**, 59–62.

Sun, S. & Wang, R. (1995b) Study on the relationship between the crystalline cone number and the growth of the Antarctic krill. *Antarctic Res.* **7**, 1–6.

Sun, S. & Wang, R. (1996) Study on the relationship between the diameter of the compound eye and the growth of the Antarctic krill. *Antarctic Res.* **7**, 87–93.

Sun, S., de la Mare, W. & Nicol, S. (1995) The compound eye as an indicator of age and shrinkage in Antarctic krill. *Antarctic Sci.* **7**, 387–92.

Taki, K. & Ogishima, T. (1997) Distribution of some developmental stages and growth of *Euphausia pacifica* Hansen in the northwestern Pacific on the basis of Norpac Net Samples. *Bull. Tohoku Nat. Fish. Res. Inst.* **59**, 95–117.

Tanasichuk, R.W. (1998) Interannual variations in the population biology and productivity of *Euphausia pacifica* in Barkley Sound, Canada, with special reference to the 1992 and 1993 warm ocean years. *Mar. Ecol. Prog. Ser.* **173**, 163–80.

Thomas, P.G. & Ikeda, T. (1987) Sexual regression, shrinkage, re-maturation and growth in female *Euphausia superba* in the laboratory. *Mar. Biol.* **95**, 357–63.

Vernet, M., Hunter, J.R. & Vetter, R.D. (1988) Accumulation of age pigments (lipofuscin) in two cold-water fishes. *Fish. Bull.* **86**, 401–407.

Vidal, J. & Smith, S.L. (1986) Biomass, growth and development of populations of herbivorous zooplankton in the southeastern Bering Sea during spring. *Deep Sea Research* **33**, 523–56.

Wahle, R.A., Tully, O. & O'Donovan, V. (1996) Lipofuscin as an indicator of age in crustaceans: analysis of the pigment in the American lobster *Homarus americanus*. *Mar. Ecol. Prog. Ser.* **138**, 117–23.

Wang, R., Lu, B., Li, C., Wang, W. & Ji, P. (1995) Age-groups of Antarctic krill (*Euphausia superba* Dana) by distribution mixture analysis from length-frequency data. *Oceanologica et Limnologica Sinica* **26**, 598–605.

Weigmann-Haass, R. (1997) Verbreitung von Makrozooplankton in der Gronlandsee im Spatherbst 1988 (Crustacea: Ostracoda, Hyperiidea (Amphipoda), Euphausiacea). *Helgolaender Meeresuntersuchungen* **51**, 69–82.

Wiborg, K.F. (1968) Atlantic euphausiids in the fjords of western Norway. *Sarsia* **33**, 35–42.

Wiborg, K.F. (1971) Investigations on euphausiids in some fjords on the west coast of Norway 1966–1969. *Fisk. Skr. Ser. Hav.* **16**, 10–33.

Yamamura, O., Inada, T. & Shimazake, K. (1998) Predation on *Euphausia pacifica* by demersal fishes: predation impact and influence of physical variability. *Mar. Biol.* **132**, 195–208.

Zikov, L.A. & Slepokurov, V.A. (1982) The equation for the estimation of the natural mortality of fishes (for example fish from Lake Endir) (in Russian). *Rybn chozyaistvo* **3**, 36–37.

Chapter 6
Reproduction in Euphausiacea

Robin Ross and Langdon Quetin

6.1 Introduction

In the group Euphausiacea, females either retain mature eggs in a brood pouch until they hatch, or release the eggs directly into the water column. All the species currently being commercially harvested, however, are broadcast spawners (Nicol & Endo, 1997). The focus of our discussion of the reproduction patterns and strategies of euphausiids is on those species that are of commercial interest, e.g. *Thysanöessa inermis*, *T. raschii*, *Meganyctiphanes norvegica*, *Euphausia pacifica*, *E. nana* and *E. superba*. However, we assume that certain reproductive characteristics are common to the entire taxonomic group. Our understanding of the reproductive cycle will be enhanced by including studies of non-harvested species, e.g. *E. lucens*, *Nyctiphanes australis*, *N. couchii* and four additional Antarctic species, *E. crystallorophias*, *T. macrura*, *E. frigida* and *E. triacantha*.

The emphasis in this chapter is on reproductive patterns and population fecundity in euphausiids. Both are critical to our understanding of the pattern and level of fishing effort a population can withstand over the long term. Reproductive patterns include topics such as age and size at first spawning, initiation and duration of the spawning season, and the distribution of spawning populations. Oöcyte maturation and the use of sexual development stages to identify seasonal patterns of ovarian development are emphasised. We address also the topics of multiple spawning, batch size, and embryo size. The implications of these reproductive characteristics and strategies for fisheries management are briefly discussed.

6.2 Reproductive cycle

The patterns of reproduction in euphausiids show trends with latitude in timing and duration of the spawning season. However, the exceptions yield clues about the environmental conditions that support successful reproduction.

Patterns of reproduction – age and size at first spawning

Euphausiids are relatively long-lived crustaceans, with life-spans that range from slightly more than a year to seven years (Table 6.1). Smaller species and populations

Table 6.1 Reproductive patterns in euphausiids. Size and age at maturity, maximum size and lifespan, type of reproductive season (continuous, intermittent, limited), number of months spawn during year, number of years an individual spawns; dur, duration; pk, peak.

Species	Mat size (mm)	1st Spawn (yr)	Max size (mm)	Lifespan (yr)	Reproductive season	Months spawning	Years spawning
E. crystallorophias	21	2+	33	4–5	limited	1	3–4
E. lucens	11	0.5	~18	1–1.2	continuous	8–10	1
E. nana	–	–	12	<1	–	–	–
E. pacifica	12	1+ 0.5	25	<1.5 to >2	limited or intermittent	1.5–2	1
E. superba	33 or 35	2+ or 3+	65	5–7	limited, usually no pk	1–3	3–6
M. norvegica	25	1+	40	2.5	limited (long dur or 2 pks)	variable	$\geqslant 2$
N. australis	11	0.75–0.8	21	~1	continuous, occ regression	8–10	1
T. inermis	22	F 2+ M 1+	30	2+	limited	1–3	1
T. raschii	–	1	25	2+	limited, 2 pks poss	1–2	2

Sources of information listed by species. E. c.: Harrington & Ikeda (1986); Siegel (1987). E. l: Pillar & Stuart (1988). E. n.: Nicol & Endo (1997). E. p.: Brinton (1976); Ross *et al.*, 1982; Nicol & Endo (1997). E. s.: Cuzin-Roudy (1987); Siegel (1987); Siegel & Loeb (1994); Nicol & Endo (1997). M. n: Nicol & Endo (1997). N. a.: Hosie & Ritz (1983); Nicol & Endo (1997). T.i.: Nicol & Endo (1997). T. r.: Nicol & Endo (1997).

of a species living in lower latitudes, e.g. *N. australis* and *E. pacifica*, tend to have life-spans between 1 and 1.5 years, whereas larger species and those in subarctic regions live to about 2.5 years, e.g. *M. norvegica*, *Thysanöessa* spp. The two largest polar euphausiids, *E. crystallorophias* and *E. superba*, have life-spans of over 4 years. Euphausiids usually reach maturity when they measure about half of their maximum total length, and have reached 30 to 60% of their life-span (Table 6.1). In several species of euphausiids, female maturity may be delayed beyond the point of 60% of the life-span. In *T. inermis* near Iceland, females mature at two years of age, towards the end of their life-span, and a small percentage survive to breed in a second reproductive season (Einarsson, 1945; Astthorsson, 1990). Siegel & Loeb (1994) suggested that Antarctic krill, *E. superba,* shows 'knife edge' maturity, i.e. reaches maturity within a single age group. However, the high percentage of immature females of this species in size classes > 40 mm in some summers suggests that under some conditions reproduction may be postponed until the fourth summer (Shaw, 1997).

Life-span and the age of first reproduction are not always the same for male and female euphausiids. When there is a difference, males have a shorter life-span. In the case of *E. pacifica* off Japan, females live 2+ to 3 years, whereas males live less than 2 years (Nicol & Endo, 1997). A similar difference appears to hold for *E. crystallorophias* in the Indian sector. Males live for 3+ years compared to the females' 4+ years (Pakhomov & Perissinotto, 1996). Although the life-spans are different, the growth rates and age of first reproduction (2+) of male and female *E. crystallorophias* are the same (Siegel, 1987). The implication for both species is that males participate in one less reproductive season than females. In other species males and females mature at different ages. Male *T. inermis* near Iceland, for example, mature a year before the females, when they are one year old not two (Astthorsson, 1990; Einarsson, 1945). However, *T. raschii* males and females from the same area mature at the same age (Astthorsson, 1990).

A common assumption is that the energy costs of spermatophore production are insignificant, and that female euphausiids expend more energy during the reproductive season than males. However, there are several lines of evidence that this assumption does not hold for *E. superba* (Virtue *et al.*, 1996), and thus may not hold for other species. For *E. superba*, large mature males had higher mortality rates than females, and lower total lipid and triacylglycerol stores. In addition, the sex ratio decreased with increasing size, with the percentage of males decreasing sharply in the size range 51–55 mm. Either the growth rates of males and females are significantly different, or mortality rates are different. In either case, the cost of reproduction is one possible explanation for the differences observed. Little is known about differences in male versus female growth or mortality rates for other species, although Pillar & Stuart (1988) suggested that the retarded growth rates in male *E. lucens* may be due to the allocation of energy to spermatophore production throughout most of the year.

Reproductive patterns

Patterns of reproduction in euphausiids can be classified as continuous, intermittent or limited (Table 6.1). With continuous reproduction there is no regression to a spent stage once maturity is reached, although the intensity of spawning is higher during some parts of the year. Continuous reproduction is primarily confined to species inhabiting the mid-latitudes and equatorial regions. *E. lucens* is characterised as a continuous spawner, with spawning most intense in late winter and early spring, just prior to upwelling and bloom (Pillar & Stuart, 1988). In a few species, reproduction is intermittent, with a cycle of maturation and regression more than once in one season, resulting in multiple cohorts (generations) per year. Intermittent reproduction was confirmed in *N. australis*, a species which produces three cohorts per year (Ritz & Hosie, 1982), and *E. pacifica* in some regions (Brinton, 1976). Intermittent reproduction was inferred in two other species in the north-east Atlantic which produce more than one generation per year, e.g. *N. couchii*, three to four per year (Lindley, 1982), and *T. longicaudata*, two per year (Lindley, 1978).

Limited reproduction means that there is a limited spawning season, with a cycle of ovarian maturation and regression once a year. Egg production is limited to a 1.5 to 3 month season, alternating with a longer period of gonadal rest. Several peaks in egg production can occur within this limited season, sometimes separated by periods of weeks; such a pattern could lead to multiple cohorts in a single year class. Limited reproduction is usually found in polar and subpolar species, e.g. the Antarctic euphausiids, *Thysanoessa* spp., *M. norvegica* (Table 6.1). However, the pattern of reproduction in a species is flexible. The reproductive season is limited over most of the range of *E. pacifica*, a temperate and subarctic species, but intermittent in the upwelling regions off Southern California (Brinton, 1976) and in some parts of Puget Sound, Washington (Ross *et al.*, 1982).

Timing and duration of spawning

From the early review of Mauchline & Fisher (1969), the timing and duration of spawning were known to vary, both between populations of the same species living in regions at similar latitudes, and among species at the same latitude (Fig. 6.1 and Fig. 6.2). As a general rule, earlier and longer spawning occurs at lower latitudes. These observations led to the hypothesis that timing and duration of spawning are keyed to the period of elevated food production, i.e. earlier in the year and longer at lower latitudes. In addition, spawning has been observed to coincide with the spring bloom, e.g. *T. inermis* and *T. longicaudata* in the Barents Sea (Dalpadado & Skjoldal, 1996), or times of high phytoplankton concentrations, e.g. *E. pacifica* in areas of intermittent reproduction (Nicol & Endo, 1997).

The premise underlying the hypothesised link between food production and spawning is that food must be available at the right time and in the right amounts to satisfy the energetic demands of reproduction. Initiation of ovarian development

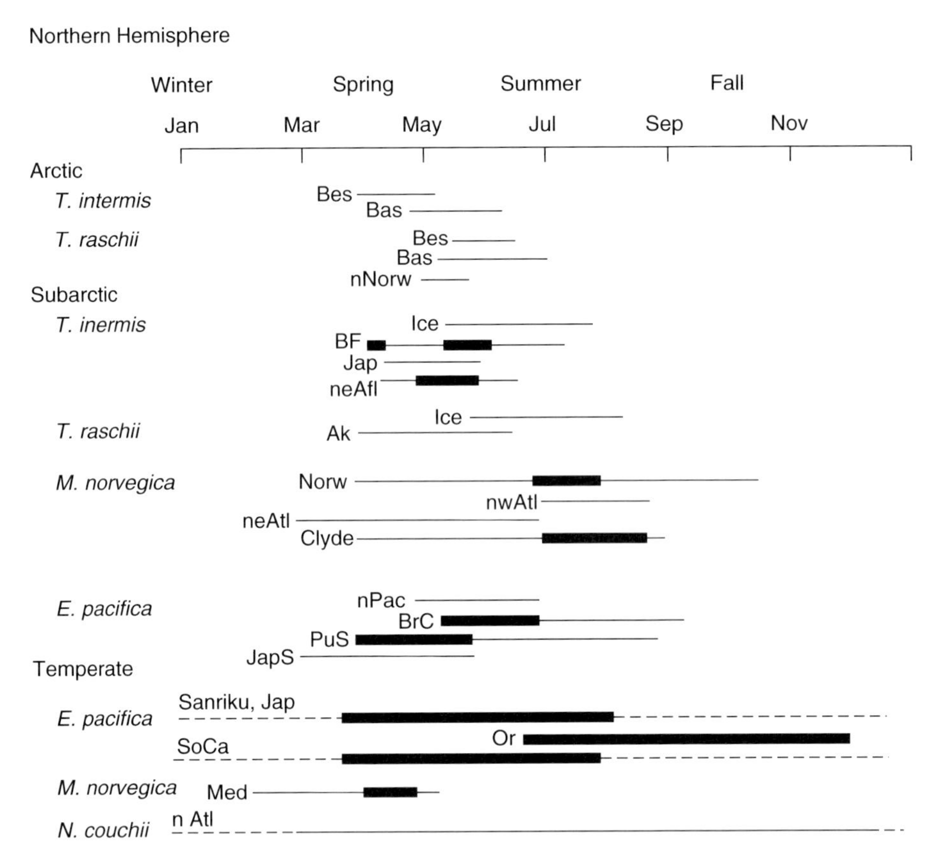

Fig. 6.1 Initiation, duration and intensity of spawning season for euphausiid species in different regions in the Northern Hemisphere. BeS, Bering Sea; BaS, Barents Sea; nNorw, northern Norwegian fjords; Ice, Icelandic fjords; BF, Bay of Fundy, Canada; Jap, Japan; ne- or nw, northeast or northwest: Atl, Atlantic Ocean; Ak, Auke Bay, Alaska; Clyde, Firth of Clyde, Scotland; nPac, north Pacific Ocean; BrC, British Columbia; PuS, Puget Sound, Washington; JapS, Japan Sea; Or, Oregon coast; SoCa, southern California coast; Med, Mediterranean Sea. Sources of information listed by species. T. i.: Kulka & Corey (1978); Hanamura *et al.* (1989); Astthorsson (1990); Einarsson (1945); Smith (1991); Nicol & Endo (1997). T. r.: Falk-Petersen & Hopkins (1981); Astthorsson (1990); Paul *et al.* (1990); Einarsson (1945); Timofeyev (1994); Smith (1991); Nicol & Endo (1997). M. n.: Lindley (1982); Boysen & Buchholz (1984); Nicol & Endo (1997); Cuzin-Roudy (1993); Cuzin-Roudy & Buchholz (1999). E. p.: Ponomareva (1966); Smiles & Pearcy (1971); Brinton (1976); Heath (1977); Ross *et al.* (1982); Bollens *et al.* (1992); Iguchi *et al.* (1993); Nicol & Endo (1997). N. c.: Lindley (1982).

may be keyed to a seasonal cue such as photoperiod, but actual spawning (release of eggs) will depend on the rate of ova maturation, and thus the rate of energy input prior to spawning. In some species energy is stored from the previous season for ovarian development, in others ovarian development is dependent on immediate food supplies. If reproduction is iteroparous (Section Population fecundity), the rate of production and number of subsequent batches of eggs is likely to depend on both energy input prior to spawning, and on food availability during the spawning season.

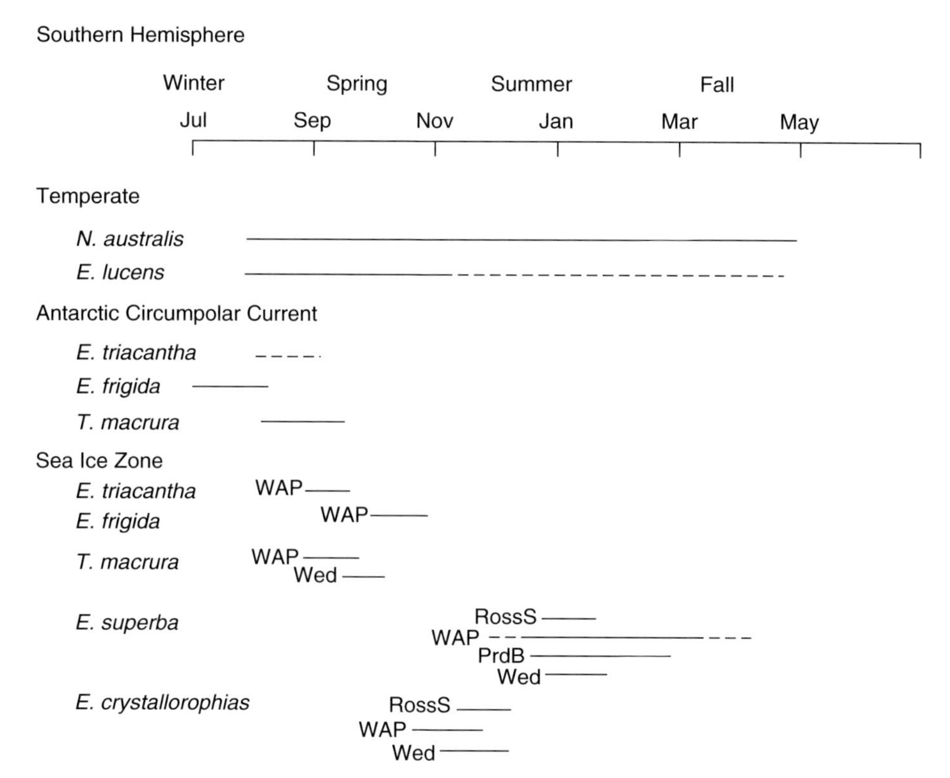

Fig. 6.2 Initiation, duration and intensity of spawning season for euphausiid species in different regions in the Southern Hemisphere. Wed, Weddell gyre; WAP, western Antarctic Peninsula; RossS, Ross Sea; PrdB, Prydz Bay and Lasarev Sea. Sources of information listed by species. N. a.: Blackburn (1980); Ritz & Hosie (1982). E. l.: Pillar & Stuart (1988); Stuart & Pillar (1988). E. t.: Baker (1959). E. f.: Makarov (1977); Siegel (1986); Makarov *et al.* (1990); Men'shenina (1992). T. m.: Stepnik (1982); Makarov *et al.* (1990); Men'shenina (1988, 1992); Makarov & Men'shenina (1992). E. s.: Makarov *et al.* (1985, 1990, 1991); Hosie & Cochran (1994); Makarov & Men'shenina (1992); Ross & Quetin (1983); Quetin *et al.* (1994). E. c.: Fevolden (1980); Makarov *et al.* (1990, 1991); Brinton & Townsend (1991); Makarov & Men'shenina (1992); Harrington & Thomas (1987).

The sequence of spawning for Antarctic euphausiids illustrates the latitudinal trend. In general, production cycles in the northern areas start earlier, and finish later, than further south (Hart, 1942; Foxton, 1956). Matching this cycle, oceanic species at lower latitudes reproduce in the early spring, whereas the shelf and coastal species at higher latitudes spawn one to two months later (Fig. 6.2). The exception is the coastal species, *E. crystallorophias*, which reproduces before *E. superba* (Fig. 6.2). Also, at higher latitudes where the period of elevated production is longer, spawning seasons are longer, e.g. west of the Antarctic Peninsula compared to the Ross Sea, and the Scotia Sea compared to the Lasarev Sea. Spiridonov (1995) confirmed this pattern for *E. superba* in a study of reproduction with data from both the Weddell Sea and waters west of the Antarctic Peninsula. He identified five habitats based on the timing of initiation, duration and intensity of spawning. In the northern most habitat, at the fringe of the extent of seasonal sea-ice, spawning was early (late November–early December), and tended to be long (3–3.5 month) but variable in duration. In contrast, in coastal areas at higher latitudes, spawning began later and was of shorter duration (~ 1.5 month).

Understanding this link between food production and spawning allows us to reconcile what appear to be conflicting views of the influence of seasonal sea-ice dynamics on the spawning season of *E. superba* (Spiridonov, 1995; Siegel & Loeb, 1995; Makarov & Men'shenina, 1992). For Antarctic krill living at the tip of the Antarctic Peninsula, Siegel & Loeb (1995) found that low sea-ice extent in winter and/or early sea-ice retreat led to a delay in spawning. In contrast, Spiridonov (1995) concluded that low sea-ice extent resulted in early spawning, and a slow (late) retreat resulted in a delayed spawning season that was not intensive. In coastal waters in the Weddell Sea, when polynas open early in the spring, i.e. simulating an early retreat, reproduction starts earlier than in waters to the north which are still covered with ice. Makarov & Men'shenina (1992) called this phenomenon the polyna or 'oasis' effect, and it can reverse the normal north–south gradient in timing of reproduction for both *E. superba* and *T. macrura*.

Can we reconcile these disparate views of the influence of seasonal sea-ice on the timing and intensity of spawning of Antarctic krill? The retreat of sea-ice impacts food availability in spring because the melting sea-ice conditions the water column for ice edge blooms. Without this conditioning, phytoplankton blooms are delayed until late spring or early summer. Since Antarctic krill do not come into reproductive condition under the ice (Cuzin-Roudy & Labat, 1992), the extent and timing of these ice edge blooms will play an important role in the rate of ovarian development and thus the timing of spawning. Sea-ice extent represents the areal extent of conditioning and thus enhanced food availability in spring; the timing of retreat represents when conditioning occurs and thus the timing of the blooms. The timing of sea-ice retreat is region-specific, that is, 'early' retreat for northern regions is likely to be earlier in the spring than in southern regions. 'Early' in one region may lead to blooms prior to the peak demand for food for ovarian development, whereas in another region 'early' retreat may create ideal conditions. In the Elephant Island area (Siegel & Loeb, 1995), at the northern edge of the influence of seasonal sea-ice, if sea-ice in winter is low little of the region will be conditioned for an ice edge bloom, and the area of enhanced food availability will be small. With an early retreat, bloom conditions may occur prior to the time of maximum need for food for ovarian development. Thus low winter sea-ice extent and/or early retreat mean lower food availability in the spring, and these environmental conditions will be associated with slow ovarian maturation, and delayed spawning. However, when polynas open early in the spring in the higher latitude coastal waters of the Weddell sea, ice edge blooms create a source of food at a time of prime need. Polynas allow the cycle of food production to start earlier in the spring at higher latitudes than in the ice-covered seas at lower latitudes; food is available for ovarian maturation, and thus spawning can start earlier. The reverse will be found with a 'late' retreat. The two viewpoints can be reconciled by assuming that sea-ice mediates the areal extent and timing of spring food for ovarian development.

The Antarctic species are not the only euphausiids to show inter-annual and geographical variability in reproduction within a species, even among locations that differ only in food availability and not in either temperature or photoperiod. The reproductive season for the Northern krill, *M. norvegica*, begins in the spring in all areas but the north-west Atlantic off Canada where the influence of the cold Laborador Current delays the onset of net primary production, and spawning does not begin until summer (Lindley, 1982). For the smaller *E. pacifica*, the population in Puget Sound begins to spawn in the spring, continuing through summer in some areas of Puget Sound but not others (Ross *et al.*, 1982; Bollens *et al.*, 1992). However, *E. pacifica* off the coast of Oregon does not spawn until late summer and early fall (Smiles & Pearcy, 1971) (Fig. 6.1). Peaks in chlorophyll a occur in April and May in Puget Sound, but off Oregon appear later in the year and are associated with the summer upwelling regime. A link between the phytoplankton bloom and the length and intensity of spawning has also been suggested as the source of the inter-annual differences in the duration of the period of high egg concentrations of *T. raschii* in Auke Bay, Alaska (Paul *et al.*, 1990).

Spawning is not usually synchronous among species where their ranges overlap. In the arctic and subarctic North Atlantic, *T. inermis* spawns earlier than *T. raschii*, which in turn spawns earlier than *M. norvegica* when that species is present (Astthorsson, 1990). Such differences may reflect the extent to which reproduction depends on that season's primary production, whether directly or indirectly, versus stored reserves from the previous season. Omnivores or carnivores may show an indirect dependence on primary production through food web links. If the populations of prey increase in response to seasonal phytoplankton blooms, food availability for the omnivores and carnivores is indirectly dependent on periods of primary productivity. In some species, initiation of ovarian maturation may be keyed so spawning occurs before the spring bloom or period of upwelling. With this strategy, stored reserves provide the energy to initiate reproduction, and seasonal production benefits the larvae. If reproduction is in response to the spring bloom or period of upwelling, the implication is that stored reserves are not adequate to fuel reproduction. The degree of inter-annual variability in the initiation of the spawning season is thus apt to vary with the degree of reliance on stored reserves versus that season's food production. In a recent comparison of the lipid biochemistry of the six most abundant species of Arctic and Antarctic euphausiids, Falk-Petersen *et al.* (August 1999) found that neutral lipid deposits are primarily accumulated for reproduction. Four species, *T. inermis*, *T. longicaudata*, *T. macrura* and *E. crystallorophias*, depend primarily on lipid reserves for spring reproduction. *T. raschii* and *E. superba* rely on food availability in the spring for final gonad maturation. *E. superba* is the single Antarctic euphausiid with a reproduction season in summer, and a recent study (Shaw, 1997) suggests that the initiation and duration of the spawning season is dependent on both spring and summer food sources.

Mating and spawning

In euphausiids, males tend to mature several months earlier in the season than females. For example, in the northern Baltic, immature male *M. norvegica* are distinguishable in October whereas immature females cannot be distinguished until January; the males also have fully developed spermatophores at nine months (Boysen & Buchholz, 1984). Mating behaviour begins up to 1–2 months prior to spawning and continues throughout the spawning season (Table 6.2). Thus transfer of the spermatophore occurs a considerable time before spawning, and one cannot assume that the presence of spermatophores on a female indicates that the female is gravid. Euphausiids lose the sperm mass held in the thelycum when they moult. Since euphausiids moult on a time-scale of weeks throughout their lives, mating must continue throughout the reproductive season to ensure that eggs are fertilised. Males with mature spermatophores in the ejaculatory ducts are found throughout the spawning season (Bargmann, 1937; Boysen & Buchholz, 1984; Astthorsson, 1990).

Table 6.2 Relative timing of mating behaviour and initiation of spawning.

Species	Mating begin	Spawning begin	Reference
E. superba	Nov	late Dec	Bargmann (1945)
M. norvegica	Jan	Apr–Oct	Falk-Petersen & Hopkins (1981) Boysen & Buchholz (1984)
T. inermis	late Feb	mid–late Apr	Hanamura *et al.* (1989)
T. inermis	Feb	Apr	Astthorsson (1990)
T. raschii	Apr	May	

Mating behaviour has been observed in *E. superba* in the ice edge zone in December (Naito *et al.*, 1986). Bargmann (1937) described the exchange of spermatophores in *E. superba*. A sequence of chase and contact behaviour between male and female *E. superba* believed to be part of mating behaviour has also been observed in large aquaria in the laboratory (Fig. 6.3) (Ross *et al.*, 1987). Durations of chases were about 5 sec, and the contact or close times 10–30 sec. In the tanks males appeared to perceive a chemical trail within 8–10 cm of the female, and turn in pursuit.

Surface aggregations have often been described as facilitating mating behaviour (Ritz, 1994). These aggregations tend to occur in the late winter or early spring, prior to reproduction. In five species, mature breeding individuals predominated in daytime surface aggregations (Table 6.3). The aggregations thus contain mature euphausiids in the close proximity necessary to initiate a pursuit sequence at a time in the cycle when finding a mate is critical. The sex ratios in these surface swarms were sometimes (Nemoto *et al.*, 1981, *E. superba*; Nicol, 1984, *M. norvegica*), but not

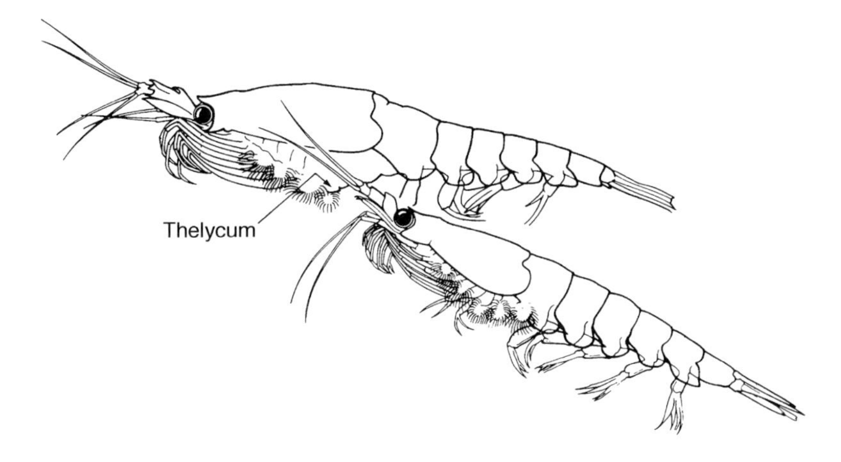

Fig. 6.3 Relative positions of the male (on the bottom) and female adult *Euphausia superba* during the chase and contact sequence of mating behaviours.

Table 6.3 Observations of daytime surface aggregations dominated by mature individuals. See Fig. 6.1 for the key.

Species	Months, or Location	Dom sex or Size	Reference
E. pacifica	Mar–Apr, Japan	female	Endo (1984)
	nw	>14 mm	Odate (1991) cited in Nicol & Endo (1997)
	May–June, nPac		Kotori (1995)
E. superba	Dec in ice zone	mating	Naito *et al.* (1986) Nicol *et al.* (1987) O'Brien (1988)
M. norvegica		female	Nicol (1984)
T. spinifera		–	Smith & Adams (1988)
T. inermis	Feb–Apr	female	Hanamura *et al.* (1989)
T. raschii	Barents Sea	female	Timoteyev (1994)

always extreme (O'Brien, 1988, *N. australis*). In another study of aggregation characteristics in *E. superba*, population composition was often skewed, with mature males and gravid females dominating some aggregations, and immature individuals dominating others (Watkins *et al.*, 1992). However, there was a positive correlation between the presence of mature males and mature females with developing ovaries (Watkins *et al.*, 1992). Such an association would be necessary for the continuing transfer of spermatophores to females throughout the reproductive season. There also appears to be spatial segregation on small scales. Off Japan, male and female *E. pacifica* may separate after mating. Males dominate the surface swarms during the day, with females at depths below 15 m (Nicol & Endo,

1997). Sex ratios in catches of *M. norvegica* also vary (Hollingshead & Corey, 1974), with surface and deep aggregrations showing different population structures that suggest that mated and ready-to-spawn females separate from the rest of the population (Nicol, 1984; Cuzin-Roudy & Buchholz, 1999).

Although information is scarce, the distribution of spawning populations of euphausiids does not appear to be homogeneous. The concept of spatial and geographical succession in Antarctic krill developmental stages has been described by several authors (Makarov & Sysoyeva, 1985; Siegel *et al.*, 1990; Lascara *et al.*, 1999). Gravid female *E. superba* are generally found in the vicinity of the continental slope (Siegel, 1992), and in waters overlying Circum-Polar Deep Water (CDW) (Hofmann *et al.*, 1992). Release of eggs over CDW is thought to be part of a reproductive strategy to enhance survival of the larvae. During the season, as the females progressively mature, their distribution changes. Less advanced stages of female *E. superba* are found on the inner shelf (Shaw, 1997) or closer to the ice edge (Cuzin-Roudy & Labat, 1992; Quetin *et al.*, 1992). Females in vitellogenesis and spawning are first found in open waters. Spawning in several other species appears to be spatially restricted. Spawning zones for *T. macrura* are found in the Antarctic Convergence, and the Secondary Frontal Zone (Men'shenina, 1988). There is a single spawning centre for *T. inermis* in the waters south of Spitzbergen, with eggs and larvae transported elsewhere (Timofeyev, 1994). And lastly, the percentage of gravid female *E. lucens* increases offshore (Pillar & Stuart, 1988). Thus, in these four cases, spawning is not randomly distributed within the species' range. The advantages to this heterogeneity are not apparent for all species, but the patterns are clear. Such distributions may enhance retention of larvae in favourable regions as has been found for fishes or enhance hatching success or reduce predation on the vulnerable early life history stages.

The concept of 'pseudopopulations' or non-breeding populations should also be mentioned in the context of the distribution of maturity stages of euphausiids. Although most species breed throughout their range of occurrence, there are exceptions. *T. inermis*, for example, does not breed north of 65–70°N (Nicol & Endo, 1997), thus the population of *T. inermis* in the southern Barents Sea (Zelikman *et al.*, 1980) is believed to be non-reproducing. There is also some debate over whether populations of *E. superba* around South Georgia are non-reproducing. Although abundances can be high, the lack of eggs and young larvae in these regions is cited as evidence for populations that are not successfully reproducing.

Ovarian maturation cycle

Ovarian development and maturation occurs over a period of several months prior to spawning. In *T. inermis*, ovaries mature over the three month period between mid-January and mid-April, prior to spawning in the spring. Ovarian development in *T. raschii* in the north-east Pacific (Zelikman, 1958), *M. norvegica* in the north-east Atlantic (Zelikman, 1958), and *E. superba* in the Antarctic (Bargmann, 1945) also takes about three months. Maturation of testes in males generally is more rapid, about 1 month in *T. inermis* (Kulka & Corey, 1978) and 2+ months in *M. norvegica* (Mauchline, 1968; Hollingshead & Corey, 1974).

General progression

The details of ovarian maturation help us to understand where environmental variability is most likely to impact rates of ovarian maturation. Ovarian maturation in crustaceans is characterised by a progression in ova development that is found in all species (Nelson, 1991). First, during oogenesis the oögonia (og) multiply in the germinal zones (GZ) of the ovary. When the oögonia leave the germinal zone, they are characterised as primary or young oöcytes (yoc). The yocs represent multiple batches of eggs in the sense that groups of eggs will develop in sequence, not simultaneously. At this point the female may enter a resting phase. Upon leaving the resting phase, groups of yoc accumulate glycoproteic yolk synthesised within the oöcyte in a process called primary or pre-vitellogenesis. At this point the ovary is increasing in size and mating may occur, but the female euphausiid is not swollen. Mating has occurred when either the thelycum is full or spermatophores are attached. The growing type 1 oöcytes (oc1) may accumulate at a second plateau before they enter vitellogenesis. During vitellogenesis, or lipidic yolk accumulation in type 2 (oc2) and type 3 (oc3) oöcytes, the oöcytes rapidly enlarge, increasing the cytoplasm to nucleus ratio. The source of the lipidic yolk is outside the oöcyte, likely the 'fat body' (Cuzin-Roudy & Buchholz, 1999). Final egg maturation occurs after vitellogenesis is complete, the nucleus moves to the periphery and loses its membrane, and meiosis occurs in the type 4 oöcyte (oc4). Another plateau can occur at this point. Ovulation, fertilisation and release of the embryos are the final steps.

The plateaus are assumed to occur at points during ovarian maturation when additional resources or environmental cues such as photoperiod are needed for further development. The cue for the second plateau is suggested to be a threshold accumulation of 'lipid' in the 'fat body', enough to complete the development of one group of oöcytes. Evidence for the existence of a 'fat body' in euphausiids is emerging for two species, *E. superba* and *M. norvegica* (Cuzin-Roudy & Buchholz, 1999).

Maturity keys based on physiological state

As oöcytes move through the succession of physiological steps of ovarian maturation as described above, they show characteristic morphological (Fig. 6.4) and biochemical changes. Keys for stages of female sexual development for two euphausiids, *E. superba* (Table 6.4, Cuzin-Roudy & Amsler, 1991) and *M. norvegica* (Cuzin-Roudy, 1993) have been developed based on both external secondary sex characteristics and the types of oöcytes in the ovary as characterised with the 'squash' technique. External secondary sex characteristics include the development

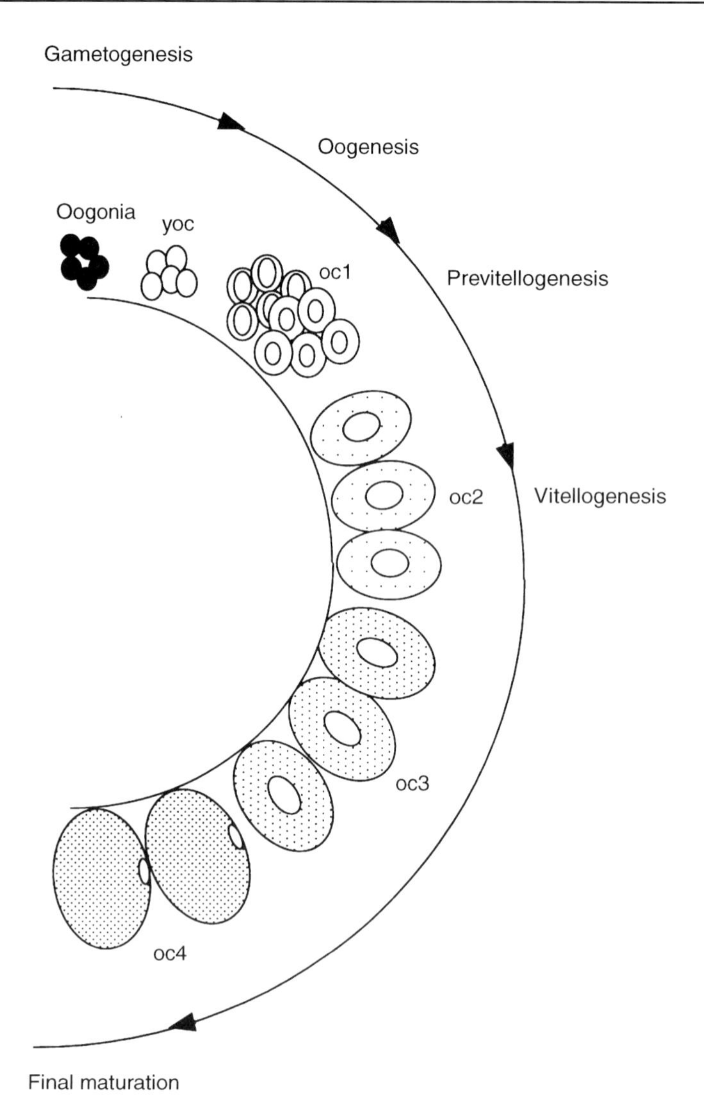

Fig. 6.4 Relative size of oöcytes and nucleus/cytoplasm ratio in the cycle from gametogenesis to final maturation. yoc, young oöcytes or primary oöcytes; oc1, stage 1 oöcytes; oc2, stage 2 oöcytes; oc3, stage 3 oöcytes; oc4, stage 4 oöcytes. Nucleus is clear. Cytoplasm becomes increasingly speckled as lipidic yolk accumulation proceeds.

of the thelycum, the size of the ovary as seen through the carapace, and the degree of swelling of the thorax. During gametogenesis the thelycum is not yet present, and the ovary is composed of germinal zones and oögonia, small non-descript cells. During oogenesis the thelycum is immature, and although the ovary is still small, some of the oögonia have transformed into the slightly larger yoc. During previtellogenesis the thelycum is completely developed. The ovary grows considerably during this phase, but the thorax is not swollen. The oöcytes (oc1) accumulate

Table 6.4 Definitions of sexual development stages (SDS) for female *Euphausia superba* (Cuzin-Roudy & Amsler, 1991).

SDS	Dev phase	Thelycum	Ovary and oöcytes
1	Gametogenesis	Absent	Ovary small. Germinal zone (GZ) and primary and secondary oögonia (og 1, og 2)
2	Oogenesis	Present, sternal and coxal plates not fully grown	Ovary small. Og 2 developed into yoc
3	Previtellogenesis	Mature thelycum	Thorax not swollen. Oc1 up to 200 μm
4	Previtellogenesis	Mature thelycum, may be red. May be mated	Thorax not swollen. Developing ovary visible. Oc1 up to 450 μm
5	Vitellogenesis	Mature thelycum, red. Mated	Ovary obvious. Oc2 present
6	Vitellogenesis	Mature thelycum, red. Mated	Thorax conspicuously swollen. Ovary full of oc3; may have various combinations of GZ, oc1, oc2 and oc3
7	Oöcyte maturation	Mature thelycum, red. Mated	Ovary in live krill opaque or bluish, full of oc4; may have various combinations of GZ, oc1, oc2, and oc4
8	Oviposition	Mature thelycum, red	Thorax swollen, but ovary reduced in size and lobes contracted. May have various combinations of GZ, oc1, oc2 and residual oc4 (continuing) or only GZ and residual oc4 (final or post spawn)
9	Ovarian reorganization	Mature thelycum, red	Thorax not swollen. Ovarian lobes irregular and lumpy, contain GZ with og1, og2 and yoc, plus residual oc4
10	Ovarian reorganization	Thelycum regressing to more juvenile form	Thorax not swollen. Ovarian lobes irregular and lumpy, contain GZ with og1, og2 and yoc, possibly residual oc4

glycoproteic yolk, and are translucent with a nucleus/cyctoplasm ratio of about 60%. Mating occurs during previtellogenesis, and a sperm mass often fills the thelycum. During vitellogenesis, the oöcyte grows, and the accumulation of lipidic yolk gives the cytoplasm in the oöcyte a cloudy or grainy appearance (oc2s and oc3s). Opaque oöcytes overlapping the size of oc1s but with a cloudy appearance are oc2s; larger oöcytes with a lower nucleus/cytoplasm ratio and a more granular appearance are oc3s. The nucleus is large and distinct in both oc2s and oc3s. Externally the ovary is now conspicuous, and the thorax swollen. Ovaries with mature oöcytes (oc4) appear opaque or even bluish gray, e.g. *E. crystallorophias* (Harrington & Thomas, 1987), *E. pacifica* (Ross *et al.*, 1982), *E. superba* (Ross & Quetin, 1983), and *M. norvegica* (Cuzin-Roudy & Buchholz, 1999). Oc4s have completed vitellogenesis, lack a large central nucleus, and meiotic figures can be seen under high magnification. The small nucleus without a membrane is in the peripheral cytoplasm (Fig. 6.4). After the mature oöcytes (oc4s) are released, the thorax remains swollen, and the female is considered post-spawn. However, the female may not have completed the reproductive season. In both *E. superba* (stage $8_c = 8$ continuing) and *M. norvegica* (stage 5_1) the presence of oöcytes in earlier stages indicates that another group or batch of oöcytes will mature and another spawning will occur that season. When the ovary in a post-spawn female contains only the germinal zone and oogonia, the reproductive season is complete for that individual, and she enters a period of reorganisation and reproductive rest.

The sexual development stages (SDS) for *E. superba* (Table 6.4) have been slightly simplified for use with *M. norvegica* (Cuzin-Roudy, 1993). For *M. norvegica*, no stage is defined for gametogenesis, and there is only one stage each for previtellogenesis and vitellogenesis. The distinction between post-spawn females that will recycle and those that have completed egg production for that season is retained (see comparison in Table 6.5). Similar keys could be used for other species, since all euphausiids move through the same stages during ovarian maturation.

Table 6.5 Sexual development stages of euphausiids and the relationship to known physiological phases of ovarian development.

Physiological phases of ovarian development	*Euphausia superba* (Cuzin-Roudy & Amsler, 1991)	*Meganyctiphanes norvegica* (Cuzin-Roudy, 1993)
Immature or resting	2 and 10	1
Oogenesis		
ONSET		
Oöcyte development		
• previtellogenesis; mating activity	3 and 4	2
• vitellogenesis	5 and 6	3
Oöcyte maturation and egg release		
• ready to spawn	7	4
• post-spawn, may recycle	8_c and 8_f	5_1 and 5_2
END		
Spent females, ovary in regression	9	6

Unlike earlier keys, these two keys distinguish physiological stages of ovarian development, not just external secondary sex characteristics. The difference allows for more detailed analysis of the reproductive cycle. For example, the key of Makarov & Denys (1980), commonly used for *E. superba*, does not distinguish between females in previtellogenesis and those undergoing ovarian reorganisation, i.e. females at the beginning and end of their reproductive cycle. Nor do they separate females in late vitellogenesis from females in the final stages of egg maturation. Finally the Makarov & Denys key does not distinguish females continuing to produce eggs after a spawn from those at the end of their reproductive season.

Analysis of sexual development stages in the populations of *E. superba* west of the Antarctic Peninsula and *M. norvegica* in the Ligurian Sea in the Mediterranean has led to an understanding of the timing and duration of the different stages of ovarian maturation (Fig. 6.5). The pattern of ovarian development in the two species is basically similar. In both species previtellogenesis occurs in the whole ovary prior to the start of the reproductive season, January to late February in *M. norvegica*, and for a longer period from September through November in *E. superba*. Vitellogenesis in both species is cyclical, with successive batches of oöcytes brought to maturity and released throughout the reproductive cycle in the spring for *M. norvegica* or summer for *E. superba*. The energy source appears to be different during the two phases of yolk accumulation. In *E. superba*, previtellogenesis cannot continue without an outside source of energy because this species does not store enough lipid for this process. In both species, the fat body in previtellogenic krill becomes progressively better developed (Cuzin-Roudy, 1993). Since vitellogenesis and egg maturation continue even with poor food, the 'fat body' may be a source of lipid during this time and reduce dependence on immediately available food during vitellogenesis. The limited period of egg production alternates with a longer period of gonadal rest.

However, there are some differences between the two species. In *M. norvegica*, the ovary continually produces new yoc throughout the season, and oöcyte development is progressive; this strategy leads to rapid successive cycles of small batches of embryos. In *E. superba*, production of yoc is not constant. The yoc at the start of ovarian maturation mature in successive batches, with the possibility of four different stages of oöcytes (mature, vitellogenic, previtellogenic, and yoc) simultaneously in the ovary. The strategy is one of pulses of oöcyte production in spawning episodes with large batches of eggs at the individual level. At the population level, both strategies lead to more even egg production over a 2 to 4 month period. This dispersal of eggs and larvae in time and space will enhance survival of the young in variable environments. A similar cycle was inferred for *T. inermis* in the Bay of Fundy from descriptions of oöcyte types and combinations in the ovary throughout the season (Kulka & Corey, 1978). We suspect that this general pattern is common to all euphausiids, but additional studies of the sexual development stages of other species will be necessary.

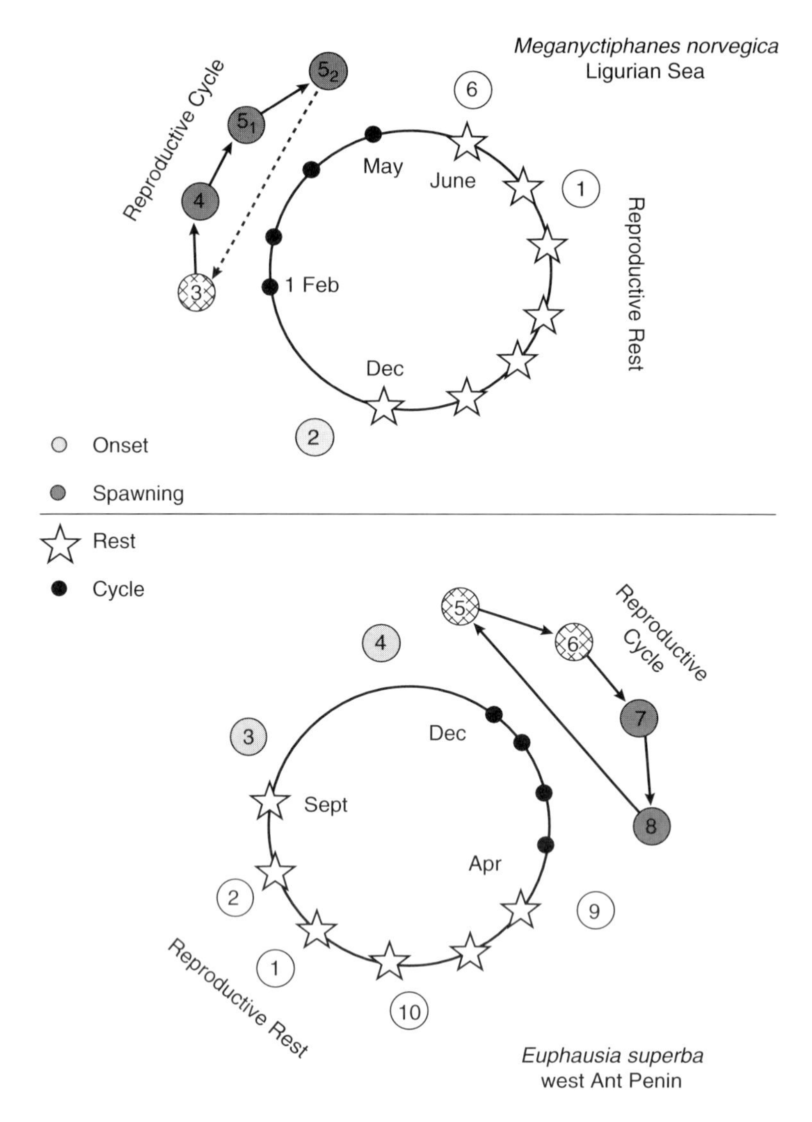

Fig. 6.5 Alternation of reproductive cycle (black circles) and resting phases (pale grey stars) throughout the year for (a) the Northern krill, *Meganyctiphanes norvegica*, in the Ligurian Sea of the Mediterranean, redrawn from Cuzin-Roudy & Buchholz (1999); and (b) the Antarctic krill, *Euphausia superba*, from samples west of the Antarctic Peninsula (data of Quetin & Ross). Numbers in circles refer to sexual development stages as described in Tables 6.4 and 6.5. Onset means onset of ovarian maturation or previtellogenesis. Spawning is period when eggs are released. The arrows indicate stages involved in ovarian cycling.

Population fecundity

Population fecundity, i.e. total rate of egg production of the population in a region, and early larval survivorship are the two main factors affecting recruitment in a region. Population fecundity in euphausiids is a product of female abundance, the percentage of females reproducing that season, the number of spawning episodes, and the number of eggs per spawning episode. The number of spawning episodes is a function of the duration of the spawning season and how often the ovary recycles during that period, i.e. the length of time between spawning episodes.

Percentage population reproducing

The assumption has been that all females, once they are the size of a mature individual, will reproduce during that reproductive season, i.e. Siegel & Loeb (1994) for *E. superba*. An examination of the sexual development stages present in the population is a test of that assumption. If all females larger than the size at maturity (Table 6.1) are in late previtellogenesis, vitellogenesis or maturation during the reproductive season, then the entire population is reproducing. However, if a percentage of the population is in a resting stage, then not all females will reproduce that season. The assumption has been verified for *M. norvegica* in the Ligurian Sea (Cuzin-Roudy & Buchholz, 1999), but refuted for *E. superba* west of the Antarctic Peninsula (Shaw, 1997). In the case of *E. superba*, the percentage of the population reproducing varies from less than 20% to nearly 100%. The percentage is the proportion of females in the reproductive cycle or having just completed the reproductive cycle of the total population of subadult and adult females in January. Thus for *E. superba* the number of seasons that an individual reproduces is an important variable. For a species that is long-lived, and which can reproduce for several seasons, the strategy of delayed reproduction or putting energy into growth instead of reproduction for a season can potentially enhance the total number of offspring produced. However, most euphausiid species only reproduce during one year (Table 6.1), with a small percentage surviving to reproduce over a second season, so this phenomenon is probably rare.

Individual fecundity

Based on the results of spawning frequency experiments with *E. pacifica* in Puget Sound, Ross *et al.* (1982) first proposed multiple spawning episodes in an individual euphausiid. Subsequently, studies of ovarian cell structure and development, and experiments with live female euphausiids of many species have established that multiple spawning episodes or batches of eggs are common in euphausiids (Table 6.6). Multiple spawning episodes in a single season theoretically increases the probability of reproductive success in a fluctuating environment.

Table 6.6 Euphausiid species with multiple broods in a reproductive season, confirmed or inferred from experiments or histological examination of the ovary or for *N. couchii* inferred from number of cohorts per year.

Species	Expt	Histology	Reference
E. crystallorophias	X		Harrington & Thomas (1987)
E. hanseni	X	X	Stuart & Nicol (1986); Stuart (1992)
E. lucens	X	X	Stuart & Nicol (1986)
E. pacifica	X		Ross *et al.* (1982)
E. superba	X	X	Ross & Quetin (1983); Cuzin-Roudy (1987)
M. norvegica	X	X	Cuzin-Roudy (1993); Cuzin-Roudy & Buchholz (1999)
N. australis	X	X	Hosie & Ritz (1983)
N. capensis		X	Stuart & Nicol (1986)
N. couchii		infer	Lindley (1982)
T. inermis		X	Kulka & Corey (1978)

The number of spawning episodes per female is a function of the duration of the spawning season and the interbrood period, itself a function of the rate of oöcyte development. For those species for which we have estimates, the interbrood period ranges from a few days to 30 days (Table 6.7). Because of its dependence on oöcyte development rate, the interbrood period (number of spawning episodes) is likely to be a function of food availability, and thus will vary inter-annually and geographically. For *E. superba*, the interbrood period varies with location and year (Quetin *et al.*, 1994), with a minimum of 6 days and a maximum of 50 days in mid-summer. With data on the duration of the spawning season, and estimates of interbrood periods from spawning frequency experiments, estimates of the number of spawning episodes per season for a reproducing *E. superba* range from 3 to 9 per season (Table 6.7). Observations of individual spawning females held in the laboratory for more than several days must be qualified with the possibility that spawning frequency is impacted by the nutritional conditions in the laboratory. However, these observations do suggest that individuals and populations may not show the same patterns. For *E. superba* (Harrington & Ikeda, 1986; Nicol, 1989), not all females produced second or third batches of embryos. Also the interval between broods was often shorter than predicted from the spawning frequency of a group of females (6–12 days, Harrington & Ikeda, 1986). In *M. norvegica* (Cuzin-Roudy & Buchholz, 1999), spawning events in individuals were only separated by 1–2 days, but spawning was confined to a restricted part of the moult cycle, which would increase estimates of population interbrood periods. In a detailed study of the relationship between spawning and moulting cycles, Cuzin-Roudy & Buchholz (1999) showed that *M. norvegica* released a complete batch of eggs (one vitellogenic

Table 6.7 Size of egg and individual fecundity estimates for euphausiid species: number of spawning episodes (SPE) per season, interbatch period, average number of eggs per female per SPE or ovarian eggs (OvEggs, E.s.) or vitellogenic cycle (M.n.), range in number of eggs per SPE or equation relating number of eggs (either per SPE or OvEggs) and total length (TL). I = intermittent range, L = limited range for E. p.

Species	Size egg μm	No. SPE per season	InterBat period	Av no. egg per fem per SPE	Range no. egg per SPE
E. crystalloophias	673, 650		6–10 d	187	
E. lucens	728	3–10 per life vs 135 per life	– 2 d	17–138 14–28 d^{-1} fem^{-1}	No = 0.00064TL^4.25 2–80
E. pacifica	580, 460	I: 2–6 L: 11–30	 2–10 d	< 16 mm, 60; > 19 mm 132	12–296 No = 39.07 + 7.87 DWT
E. superba	600 (555–630) 556–803 3 sites [560–590]	if repro, 3–9	5–33 d	1309 1417 2809 1372 45 mm–2469; 50 mm–3588 45 mm–2900 OvEggs ’82 45 mm–1700 OvEggs ’78	1140–1688 627–3115 89–6167 263–3662 500–8000 OvEggs = −15770 + 415.1 (TL) ’82 OvEggs = −7224 + 198.5 (TL) ’78
M. norvegica	722, 740	2 to 4 vit cycle	2–3/IMP	650 per vit cycle	3 loc: 513, 681, 735
*N. australis**	378–405	8–10 per life	30 d		49–193

* egg size in ovisac

Sources of information listed by species initials. E. c.: Harrington & Thomas (1987), Ideda (1986). E. l.: Pillar & Stuart (1988) lower estimate of SPE per lifespan, Stuart & Nicol (1986), Stuart (1992) higher estimate of SPE per lifespan. E. p.: Brinton (1976), Ross *et al.* (1982), Suh *et al.* (1993), Iguchi & Ikeda (1994). E. s.: Cuzin-Roudy (1987), Kikuno (1981), Kikuno & Kawamura (1983), Siegel (1985), Harrington & Ikeda (1986), Ross & Quetin (1983), Nicol (1989). M. n.: Marschall (1983), Cuzin-Roudy & Buchholz (1999). N. a.: Hosie & Ritz (1982, 1983).

cycle) in two main and usually one residual spawning event during the premoult period. In the Ligurian Sea, spawning moult cycles alternate with vitellogenic moult cycles. This pattern may not hold throughout its distribution range; in areas of low food availability, an extended period of reorganisation of the ovary may be necessary before a new cycle of egg production begins. Furthering our understanding of what drives this variability is critical to understanding reproductive success in euphausiids.

The number of eggs released per spawning episode varies significantly both within and between species (Table 6.7). Although an increase in fecundity with female size is predicted, the correlation with total length is usually quite low (i.e. *E. superba* as cited in Ross & Quetin, 1983; Harrington & Ikeda, 1986; Nicol *et al.*, 1995). Most investigators have inferred that the number of eggs released per spawning episode is influenced by factors other than body size. Somers (1991) examined five different regression models for the relationship between fecundity and female size for crustaceans. He concluded that the allometric model (log-transformed fecundity versus log-transformed female length) was the better choice, primarily because of the relative scale independence and the ability to make comparisons among taxa. From general allometric concepts, the relation between fecundity and female volume (expressed as length cubed) should be linear (i.e. two volume measures), as has been shown for the relationship between wet weight and ovarian oöcytes in *E. superba* and *M. norvegica* (Xuefeng & Rong, 1995; Cuzin-Roudy, 2000). The relationship between fecundity and size may also show high variability because the numbers of eggs released by the same female in successive spawning episodes are not equal, as seen in both *M. norvegica* and *E. superba* (Harrington & Ikeda, 1986; Nicol, 1989; Cuzin-Roudy & Buchholz, 1999). In fact, if all mature or maturing oöcytes (oc4) in the ovary are counted, the relationship between the number of eggs and the size of the euphausiid improves significantly for both *E. superba* (Siegel, 1985; Xeufeng & Rong, 1995; Cuzin-Roudy, 2000) and *M. norvegica* (Cuzin-Roudy, 2000). The implication is that a batch of eggs is released over a period of days in several spawning episodes, and that the sum of the eggs released in this suite of successive episodes correlates well with body size. Batch size or number of eggs per spawning episode can also vary among geographical sites (Harrington & Ikeda, 1986) or between years (Siegel, 1985; unpublished data, Ross & Quetin). Presumably the nutritional conditions prior to and during the spawning season also affect the size of the batches.

Mauchline (1988) found that the relationship between brood volume and body volume for 13 species of brooding euphausiids was logarithmic, with a correlation of 0.871. The slope was slightly greater than 1, implying that larger species tended to produce larger broods, but the increase was small, from 10 to 15% of body volume. For six species of epipelagic euphausiids, the relationship between the average number of eggs per spawning episode or vitellogenic cycle and the volume (maximum total length cubed) was linear, with $r^2 = 0.987$ (Fig. 6.6a).

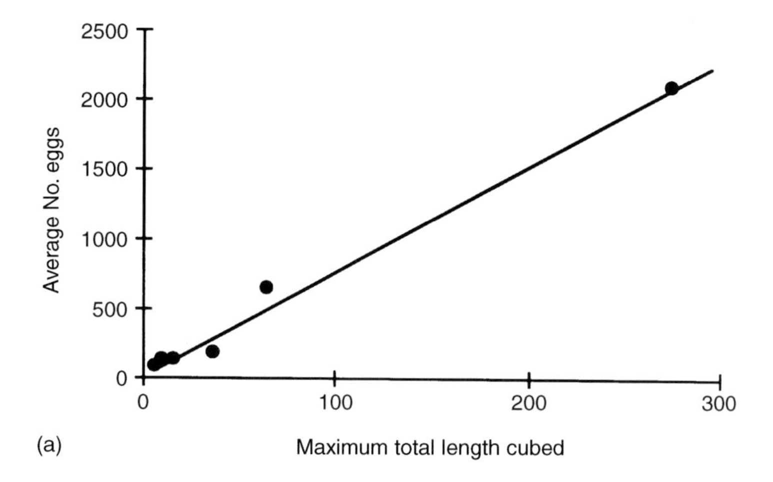

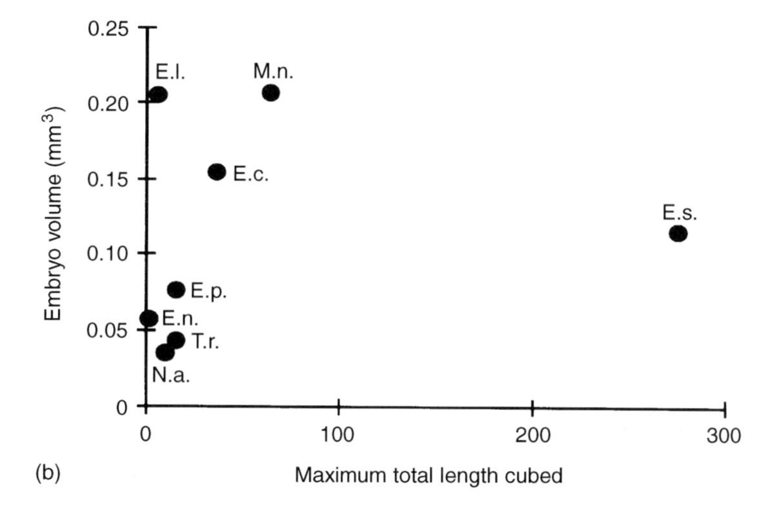

Fig. 6.6 Intraspecific comparison of euphausiid volume (expressed as the cube of total maximum length, Table 6.1) and (a) average number of eggs per spawning episode, Table 6.7; and (b) average diameter of the egg, average of values shown in Table 6.7 plus E. n. from Hirota *et al.* (1984). Initials on points are the first letter of the genus and species of euphausiid. The line shown in (a) is average number of eggs = 32.4 + 7.58 • TL^3; n = 6, $r^2 = 0.987$.

Embryo size

Embryo size also varies within and between species. Inter-annual and geographical differences in embryo size are well documented for euphausiids (Mauchline & Fisher, 1969). Such differences have implications for the energy reserves available for the non-feeding larvae, and may impact survivorship of early life history stages.

No studies have explicitly addressed this question, but variation in parental input into the oöcyte before release and its impact on inter-annual variability in recruitment should be investigated.

Mauchline (1988) compared average egg volume and body volume for a suite of 16 epipelagic euphausiids, 7 brooders and 9 broadcast spawners. The relationship was allometric, with r = 0.833:

$$\log \text{egg volume (ml)} = 0.301 \log \text{body volume (ml)} = 4.165$$

The exponent less than unity suggests that large euphausiids have larger eggs than small euphausiids, but that egg size relative to body size decreases. From a slightly different perspective, however, the conclusions are somewhat different. A plot of egg volume against body volume (maximum TL cubed) shows that egg volume does increase with body volume for six of the eight species of euphausiids plotted. However, the egg of *E. lucens* is larger than would be predicted, and that of *E. superba* much smaller than would be predicted. Some interesting differences were obscured by the use of a log-log plot.

Mauchline (1988) suggested that there was a minimum viable size of egg to explain the decrease in relative egg size with increasing body size. The hypothetical minimum egg size also has implications for individual fecundity. Euphausiids can increase their total fecundity either by producing successive broods, as has been shown to be true for most euphausiids, or by decreasing egg size relative to body size in order to increase the number of eggs per spawning episode. In the case of *E. superba*, it appears as if the overall reproductive strategy includes both mechanisms.

6.3 Development

Eggs from different species vary not only in absolute size but also in the size of the perivitelline space (PVS), the space between the vitelline membrane and the capsule. The duration and extent of the ontogenetic migration during the early life history of euphausiids is influenced by the size of the PVS, and sinking rate and development rate of the embryo exposed to characteristic hydrographic conditions (Ross & Quetin, 1991; Hofmann *et al.*, 1992). Many investigators have suggested that a strong developmental descent and ascent is characteristic of oceanic euphausiids (Makarov, 1979, 1983; Mauchline, 1980; Marr, 1962; Hempel *et al.*, 1979; Williams & Lindley, 1982). These species would have small PVS. Neritic species, on the other hand, would have a large PVS, and would be either neutrally buoyant or have slow sinking rates. The vertical extent of the developmental descent and ascent would be small. The larger vertical extent in oceanic euphausiids would reduce intraspecific predation, whereas the small vertical extent in neritic euphausiids would reduce the extent of benthic predation. The data to date (Table 6.8) do not clearly support this hypothesis. For *M. norvegica* and *T. raschii*, both

Table 6.8 Perivitelline space (PVS) as percentage volume of embryo and qualitative characterisation, sinking rate and reference.

Species	PVS % vol	Qual char of PSV	Sink rate (m d^{-1})	Reference
Neritic				
E. crystallorophias	–	'larger than *E. superba*	'neutral'	Harrington & Thomas (1987)
M. norvegica	78		61–104	Marschall (1983)
		'med-large'	132–180	Mauchline & Fisher (1969)
T. raschii	52		42–67	Marschall (1983)
		'large'	132–180	Mauchline & Fisher (1969)
Oceanic				
E. pacifica	–	small	Sink	Bollens *et al.* (1992)
E. superba	9	small	203–231	Marschall (1983)
			175	Quetin & Ross (1984)
E. nana	–	small	–	
T. longicaudata	–	'very small'	–	Williams & Lindley (1982)

slow and fast sinking rates have been measured (Mauchline & Fisher, 1969; Marschall, 1983). For some species with a small PVS, high sinking rates but fast development rates still yield a small vertical excursion, as seen in *E. pacifica* in Dabob Bay (Bollens *et al.*, 1992). More research is clearly warranted.

6.4 Implications for fisheries management

The goal of management of euphausiid fisheries, as for all fisheries, is to ensure that recruitment to the adult population is adequate to compensate for the stock that is harvested. The harvest rate must depend on the rate of recruitment into the adult population that in turn depends on the population fecundity. Thus reproduction (egg production) and its impact on recruitment should be a critical aspect of management of crustacean fisheries (Botsford, 1991). However, although recruitment depends on egg production through the relationship between stock and recruitment, the impact of variation in egg production is subsequently modified by the impact of the fluctuating pre-recruitment environment on the survival of the early life history stages.

At present we know very little about stock-recruitment relationships in crustaceans (Botsford, 1991). Data necessary to establish the relationship between spawning stock and reproductive output in euphausiids are scarce. The time series needs to be long, and both stock and recruitment must be measured on relevant time and space scales. One critical issue is whether the stock measured is the one that produced the larvae recruited to that adult population, or if the eggs and larvae were transported by currents prior to recruitment to the adult population.

Zelikman *et al.* (1980) suggested that the lack of correlation between stock and recruitment in euphausiids supported the concept of a 'self regulating' population, i.e. density dependent mechanisms regulating recruitment. The amount of uncertainty about the relative importance of egg production and early larval mortality on recruitment, however, suggests that this conclusion may be premature. The effects of egg production on recruitment may be hidden by processes occurring during the early life history stages, i.e. the effects of abiotic and biotic factors in the pre-recruitment environment. In the case of *E. superba*, we know that interannual variability in recruitment success of *E. superba* is high (Siegel & Loeb, 1995). Although the mechanism proposed is the over-winter survival of the larvae (Ross & Quetin, 1991; Quetin *et al.*, 1996; Siegel & Loeb, 1995), the role of reproductive output has not been thoroughly evaluated.

An alternative explanation for the lack of correlation between stock and recruitment is that population fecundity is complex and varies greatly with stock. Emerging evidence for euphausiids suggests that the number of eggs produced per female may show large inter-annual variation due to the effect of nutritional history on factors such as the percentage of the population reproducing, variation in numbers of spawning episodes and batch size. Early larval survival may also be affected by the amount of reserves in the embryo. Variation in embryo size for euphausiids is well documented, and this variation implies that the amount of reserves also varies.

Given our uncertainties about the relative roles of egg production and larval survivorship in recruitment, the conservative management tactic would be to guard egg production. Botsford (1991) presents the argument that such a management tactic does not always create the highest yield, but that the population will be maintained. To achieve this goal of guarding egg production, restrictions may be placed on the timing or spatial occurrence of commercial harvesting, or on the part of the population that can be fished. One common tactic when reproduction is concentrated in one part of the year is a limited season for harvesting after the reproductive season. With the species of euphausiids currently harvested, the limited reproductive seasons would allow this tactic.

The restricted distribution of spawning populations would allow the use of a somewhat different policy, basically to prohibit the harvesting of gravid females, analogous to prohibiting 'berried' lobsters and crabs from being kept. Fishing pseudopopulations, for example, would have no immediate impact on population recruitment. We also know that swarming is often correlated with reproductive activity, so fishing surface swarms may selectively harvest the portion of the population that we most want to guard. On the other hand, if we understand the behaviour of individuals in the surface aggregations better, we may be able to take advantage of differences in the behaviour of mated females in terms of timing of spawning or diel distribution.

A more complex management tactic would be based on the fact that population fecundity will depend on the size–age structure of the population, which in turn depends on harvest policies. Larger females produce the greater number of eggs.

The goal would be to prevent the overfishing of the large egg-producing females and the males needed to mate with those females throughout the season. In some crustacean fisheries, a size limit is set for a single sex fishery (e.g. males below a certain size, or females above a certain size), but this tactic would not be possible with most euphausiids. Even though catches often show a skewed sex ratio (Watkins *et al.*, 1992) throwing back the 'wrong' sex would not be feasible. Once caught, large individuals would not survive if thrown back. In summary the best options for management tactics based on our knowledge of euphausiid reproduction appear to be restrictions of the fishery in time and space to guard the large egg-producing females.

Acknowledgments

We thank our many colleagues in the field of euphausiid research for stimulating and enlightening discussions. Those with J. Cuzin-Roudy and S. Nicol have been particularly valuable. An anonymous reviewer asked questions that helped clarify the text. This publication was supported by NSF Grant OPP96-32763, with additional support from the University of California at Santa Barbara. This is Palmer LTER publication No. 182.

References

Astthorsson, O.S. (1990) Ecology of the Euphausiids *Thysanoessa raschi*, *T. inermis* and *Meganyctiphanes norvegica* in Isafjord-deep, northwest-Iceland. *Mar. Biol.* **107**, 147–57.

Baker, A.C. (1959) The distribution of *Euphausia triacantha* Holt and Tattersall. *Discovery Rep.* **29**, 309–339.

Bargmann, H.E. (1937) The reproductive system of *Euphausia superba*. *Discovery Rep.* **14**, 325–50.

Bargmann, H.E. (1945) The development and life-history of adolescent and adult krill, *Euphausia superba*. *Discovery Rep.* **23**, 103–176.

Blackburn, M. (1980) *Observations on the distribution of* Nyctiphanes australis *Sars (Crustacea, Euphausiidae) in Australian waters.* Commonwealth Scientific and Industrial Research Organization Division of Fisheries and Oceanography, Cronulla, N.S.W.

Bollens, S.M., Frost, B.W. & Lin, T.S. (1992) Recruitment, growth, and diel vertical migration of *Euphausia pacifica* in a temperate fjord. *Mar. Biol.* **114**, 219–28.

Botsford, L.W. (1991) Crustacean egg production and fisheries management. In *Crustacean Egg Production*, Vol. 7 (A. Wenner & A. Kuris, eds), pp. 379–94. Balkema, Rotterdam.

Boysen, E. & Buchholz, F. (1984) *Meganyctiphanes norvegica* in the Kattegat. Studies on the annual development of a pelagic population. *Mar. Biol.* **79**, 195–207.

Brinton, E. (1976) Population biology of *Euphausia pacifica* off southern California. *Fish. Bull.* **74**, 733–62.

Brinton, E. & Townsend, A. W. (1991) Developmental rates and habitat shifts in the Antarctic neritic euphausiid *Euphausia crystallorophias*, 1986–87. *Deep-Sea Res.* **38**, 1195–211.

Cuzin-Roudy, J. (1987) Sexual differentiation in the Antarctic krill *Euphausia superba* Dana (Crustacea: Euphausiacea). *J. Crust. Biol.* **7**, 518–24.

Cuzin-Roudy, J. (1993) Reproductive strategies of the Mediterranean krill, *Meganyctiphanes norvegica* and the Antarctic krill, *Euphausia superba* (Crustacea, Euphausiacea). *Invertebrate Repro Devel.* **23**, 105–114.

Cuzin-Roudy, J. (2000) Seasonal reproduction, multiple spawning and fecundity in Northern krill, *Meganyctiphanes norvegica*, and Antarctic krill, *Euphausia superba*. In Proceedings of the Second International Krill Symposium, Santa Cruz, California, August 1999. *Can. J. Fish. Aquat. Sci.* (in press).

Cuzin-Roudy, J. & Amsler, M.O. (1991) Ovarian development and sexual maturity staging in antarctic krill, *Euphausia superba* Dana (Euphausiacea). *J. Crust Biol.* **11**, 236–49.

Cuzin-Roudy, J. & Buchholz, F. (1999) Ovarian development and spawning in relation to the moult cycle in Northern krill, *Meganyctiphanes norvegica* (Crustacea: Euphausiacea), along a climatic gradient. *Mar. Biol.* **133**, 267–81.

Cuzin-Roudy, J. & Labat, J.P. (1992) Early summer distribution of Antarctic krill sexual development in the Scotia-Weddell region: a multivariate approach. *Polar Biol.* **12**, 65–74.

Dalpadado, P. & Skjoldal, H.R. (1996) Abundance, maturity and growth of the krill species *Thysanoessa inermis* and *T. longicaudata* in the Barents Sea. *Mar. Ecol. Prog. Ser.* **144**, 175–83.

Einarsson, H. (1945) Euphausiacea. I. Northern Atlantic species. *Dana Report* **27**, 1–175.

Endo, Y. (1984) Daytime surface swarming of *Euphausia pacifica* (Crustacea: Euphausiacea) in the Sanriku coastal waters off northeastern Japan. *Mar. Biol.* **79**, 269–76.

Falk-Petersen, S. & Hopkins, C. C. E. (1981) Ecological investigations on the zooplankton community of Balsfjorden, northern Norway: population dynamics of the euphausiids *Thysanoessa inermis* (Kröyer), *T. raschii* (M. Sars) and *Meganyctiphanes norvegica* (M. Sars) in 1976 and 1977. *J. Plank. Res.* **3**, 177–92.

Falk-Petersen, S., Hagen, W., Kattner, G., Clarke, A. & Sargent, J. (1999) *Lipids – Key components of dominant species of Arctic and Antarctic krill.* Paper presented at the Second International Symposium on *Krill*, Santa Cruz CA, August 23–27, 1999.

Fevolden, S.E. (1980) Krill off Bouvetoya and in the southern Weddell Sea with a description of larval stages of *Euphausia crystallorophias*. *Sarsia* **65**, 149–62.

Foxton, P. (1956) The distribution of the standing crop of zooplankton in the Southern Ocean. *Discovery Rep.* **28**, 191–236.

Hanamura, Y., Kotori, M. & Hamaoka, S. (1989) Daytime surface swarms of the Euphausiid *Thysanöessa inermis* off the west. *Mar. Biol.* **102**, 369–76.

Harrington, S.A. & Ikeda, T. (1986) Laboratory observations on spawning, brood size and egg hatchability of the antarctic krill *Euphausia superba* from Prydz Bay, Antarctica. *Mar. Biol.* **92**, 231–35.

Harrington, S.A. & Thomas, P.G. (1987) Observations on spawning by *Euphausia crystallrophias* from waters adjacent to Enderby Land (East Antarctica) and speculations on the early ontogenetic ecology of neritic euphausiids. *Polar Biol.* **7**, 93–95.

Hart, T.J. (1942) Phytoplankton periodicity in Antarctic surface waters. *Discovery Rep.* **8**, 1–268.

Heath, W.A. (1977) *The ecology and harvesting of euphausiids in the Strait of Georgia.* University of British Columbia, Vancouver, pp. 187.

Hempel, I., Hempel, G. & Baker, A.D. (1979) Early life history stages of krill (*Euphausia superba*) in Bransfield Strait and Weddell Sea. *Meereforschung* **27**, 267–81.

Hirota, Y., Nemoto, T. & Marumo, R. (1984) Larval development of *Euphausia nana* (Crustacea: Euphausiacea). *Mar. Biol.* **81**, 311–22.

Hofmann, E.E., Capella, J.E., Ross, R.M. & Quetin, L.B. (1992) Models of the early life history of *Euphausia superba* – Part I. Time and temperature dependence during the descent–ascent cycle. *Deep-Sea Res.* **39**, 1177–1200.

Hollingshead, K.W. & Corey, S. (1974) Aspects of the life history of *Meganyctiphanes norvegica* (M. Sars), Crustacea (Euphausiacea), in the Passamaquoddy Bay. *Can. J. Zool.* **52**, 495–505.

Hosie, G.W. & Cochran, T.G. (1994) Mesoscale distribution patterns of macrozooplankton communities in Prydz Bay, Antarctica – January to February 1991. *Mar. Ecol. Prog. Ser.* **106**, 21–39.

Hosie, G. & Ritz, D. (1983) Contribution of moulting and eggs to secondary production in *Nyctiphanes australis* (Crustacea: Euphausiacea). *Mar. Biol.* **77**, 215–20.

Iguchi, N. & Ikeda, T. (1994) Experimental study on brood size, egg hatchability and early development of a euphausiid *Euphausia pacifica* from Toyama Bay, southern Japan Sea. *Bull. Japan Sea Nat. Fish. Res. Inst.* **44**, 49–57.

Iguchi, N., Ikeda, T. & Imamura, A. (1993) Growth and life cycle of a euphausiid crustacean (*Euphausia pacifica*) Hansen in Toyama Bay, southern Japan Sea. *Bull. Japan Sea Nat. Fish. Res. Inst.* **43**, 69–81.

Ikeda, T. (1986) Preliminary observations on the development of the larvae of *Euphausia crystallorophias* Holt and Tattersall in the laboratory. *Mem. Natl Inst. Polar Res. Tokyo*, pp. 183–86.

Kikuno, T. (1981) Spawning behaviour and early development of the antarctic krill, *Euphausia superba* Dana, observed on board R.V. Kaiyo Maru in 1979–80. *Antarctic Rec.* **73**, 97–102.

Kikuno, T. & Kawamura, A. (1983) Observation of the ovarian eggs and spawning habits in *Euphausia superba* Dana. *Mem. Natl Inst. Polar Res. Tokyo.* Special Issue 27, 104–128.

Kotori, M. (1995) An incidence of surface swarming of *Euphausia pacifica* off the coast of western Hokkaido, Japan. *Bull. Plank. Soc. Japan* **42**, 80–84.

Kulka, D.W. & Corey, S. (1978) The life history of *Thysanoessa inermis* (Kroyer) in the Bay of Fundy. *Can. J. Zool.* **56**, 492–506.

Lascara, C.M., Hofmann, E.E., Ross, R.M. & Quetin, L.B. (1999) Seasonal variability in the distribution of Antarctic krill, *Euphausia superba*, west of the Antarctic Peninsula. *Deep-Sea Res.* **46**, 951–84.

Lindley, J.A. (1978) Population dynamics and production of euphausiids. I. *Thysanoessa longicaudata* in the North Atlantic Ocean. *Mar. Biol.* **46**, 121–30.

Lindley, J.A. (1982) Population dynamics and production of Euphausiids. III. *Meganyctiphanes norvegica* and *Nyctiphanes couchii* in the North Atlantic Ocean and the North Sea. *Mar. Biol.* **66**, 37–46.

Makarov, R.R. (1977) Distribution of the larvae and some questions of the ecology of reproduction of the euphausiid *Euphausia frigida* Hansen, 1911 (Crustacea, Euphausiacea) in the southern part of the Scotia Sea. *Oceanology* **17**, 208–213.

Makarov, R.R. (1979) Larval distribution and reproductive ecology of *Thysanoessa macrura* (Crustacea: Euphausiacea) in the Scotia Sea. *Mar. Biol.* **52**, 377–86.

Makarov, R.R. (1983) Some problems in the investigation of larval euphausiids in the Antarctic. *Berichte zur Polarforschung* **4**, 58–69.

Makarov, R.R. & Denys, C.J. (1980) Stages of sexual maturity of *Euphausia superba* Dana. *BIOMASS Handbook* **11**, 1–11.

Makarov, R.R. & Men'shenina, L.L. (1992) Larvae of euphausiids off Queen Maud Land. *Polar Biol.* **11**, 515–23.

Makarov, R.R. & Sysoyeva, M.V. (1985) Biology and distribution of *Euphausia superba* in the Lazarev Sea and adjacent waters. *Legkaya i pishevaya promyshlennost', Moscow*, 110–116.

Makarov, R., Men'shenina, L. & Spiridonov, V. (1990) Distributional ecology of euphausiid larvae in the Antarctic Peninsula region and adjacent waters. Proceedings of the NIPR Symposium on Polar Biology **3**, 23–35.

Makarov, R.R., Men'shenina, L.L., Timonin, V.P. & Shurunov, N.A. (1991) Ecology of larvae and reproduction of Euphausiidae in the Ross Sea. *Soviet J. Mar. Biol.* **16**, 156–62.

Makarov, R.R., Solyankin, Y. & Shevtsov, V.V. (1985) Environmental conditions and adaptive features of the biology of *Euphausia superba* Dana in the Lazarev Sea. *Polar Geogr. Geol.* **9**, 146–64.

Marr, J.W.S. (1962) The natural history and geography of the Antarctic krill (*Euphausia superba* Dana). *Discovery Rep.* **32**, 33–464.

Marschall, H.-P. (1983) Sinking speed, density and size of Euphausiid eggs. *Meeresforschung* **30**, 1–9.

Mauchline, J. (1968) The development of the eggs in the ovaries of euphausiids and estimation of fecundity. *Crustaceana* **14**, 155–63.

Mauchline, J. (1980) The biology of mysids and euphausiids. *Adv. Mar. Biol.* **18**, 1–681.

Mauchline, J. (1988) Egg and brood sizes of oceanic pelagic crustaceans. *Mar. Ecol. Prog. Ser.* **43**, 251–58.

Mauchline, J. & Fisher, L.R. (1969) The biology of euphausiids. *Adv. Mar. Biol.* **7**, 1–454.

Men'shenina, L.L. (1988) Ecology of spawning and larval development of *Thysanoessa macrura* in the southwestern sector of the Antarctic Ocean (with remarks on *T. vicina*). *Oceanology* **28**, 774–80.

Men'shenina, L.L. (1992) Distribution of Euphausiid larvae in the Weddell Gyre in Sep.–Oct. 1989. Proceedings of the NIPR Symposium on *Polar Biology* **5**, 44–54. Tokyo, National Institute of Polar Research.

Naito, Y., Taniguchi, A. & Hamada, E. (1986) Some observations on swarms and mating behavior of antarctic krill (*Euphausia superba* Dana). *Mem. Natl Inst. Polar Res*, Special Issue 40, 178–82.

Nelson, K. (1991) Scheduling of reproduction in relation to molting and growth in malacostracan crustaceans. In *Crustacean Egg Production*, Vol. 7 (A. Wenner & A. Kuris, eds), pp. 77–133. Balkema, Rotterdam.

Nemoto, T., Doi, T. & Nasu, K. (1981) Biological characteristics of krill caught in the Southern Ocean. In *Selected Contributions to the Woods Hole Conference on Living Resources of the Southern Ocean 1976*, Biomass Vol. 2., pp. 47–64. SCAR, Cambridge.

Nicol, S. (1984) Population structure of daytime surface swarms of the euphausiid *Meganyctiphanes norvegica* in the Bay of Fundy. *Mar. Ecol. Prog. Ser.* **18**, 29–39.

Nicol, S. (1989) Apparent independence of the spawning and moulting cycles in female Antarctic krill (*Euphausia superba* Dana). *Polar Biol.* **9**, 371–75.

Nicol, S. & Endo, Y. (1997) Krill fisheries of the world. *FAO Fisheries Technical Paper* 367, 100 pp. Food and Agriculture Organization of the United Nations, Rome.

Nicol, S., de la Mare, W.K. & Stolp, M. (1995) The energetic cost of egg production in Antarctic krill (*Euphausia superba* Dana). *Antarctic Sci.* **7**, 25–30.

Nicol, S., James, A. & Pitcher, G. (1987) A first record of daytime surface swarming by *Euphausia lucens* in the Southern Benguela region. *Mar. Biol.* **94**, 7–10.

O'Brien, D.P. (1988) Surface schooling behaviour of the coastal krill *Nyctiphanes australis* (Crustacea: Euphausiacea) off Tasmania, Australia. *Mar. Ecol. Prog. Ser.* **42**, 219–33.

Odate, K. (1991) Fishery biology of the krill, *Euphausia pacifica*, in the northeastern coasts of Japan. *Suisan Kenkyu Sosho* **40**, 1–100 (in Japanese).

Pakhomov, E.A. & Perissinotto, R. (1996) Antarctic neritic krill *Euphausia crystallorophias*: spatio-temporal distribution, growth and grazing rates. *Deep-Sea Res.* **43**, 59–87.

Paul, A.J., Coyle, K.O. & Ziemann, D.A. (1990) Timing of spawning of *Thysanoessa raschii* (Euphausiacea) and occurrence of their feeding stage larvae in an Alaskan Bay. *J. Crust. Biol.* **10**, 69–78.

Pillar, S.C. & Stuart, V. (1988) Population structure, reproductive biology and maintenance of *Euphausia lucens* in the southern Benguela Current. *J. Plank. Res.* **10**, 1083–98.

Ponomareva, L.A. (1966) *The euphausiids of the North Pacific, their distribution, ecology, and mass species*. Inst. Okeanol. Akad. Sci. USSR, Moscow.

Quetin, L.B. & Ross, R.M. (1984) Depth distribution of developing *Euphausia superba* embryos, predicted from sinking rates. *Mar. Biol.* **79**, 47–53.

Quetin, L.B., Ross, R.M. & Clarke, A. (1994) Krill energetics: seasonal and environmental aspects of the physiology of *Euphausia superba*. In *Southern Ocean Ecology: the BIOMASS Perspective* (S. El-Sayed, ed.), pp. 165–84. Cambridge University Press, Cambridge.

Quetin, L.B., Ross, R.M., Frazer, T.K. & Haberman, K.L. (1996) Factors affecting distribution and abundance of zooplankton, with an emphasis on Antarctic krill, *Euphausia superba*. In *Foundations for Ecological Research West of the Antarctic Peninsula*, Vol. 70 (R.M. Ross, E.E. Hofmann, & L.B. Quetin, eds), pp. 357–71. American Geophysical Union, Washington, D.C.

Quetin, L.B., Ross, R.M., Prézelin, B.B., Haberman, K.L., Hacecky, K.L. & Newberger, T. (1992) Palmer LTER program: Biomass and community composition of euphausiids within the peninsula grid, November 1991 cruise. *Antarctic J. US* **27**, 244–45.

Ritz, D. & Hosie, G. (1982) Production of the euphausiid *Nyctiphanes australis* in Storm Bay, south-eastern Tasmania. *Mar. Biol.* **68**, 103–108.

Ritz, D.A. (1994) Social aggregation in pelagic invertebrates. In *Advances in Marine Biology*, Vol. 30 (J.H.S. Blaxter & A.J. South, eds), pp. 155–207. Academic Press, London.

Ross, R.M. & Quetin, L.B. (1983) Spawning frequency and fecundity of the antarctic krill *Euphausia superba*. *Mar. Biol.* **77**, 201–205.

Ross, R.M. & Quetin, L.B. (1991) Ecological physiology of larval euphausiids, *Euphausia superba* (Euphausiacea). *Memoirs of the Queensland Museum* **31**, 321–33.

Ross, R.M., Daly, K.L. & English, T.S. (1982) Reproductive cycle and fecundity of *Euphausia pacifica* in Puget Sound, Washington. *Limn. Oceanogr.* **27**, 304–314.

Ross, R.M., Quetin, L.B., Amsler, M.O. & Elias, M.C. (1987) Larval and adult Antarctic krill, *Euphausia superba*, winter-over at Palmer Station. *Antarctic J. US* **22**, 205–206.

Shaw, C.T. (1997) Effect of sea ice conditions on physiological maturity of female Antarctic krill (*Euphausia superba* Dana) west of the Antarctic Peninsula. Master of Arts thesis,

Ecology, Evolution and Marine Biology, University of California at Santa Barbara, Santa Barbara, pp. 98. Available through interlibrary loan, call # QH319.C2.S25 SHAC 1997, or through the authors of this chapter.

Siegel, V. (1985) On the fecundity of Antarctic krill, *Euphausia superba* (Euphausiacea). *Archiv für FischereiWissenschaft* **36**, 185–93.

Siegel, V. (1986) Structure and composition of the Antarctic krill stock in the Bransfield Strait (Antarctic Peninsula) during the Second International BIOMASS Experiment (SIBEX). *Archiv für FischereiWissenschaft* **37**, 51–72.

Siegel, V. (1987) Age and growth of antarctic Euphausiacea (Crustacea) under natural conditions. *Mar. Biol.* **96**, 483–95.

Siegel, V. (1992) Assessment of the krill *Euphausia superba* spawning stock off the Antarctic Peninsula. *Archiv für Fischereiwissenschaft* **41**, 101–130.

Siegel, V. & Loeb, V. (1994) Length and age at maturity of Antarctic krill. *Antarctic Sci.* **6**, 479–82.

Siegel, V. & Loeb, V. (1995) Recruitment of Antarctic krill (*Euphausia superba*) and possible causes for its variability. *Mar. Ecol. Prog. Ser.* **123**, 45–56.

Siegel, V., Bergstrom, B., Stromberg, J.O. & Schalk, P.H. (1990) Distribution, size frequencies and maturity stages of krill, *Euphausia superba*, in relation to sea-ice in the northern Weddell Sea. *Polar Biol.* **10**, 549–57.

Smiles, M.C., Jr. & Pearcy, W.G. (1971) Size structure and growth rate of *Euphausia pacifica* off the Oregon coast. *Fish. Bull.* **69**, 79–86.

Smith, S.L. (1991) Growth, development and distribution of the euphausiids *Thysanoessa raschii* (M. Sars) and *Thysanoessa inermis* (Kröyer) in the southeastern Bering Sea. *Polar Res.* **10**, 461–78.

Smith, S.E. & Adams, P.B. (1988) Daytime surface swarms of *Thysanoessa spinifera* (Euphausiacea) in the Gulf of the Farallones, California. *Bulletin of Marine Science* **42**, (1), 76–84.

Somers, K.M. (1991) Characterizing size-specific fecundity in crustaceans. In *Crustacean Egg Production*, Vol. 7 (A. Wenner & A. Kuris, eds), pp. 357–78. Balkema, Rotterdam.

Spiridonov, V.A. (1995) Spatial and temporal variability in reproductive timing of Antarctic krill (*Euphausia superba* Dana). *Polar Biol.* **15**, 161–74.

Stepnik, R. (1982) All-year populational studies of Euphausiacea (Crustacea) in the Admiralty Bay (King George Island, South Shetland Island Antarctic). *Polish Polar Res.* **3**, 49–68.

Stuart, V. (1992) Fecundity of *Euphausia lucens* (Hansen) – Laboratory evidence for multiple broods. *J. Exp. Mar. Biol. Ecol.* **160**, 221–28.

Stuart, V. & Nicol, S. (1986) The reproductive potential of three euphausiid species from the southern Benguela region. *J. Exp. Mar. Biol. Ecol.* **103**, 267–74.

Stuart, V. & Pillar, S.C. (1988) Growth and production of *Euphausia lucens* in the southern Benguela Current. *J. Plank. Res.* **10**, 1099–112.

Suh, H.L., Soh, H.Y. & Hong, S.Y. (1993) Larval development of the euphausiid *Euphausia pacifica* in the Yellow Sea. *Mar. Biol.* **115**, 625–33.

Timofeyev, S.F. (1994) Population structure of summer aggregations of euphausiids *Thysanoessa raschii* (M. Sars) in the southern part of the Barents Sea. *Oceanology* **33**, 785–89.

Virtue, P., Nichols, P.D., Nicol, S. & Hosie, G. (1996) Reproductive trade-off in male Antarctic krill, *Euphausia superba*. *Mar. Biol.* **126**, 521–27.

Watkins, J.L., Buchholz, F., Priddle, J., Morris, D.J. & Ricketts, C. (1992) Variation in reproductive status of Antarctic krill swarms – Evidence for a size-related sorting mechanism. *Mar. Ecol. Prog. Ser.* **82**, 163–74.

Williams, R.L. & Lindley, J.A. (1982) Variability in abundance, vertical distribution and ontogenetic migrations of *Thysanoessa longicaudata* (Crustacea: Euphausiacea) in the North-Eastern Atlantic. *Mar. Biol.* **69**, 321–30.

Xuefeng, Z. & Rong, W. (1995) Reproductive characteristics of Antarctic krill, *Euphausia superba* Dana, in the Prydz Bay region. *Antarctic Res.* **6**, 58–72.

Zelikman, E.A. (1958) On gonad maturation and female productivity in species of euphausians abundant in the Barents Sea. *Doklady Adad Nauk S.S.S.R* **118**, 201–204.

Zelikman, E.A., Lukashevich, I.P., Drobysheva, S.S. & Degtereva, A.A. (1980) Fluctuations of the number of eggs spawned by *Thysanoessa inermis* Kr. and *T. raschii* (M. Sars) (Euphausiacea) from the Barents Sea. *Oceanology* **20**, 716–21.

Chapter 7
Role of Krill in Marine Food Webs

7.1 Japanese waters

Yoshi Endo

Off the Japanese coast *Euphausia pacifica* is preyed upon by almost all the commercially important fish species which are endemic to, or seasonally migrate through Sanriku waters (e.g., Endo, 1981; Yamamura, 1993; Kuroda, 1994; Yamamura *et al.*, 1998). It is one of the main food items of Pacific cod (*Gadus macrocephalus*), walleye pollack (*Theragra chalcogramma*), Japanese chub mackerel (*Scomber japonicus*) and sand lance (*Ammodytes personatus*). In the *E. pacifica* fishing season, sand lance, black-tailed gull (*Larus crassirostris*) and rhinoceros auklet (*Cerorhinca monocerata*) feed on them intensively (Komaki, 1967). For example, Takeuchi (1985) has reported that on average each individual sand lance in a school contained 82 individual *E. pacifica* in its stomach. In this chapter, I consider the importance of *E. pacifica* both in demersal and pelagic environments, with the former being reported more fully.

Demersal environment

Fujita (1994) suggested that *E. pacifica* is most important as food for demersal fishes in the upper slope region. There, compared with the continental shelf and deeper areas (> 300 m) in Sanriku waters, the abundance of demersal fish is highest, but the species diversity is low. Blue eye, *Chlorophthalmus albatrossis* (*borealis*), is a finfish for which the annual commercial catch is small, about 200–400 tonnes. It feeds on *E. pacifica* throughout the year in waters off the Fukushima Prefecture (Akiyama, 1996). This implies that *E. pacifica* occurs in the offshore area within the depth range from 150 to 260 m throughout the year.

Yamamura (1993) examined the resource partitioning of demersal fish in the offshore areas of Sendai Bay during May to June 1989 (see Fig. 3.1.4 for place names). Pacific cod was dominant in areas where the water was shallower than 200 m deep and they made up 76% by number of the demersal fish present there. Walleye pollack was dominant between 300–400 m accounting for 70% of the fish, and threadfin hakeling (*Laemonema longipes*) was dominant, 89% numerically, of

Plate 1 Antarctic Krill, *Euphausia superba* swimming, showing the greenness of the hepatopancreas caused by feeding on phytoplankton. (Photographer: Inigo Everson ARPS)

Plate 2 (*Below.*) Colour chart recommended for use by CCAMLR in determining the feeding status of krill in the Southern Ocean. (CCAMLR 1993)

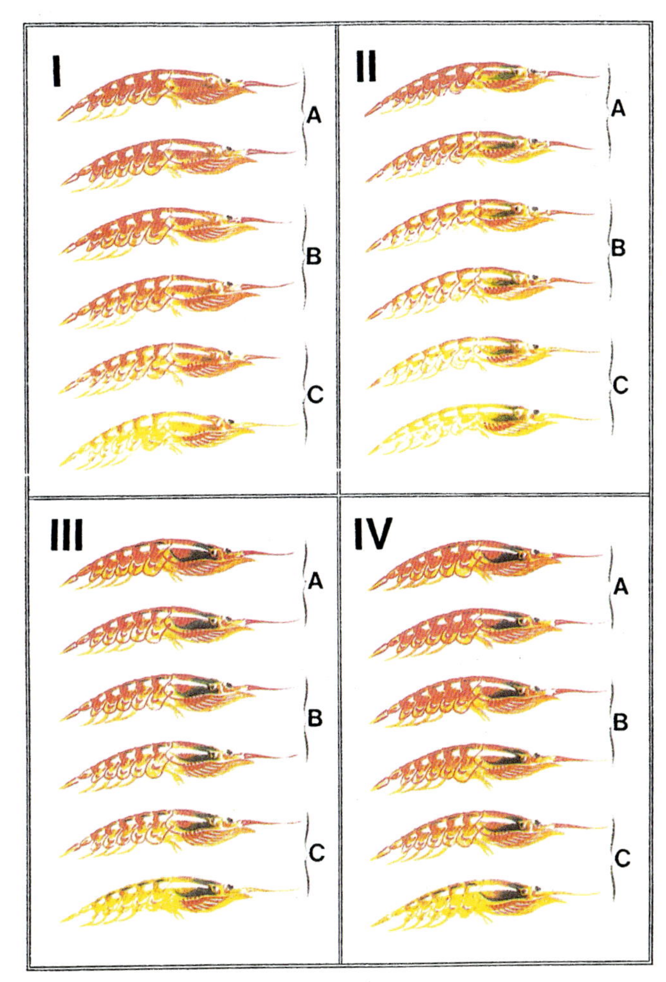

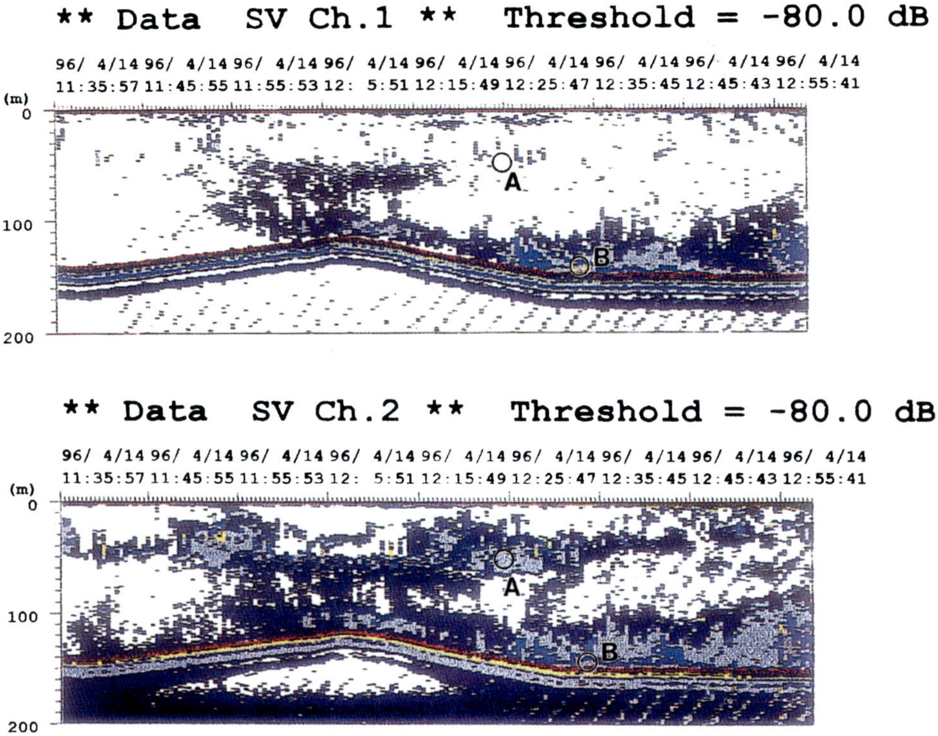

** Data SV Ch.1 ** Threshold = -80.0 dB

96/ 4/14 96/ 4/14 96/ 4/14 96/ 4/14 96/ 4/14 96/ 4/14 96/ 4/14 96/ 4/14 96/ 4/14 96/ 4/14
11:35:57 11:45:55 11:55:53 12: 5:51 12:15:49 12:25:47 12:35:45 12:45:43 12:55:41

(m)
0
A
100
B
200

** Data SV Ch.2 ** Threshold = -80.0 dB

96/ 4/14 96/ 4/14 96/ 4/14 96/ 4/14 96/ 4/14 96/ 4/14 96/ 4/14 96/ 4/14 96/ 4/14 96/ 4/14
11:35:57 11:45:55 11:55:53 12: 5:51 12:15:49 12:25:47 12:35:45 12:45:43 12:55:41

(m)
0
A
100
B
200

Plate 3 Echocharts of volume backscatter at 38 kHz (upper) and 120 kHz (lower) at a station off Iwate Prefecture. At the point A the difference ($SV_{120} - SV_{38}$) was 13 dB indicating the presence of krill (*Euphausia pacifica*) with an estimated size of 17.0 mm and numerical density of 8.3 individuals per cubic metre. At point B the difference was 1.4 dB indicating the presence of walleye pollock close to the seabed. (Miyashita *et al.* 1997)

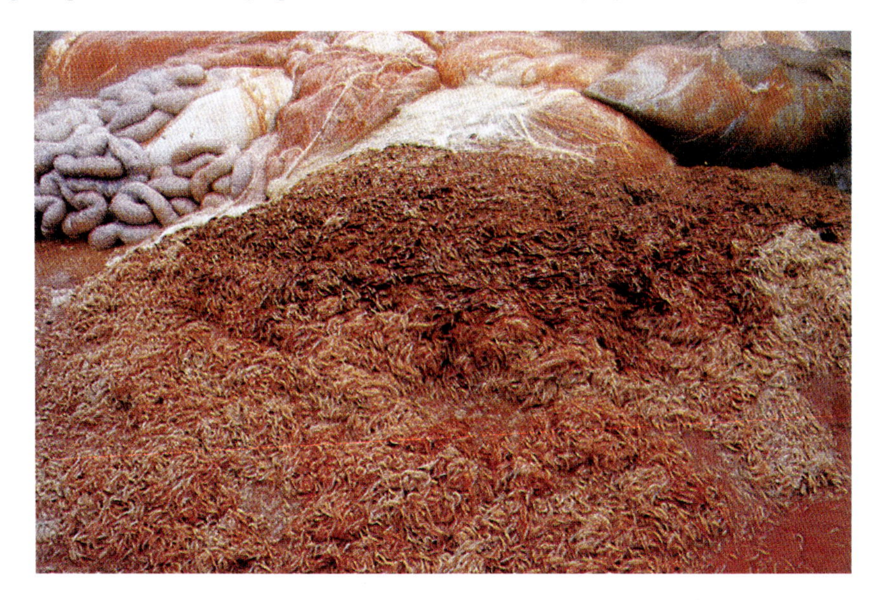

Plate 4 Stomach of a freshly flensed fin whale cut open to show the large amount of krill eaten. (Photographer: Hidehiro Kato)

Plate 5 (*Above.*) Krill trawl being hauled onboard the West German research vessel FFS Walther Herwig when undergoing trials in the Southern Ocean. Note the fine mesh netting in all parts of the net that are needed to retain the krill. (Photographer: Inigo Everson ARPS)

Plate 6 (*Right.*) Cans of krill: processed krill tails produced by different factories in the former USSR. (Photographer: Inigo Everson ARPS)

Plate 7 Japanese canned krill tails, marketed in the 1980s as 'Okiami'. (With permission from the photographer: S Nicol)

Plate 8 Advertisement outside a fish processing factory in Yalta conveying the message that 'Thanks to the composition of Ocean Paste young people will grow well together. Good Health'. (Photographer: Inigo Everson ARPS)

all fish in the area deeper than 400 m. Stomach contents analysis showed that Pacific cod with body length <40 cm eat mainly *E. pacifica* and the amphipod (*Themisto japonica*). Larger individuals eat Japanese sardine (*Sardinops melanostictus*). Walleye pollack, irrespective of their body size, eat mainly *E. pacifica*. Threadfin hakeling with body length < 40 cm eat copepods, euphausiids and mesopelagic decapods. Larger individuals eat lantern fishes. It is noteworthy that 62% of total biomass of demersal fishes were zooplankton feeders whereas benthos feeders comprised only 1.3%. This implies that the relative abundance of pelagic prey is high in this area.

Based on four years of bottom trawl surveys in a similar survey area off Sendai Bay in May and November, Yamamura *et al.* (1998) examined the importance of *E. pacifica* as food of demersal fishes. Of 87 fish species examined for stomach contents, 24 (27.6%) ingested *E. pacifica*. In terms of biomass, however, *E. pacifica* feeders comprised more than 90% in 10 out of 16 sample sets. Pacific cod and walleye pollack had the highest predation impact. The contribution of *E. pacifica* to the total diet of the fish assemblages was highest in shallow (< 300 m) regions during May, accounting for 40% by mass. Yamamura *et al.* (1998) found a smaller contribution of *E. pacifica* in warm years, and pointed out that the benthopelagic population of *E. pacifica* moved offshore in accordance with the preferred temperature range. The average predation impact on *E. pacifica* was maximal in May (4.6 kg wet wt km^{-2} d^{-1}) and minimal in November in the shallow area (0.4 kg wet wt km^{-2} d^{-1}). Annual estimates of consumption by demersal fishes in the survey area ranged from 15 to 64% of the annual commercial catch of *E. pacifica* by local fisheries. The authors suggested that *E. pacifica* fisheries have the potential to exert a considerable impact on demersal fish assemblages.

Kodama & Izumi (1994) compiled data on demersal fish landings at the Ishinomaki Fish Market, and examined stomach contents to quantify the importance of *E. pacifica* as prey. These landings were made by an offshore trawl fishery which is operated mainly in the vicinity of Kinkasan Island over water 150 to 2000 m deep. The average yearly landing of demersal fishes at the Ishinomaki Fish Market was 45 740 tons during the period from 1983 to 1988. Except for 10 127 tons of sand lance caught in the shallow Sendai Bay area, gadoid fishes predominated constituting 88% (31 355 tons) of the total landings. These were followed by Japanese common squid (*Todarodes pacificus*) 1.8% (652 tons), spinycheek rockfish (*Sebastolobus macrochir*) 1.3% (460 tons), octopuses 0.6% (210 tons) and flatfishes 0.6% (203 tons). Gadoid fish catches were dominated by walleye pollack which made up 76% of the catch. This was followed by the other gadoids, Pacific cod 7%, and deep-sea rattails 17%. This indicates the importance of walleye pollack as a demersal fish within the ecosystem of this area.

The importance of *E. pacifica* as food is linked to the body size of the fish predators (Kodama & Izumi, 1994). For instance, the percentages of walleye pollack which consumed *E. pacifica* as prey were 84% for fish of body length of 10–19 cm, 77% for fish of body length 20–29 cm, 60% for fish of body length 30–39 cm, and

42% at 40–64 cm. For Pacific cod, comparable results were 24% at 10–19 cm, 47% at 20–29 cm, 13% at 30–39 cm, 3% at 40–59 cm and 0% at 60–79 cm. For Pacific herring, *E. pacifica* were consumed by 65% of fish between 17–23 cm long and 70% of fish 24–32 cm long.

Several assumptions must be made to calculate daily consumption of *E. pacifica* by demersal fish. In developing their trophic analysis of the ecosystem, Kodama & Izumi (1994) categorised the Pacific herring as a demersal fish because, over the southern part of its range in Sanriku waters, it behaves like a demersal fish (Kodama, personal communication). Three general assumptions were made regarding the energy balance in the system. Firstly, fishing efficiency was assumed to be 0.3 and therefore fish biomass equals annual catch divided by 0.3. Secondly, the daily ration was assumed to be equal to the average weight of stomach contents. This assumes that the stomach is filled no more than once a day, an assumption that has to be made because the assimilation rate is unknown. This may cause an underestimation of the amount of *E. pacifica* consumed. Thirdly, the ratio of *E. pacifica* weight to the total weight of stomach contents was assumed to be equal to the occurrence of *E. pacifica* divided by the sum of occurrences of all the food items.

The results are shown in Table 7.1.1. Walleye pollack are estimated to consume 540 000 tons, Pacific cod 8500 tons and Pacific herring 1300 tons of *E. pacifica* per year. A total of 550 000 tons were consumed each year off Miyagi and Fukushima Prefectures. As about 52% of demersal fish landings from the whole Sanriku coast to Kashima-nada are landed at the Ishinomaki Fish Market, the consumption of *E. pacifica* by fish probably amounts to about 1 000 000 tons per annum.

As discussed by Endo in Chapter 3, the benthopelagic population of *E. pacifica* occurs in the daytime throughout the year except in spring in the Sanriku and Joban

Table 7.1.1 Estimates of *Euphausia pacifica* consumed annually by the three demersal fish species in the vicinity of Kinkasan Island (Adapted from Kodama & Izumi, 1994).

Fish species	Body length (cm)	Annual catch (t)	Biomass (t)	1-E[1]	Daily ration (%)	Krill ratio[2]	Annual Krill consumption (t)
Walleye pollack	10–25	12 265	40 883	0.904	2.3	0.74	231 594
	25–35	18 215	60 717	0.935	1.7	0.69	237 339
	>40	10 346	34 487	0.891	1.5	0.44	75 503
	total	40 826	136 087				544 436
Pacific cod	15–25	496	1 653	0.946	2.6	0.31	4542
	30–40	1814	6047	0.951	1.9	0.09	3746
	>50	1067	3557	0.901	1.9	0.01	213
	total	3377	11 257				8501
Pacific herring		180	600	0.641	1.5	0.61	1295

[1] E denotes proportion of individuals with empty stomach.
[2] Occurrence of *Euphausia pacifica* in the fish stomach relative to the occurrence of total food items.

areas. Benthopelagic *E. pacifica* support very high abundance of demersal fish population. From submersible observations on benthopelagic *E. pacifica* we have noted dense beds of brittle stars on the seabed (Endo & Kodama, in preparation). It has been reported that the seas surrounding the northern part of Japan are blanketed by dense beds of the brittle star *Ophiura sarsii* between depths of about 200 and 600 m with a mean density estimated to be 373 individuals m^{-2} (Fujita & Ohta, 1989). The brittle stars may depend on *E. pacifica* and lantern fishes for food (Stancyk *et al.*, 1998). In addition to this information, quantitative estimates of the biomass of benthopelagic *E. pacifica* are required for the entire distributional range in Japanese waters to elucidate the relative importance of the species in the pelagic as well as benthopelagic food web in Sanriku waters.

Pelagic environment

Pelagic fishes such as sardines, Japanese chub mackerel and Japanese common squid also migrate in large numbers into the Sanriku area in May to January. Their biomass is estimated to be more than 1 000 000 tons (Kodama & Izumi, 1994). Quantitative estimates of *E. pacifica* consumed by these pelagic species are also needed.

Based on 40 years of coastal whaling operations from 1948 to 1987, Kasamatsu & Tanaka (1992) reported that 46.5% of minke whale, *Balaenoptera acutorostrata*, were feeding on krill in Joban–Sanriku waters. It is highly probable that most of the krill were *E. pacifica*. Ogi (1994) has provided some information on seabirds which feed on euphausiids in Sanriku waters. In early summer, the sooty shearwater, (*Puffinus griseus*) the slender-billed shearwater (*P. tenuirostris*) and the pale-footed shearwater (*P. carneipes*) visit Sanriku waters in large flocks during their northward migration from the southern hemisphere.

Although the abundance of the rhinoceros auklet (*Cerorhinca monocerata*) and the streaked shearwater (*Calonectris leucomelas*) which breed in this area, are present in reasonable numbers, the sooty shearwater is by far the most numerous. The slender-billed shearwater is more important as a euphausiid feeder than the piscivorous sooty shearwater because the species is thought to be suited for sea surface feeding (Ogi, 1994). This was confirmed by stable isotope analyses (Minami *et al.*, 1995). In Sanriku waters the slender-billed shearwater had significantly lower $\delta^{15}N$ values than the sooty shearwater with a difference of 1.2%. This suggested that the trophic level of the slender-billed shearwater was lower than that of the sooty shearwater. In fact, as the slender-billed shearwater feed mainly on the euphausiid, *Nyctiphanes australis,* in their breeding area of Tasmanian waters, Ogi (1994) suggests that they feed on a large amount of *E. pacifica* in Sanriku waters.

Euphausiids are reported to occupy 83% of stomach contents of the slenderbilled shearwater in wet weight in the Okhotsk Sea and even higher percentages in some part of the Bering Sea (Ogi *et al.*, 1980). Ogi *et al.* (1980) suggested that the slender-billed shearwater consume 30 000 tonnes of *Thysanoessa raschi* in Bristol Bay during their stay from late April to early June assuming each bird feeds on 90 g of *T.*

raschi per day. The consumption of euphausiids by the slender-billed shearwater is almost equal to that consumed by sockeye salmon.

Although seabird observations during winter are rare in Sanriku waters, Ogi (1994) found a large number of auklets present in late February to early March. These include the thick-billed murre (*Uria lomvia*), the common murre (*U. aalge*), the rhinoceros auklet, the crested auklet (*Aethia cristatella*), the ancient murrelet (*Synthliboramphus antiquum*) and so on. These auklets seem to feed on euphausiids, almost exclusively *E. pacifica*, and copepods, because dense patches of zooplankton are always being detected on the echogram under small flocks of crested auklets. Ogi suggests that these auklets are not endemic to Sanriku waters, but come from the Okhotsk Sea to overwinter in the area. The importance of euphausiids as prey for thick-billed murres in the North Pacific ranks third following squids and fish (Ogi, 1980). The largest number of euphausiids found in a single thick-billed murre was 2246 in the East Kamchatka Current. Quantitative estimates of *E. pacifica* consumed by these seabirds in Sanriku waters are also needed.

Energy flow through **Euphausia pacifica**

Wada & Yabuki (1994) constructed a trophodynamic model for the continental shelf ecosystem of the Oyashio region. They took into account 4 physical factors

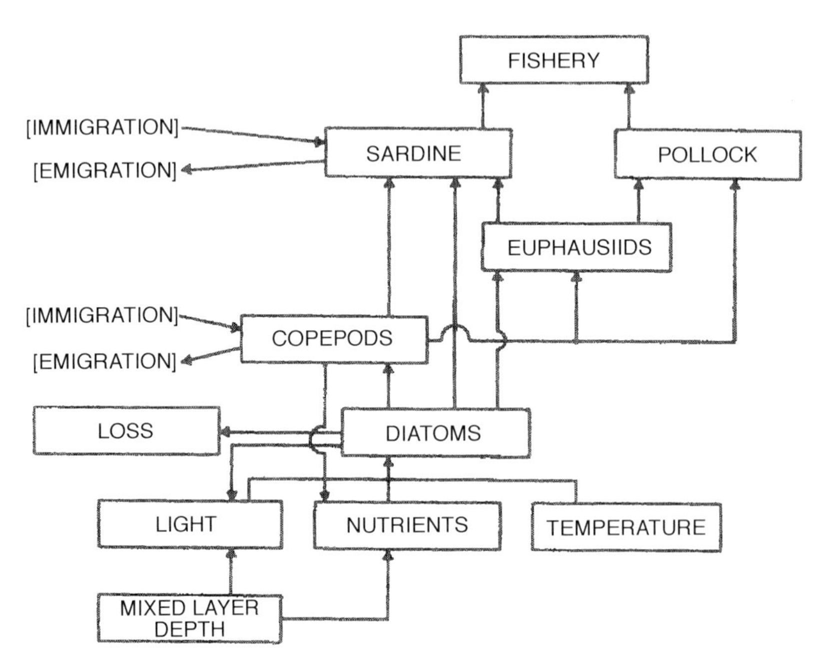

Fig. 7.1.1 Structure of the trophodynamic model in the shelf area of the Oyashio region (Wada & Yabuki, 1994).

(light, nutrients, temperature and mixed layer depth), 7 biological factors (diatoms, copepods, euphausiids, sardines, walleye pollack, fisheries and diatom sinking), and 21 processes which link these factors (Fig. 7.1.1). It is interesting to note that a deepening of the mixed layer depth and an increase in water temperature lead to

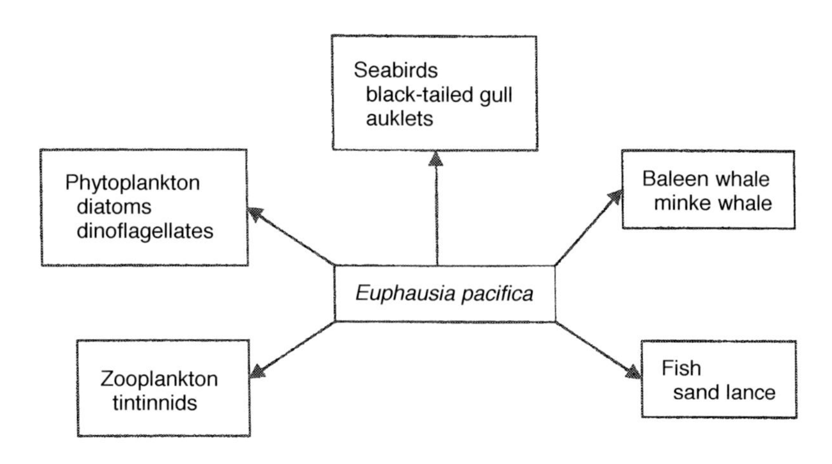

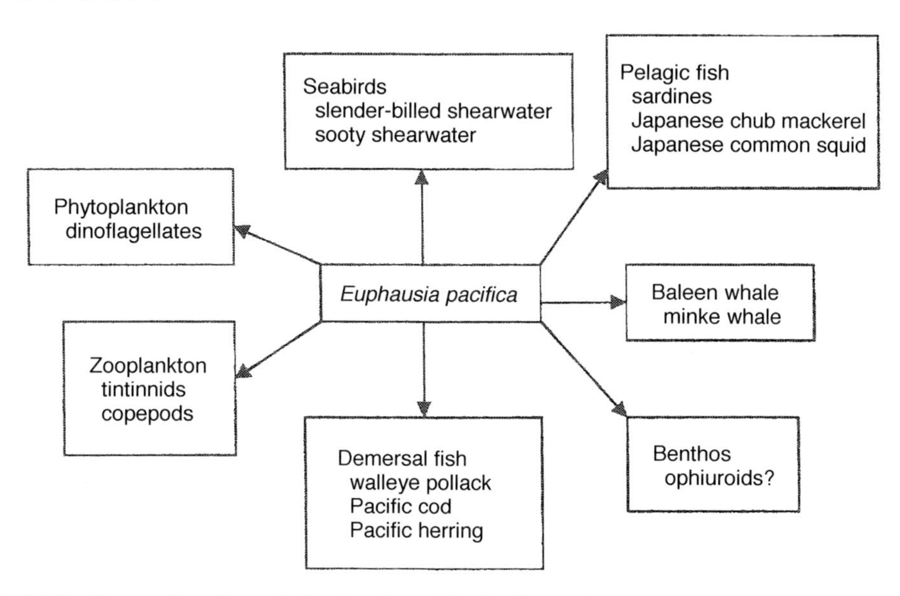

Fig. 7.1.2 Energy flow through *Euphausia pacifica* in the marine ecosystem of Sanriku and Joban waters based on various sources referred in the text.

increased euphausiid production. Conversely, a shoaling of the mixed layer depth and a decrease in solar radiation and water temperature decrease it.

The energy flow through *E. pacifica* in the marine ecosystem of Sanriku and Joban areas is summarised in Fig. 7.1.2. *E. pacifica* eats mainly diatoms in spring and copepods, dinoflagellates and tintinnids in other seasons (Endo, 1981, Nakagawa *et al.*, in preparation). Copepods, however, seem to be most important as prey for *E. pacifica* throughout the year in terms of carbon (Nakagawa *et al.*, in preparation). In spring, *E. pacifica* forms dense swarms and is pelagic both day and night. Blacktailed gull, auklets, sand lance and minke whales target the swarms. In other seasons *E. pacifica* moves to offshore waters and forms benthopelagic aggregations during daytime. Therefore, many pelagic and demersal predators eat *E. pacifica*.

References

Akiyama, M. (1996) Body length composition of *Euphausia pacifica* in the stomach contents of blue eye, *Chlorophthalmus albatrossis* (*borealis*), in Joban coastal waters. *Rep. Res. Meetings on North Pacific Krill Resources* **5**, 99–100 (in Japanese).

Endo, Y. (1981) Ecological Studies on the Euphausiids Occurring in the Sanriku Waters with Special Reference to Their Life History and Aggregated Distribution. Ph.D. thesis, Tohoku University, Sendai, 166 pp. (in Japanese with English abstract).

Fujita, T. (1994) Importance of planktonic organisms as food of demersal fish population. *Gekkan Kaiyo (Kaiyo Monthly)* **26**, 236–41 (in Japanese).

Fujita, T. & Ohta, S. (1989) Spatial structure within a dense bed of the brittle star *Ophiura sarsii* (Ophiuroidea: Echinodermata) in the bathyal zone off Otsuchi, northeastern Japan. *J. Oceanogr. Soc. Jap.* **45**, 289–300.

Kasamatsu, F. & Tanaka, S. (1992) Annual changes in prey species of minke whales taken off Japan 1948–87. *Nippon Suisan Gakkaishi* **58**, 637–51.

Kodama, J. & Izumi, Y. (1994) Factors relevant to the fishing ground formation of *Euphausia pacifica* and the relation to the demersal fish resources. *Gekkan Kaiyo (Kaiyo Monthly)* **26**, 228–35 (in Japanese).

Komaki, Y. (1967) On the surface swarming of euphausiid crustaceans. *Pacif. Sci.* **21**, 433–48.

Kuroda, K. (1994) Euphausiid fishery in the Japanese waters and related research activities. *Gekkan Kaiyo (Kaiyo Monthly)* **26**, 203–209 (in Japanese).

Minami, H., Minagawa, M. & Ogi, H. (1995) Changes in stable carbon and nitrogen isotope ratios in sooty and short-tailed shearwaters during their northward migration. *The Condor* **97**, 565–74.

Nakagawa, Y., Endo, Y. & Taki, K. (in press) Diet of *Euphasia pacifica Hansen* in Sanrika waters off northeastern Japan. *Plankton Biol. Ecol.*

Ogi, H. (1980) The pelagic feeding ecology of thick-billed murres in the North Pacific, March–June. *Bull. Fac. Fish., Hokkaido Univ.* **31**, 50–72.

Ogi, H. (1994) Euphausiids as food of sea birds. *Gekkan Kaiyo (Kaiyo Monthly)* **26**, 242–47 (in Japanese).

Ogi, H., Kubodera, T. & Nakamura, K. (1980) The pelagic feeding ecology of the short-tailed shearwater *Puffinus tenuirostris* in the subarctic Pacific region. *J. Yamashina Inst. Ornith.* **12**, 157–82.

Stancyk, S.E., Fujita, T. & Muir, C. (1998) Predation behavior on swimming organisms by *Ophiura sarsii*. In *Echinoderms: San Francisco* (R. Mooi & M. Telford, eds), pp. 425–29. Balkema, Rotterdam.

Takeuchi, I. (1985) A preliminary report on the stomach contents of sand lance in the coastal area of Miyagi Prefecture. *Bull. Tohoku Branch Jap. Soc. of Sci. Fish.* **35**, 19–21 (in Japanese).

Wada, T. & Yabuki, K. (1994) Trophodynamic model in the shelf area of the Oyashio region. *Gekkan Kaiyo (Kaiyo Monthly)* **26**, 251–55 (in Japanese).

Yamamura, O. (1993) Resource partitioning among demersal fishes off Sendai Bay. *GSK Northern Japan Demersal Fish Res. Group Rep.* **26**, 61–70 (in Japanese).

Yamamura, O., Inada, T. & Shimazaki, K. (1998) Predation on *Euphausia pacifica* by demersal fishes: predation impact and influence of physical variability. *Mar. Biol.* **132**, 195–208.

7.2 Canadian waters

Ron Tanasichuk

The Canadian west, east and Arctic coasts certainly support fish, bird and mammalian species that would depend on krill as food. Most information is descriptive. Consumption estimates exist only for pelagic fish from the south-west coast of Vancouver Island, on the Canadian west coast.

Pacific hake (*Merluccius productus*), Pacific herring (*Clupea pallasi*) and spiny dogfish (*Squalus acanthias*) dominate the pelagic fish biomass along the south-west coast of Vancouver Island. Tanasichuk (1999a) reported that euphausiids accounted for an average of 84 and 89% of the daily ration of hake and dogfish respectively over August 1985–97, and for all of the daily ration for herring over those years (Fig. 7.2.1). Tanasichuk (1999b) reported that hake persist in preferring *Thysanoessa spinifera* and euphausiids longer than 17 mm even though there has been a large reduction in euphausiid biomass, a shift to smaller animals in the population and a change in species composition following the 1992 ENSO event (Tanasichuk, 1998 a,b). Ware and McFarlane (1995) comment on how euphausiid distributions influence hake movements. Waddell *et al.* (1992) and Morris & Healey (1990) found that chinook (*Onchorynchus tschawytscha*) and coho (*O. kisutch*) salmon feed heavily on euphausiids. Other significant regions on the west coast are the Strait of Georgia and Hecate Strait. Euphausiids are important prey for the resident hake stock, young Pacific herring and coho, chinook and chum (*O. keta*) salmon in the Strait of Georgia (R.J. Beamish, Fisheries and Oceans Canada, pers. comm.) They are also taken when available to flatfishes in Hecate Strait (J. Fargo, Fisheries and Oceans Canada, pers. comm.).

Perry & Waddell (1997) summarised information on zooplankton feeding by seabirds found in west coast waters. Euphausiids were specifically identified as prey for 17 of the 24 species. *T. spinifera*, *T. longipes* and *Euphausia pacifica* were commonly observed in the diets. Euphausiids are not important food for birds in the Strait of Georgia (Mackas & Fulton, 1989).

A number of baleen whale species feed on euphausiids along the Pacific and Atlantic coasts every summer (G. Ellis, Fisheries and Oceans Canada, pers. comm.). Blue (*Balenoptera musculus*), humpback (*Megaptera novaeangliae*), fin (*Balenoptera physalus*) and minke (*Balenoptera acutorostrata*) whales rely heavily on krill. Recently, humpbacks have shifted away from the south-west coast of Vancouver Island. This is coincident with the decline in euphausiid biomass after the 1992 ENSO event.

Euphausiids are not important prey in the Canadian Arctic. Gaston & Bradstreet (1993) reported thick-billed murres (*Uria lomvia*), the most numerous nesting seabird in the eastern Canadian Arctic, fed on hyperiid amphipods. Euphausiids were a minor prey item for Arctic cod (*Boreogadus saida*) in Beaufort Sea coastal

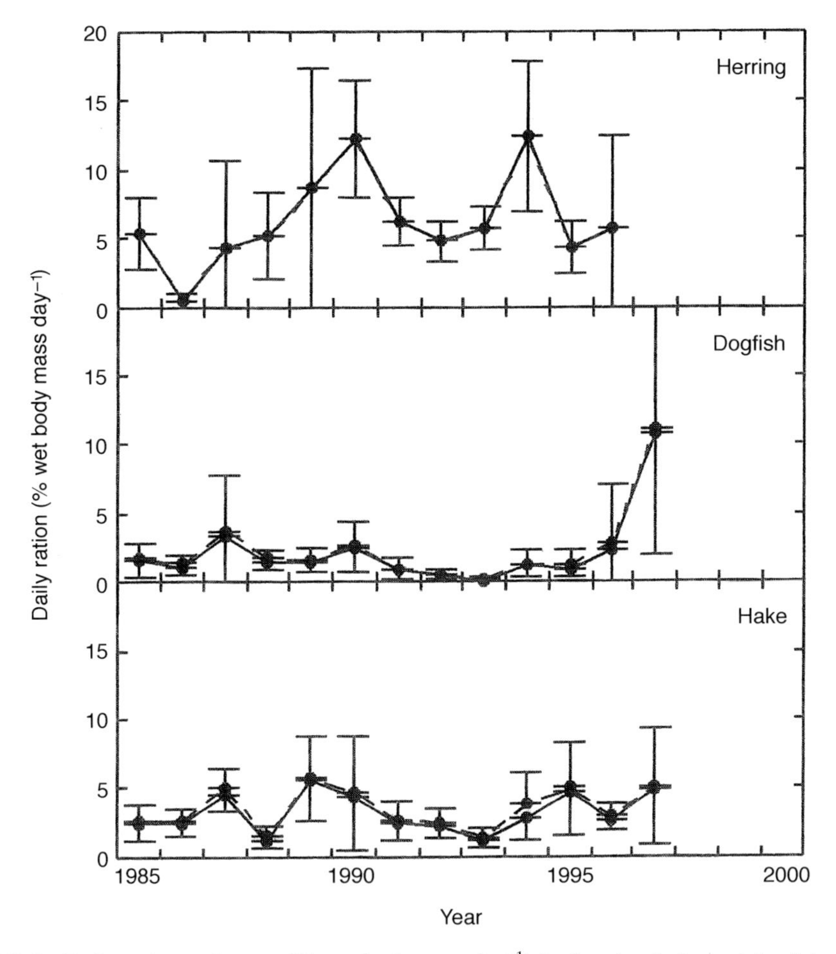

Fig. 7.2.1 Daily ration estimates (% wet body mass day^{-1}) for herring, hake and dogfish, August 1985–97. Solid lines are the euphausiid ration and dotted lines are total ration. Bars are 95% confidence limits.

waters, Alaska (Craig *et al.*, 1982). Bowhead whales (*Balaena mysticetus*) feed on euphausiids.

Euphausiids are important prey for fish species feeding along the east coast of Canada. They were important for silver hake (*Merluccius bilinearis*) and cod (*Gadus morhua* Linneaus) along western Nova Scotia and haddock (*Melanogrammus aeglefinus*) in the Georges Bank area (Maurer, 1975). Young silver hake, red hake (*Urophycis chuss* Walbaum), adult herring (*Clupea harengus* L.) and mackerel on Georges Bank fed mainly on euphausiids (*Meganytiphanes*) (Vinogradov, 1984). The dominant fish species in the Scotian shelf area feed mainly on euphausiids (Harding 1996). Vesin *et al.* (1981) found that euphausiids (*T. raschii*) dominated the diet of capelin (*Mallotus villosus*) in the western Gulf of St. Lawrence. Pedersen & Riget (1992) found that euphausiids were important prey for redfish (*Sebastes* sp.) off west Greenland. Lilly & Rice (1983) reported that *T. raschii* was an important prey item for Atlantic cod feeding on the northern Grand Bank in spring. Euphausiids were important prey for three (beaked redfish (*Sebastes mentalla*), Acadia redfish (*S. fasciatus*) and longfin hake (*U. chesteri*)) of the 14 species collected from the Flemish Cap, about 320 km off the Newfoundland Coast (Rodriguez-Marin *et al.*, 1994).

There is diet information for procellariiform and alcid marine birds from the east coast. Brown *et al.* (1981 cited in Prince & Morgan, 1987) found that, in seasons when *Meganyctiphanes* and sandlance *Ammodytes* were available, *Puffinus gravis* mainly took fish and *P. griseus* took euphausiids. *M. norvegica* was important food for Leach's storm petrel (*Oceanodroma leucorhoa*) off Nova Scotia (Linton, 1978, cited in Prince & Morgan, 1987). Bradstreet (1983, cited in Bradstreet & Brown, 1985) found that adult razorbill (*Alca torda*), one of the five alcid species occurring along the east coast, fed on sculpins and euphausiids.

References

Bradstreet, M.S.W. & Brown, R.G.B. (1985) Feeding ecology of the Atlantic Alcidae. In *The Atlantic Alcidae: The Evolution, Distribution and Biology of the Auks Inhabiting the Atlantic Ocean and Adjacent Waters* (D.N. Nettleship and T.R. Birkhead, eds), pp. 264–318. Academic Press, London.

Craig, P.C., Griffiths, W.B., Haldorson, L. & McElderry, H. (1982) Ecological studies of Arctic cod (*Boreogadus saida*) in Beaufort Sea coastal waters, Alaska. *Can. J. Fish. Aquat. Sci.* **39**, 395–406.

Gaston, A.J. & Bradstreet, M.S.W. (1993) Intercolony differences in the summer diet of thick-billed murres in the eastern Canadian Arctic. *Can. J. Zoology* **71**, 1831–40.

Harding, G.C.H. (1996) Ecological factors to be considered in establishing a new krill fishery in the Maritimes region. DFO Atlantic Fisheries Research Document 96/99. 11 pp.

Lilly, G.R. & Rice, J. C. (1983) Food of Atlantic cod (*Gadus morhua*) on the northern Grand Bank in spring. NAFO SCR Document 83/9/87. 35 pp.

Mackas, D.L. & Fulton, J.D. (1989) Distribution and aggregation of zooplankton in the Strait of Georgia and their potential availability to marine birds. In *The Ecology and Status of Marine and Shoreline Birds in the Strait of Georgia, British Columbia* (K. Vermeer & R.W. Butler eds). Proceedings of a symposium sponsored by the Pacific Northwest Bird and Mammal Society and the Canadian Wildlife Service, held in Sidney, B. C. 11 December 1987, pp. 19–25.

Maurer, R. (1975) A preliminary description of some important feeding relationships. ICNAF Research Document 75/IX/130. 15 pp.

Morris, J.F.T. & Healey, M.C. (1990) The distribution, abundance and feeding habits of chinook and coho salmon on the fishing banks. *Canadian Technical Report of Fisheries and Aquatic Sciences 1759*, 75 pp.

Pedersen, S.A. & Riget, F. (1992) Feeding habits of redfish, *Sebastes* sp., in West Greenland waters with special emphasis on predation on shrimp. ICES Document C.M. 1992/G:24.

Perry, R.I. & Waddell, B.J. (1997) Zooplankton in Queen Charlotte Island waters: distribution and availability to marine birds. In *The Ecology, Status, and Conservation of*

Marine and Shoreline Birds of the Queen Charlotte Islands (K. Vermeer & K.H. Morgan, eds). Canadian Wildlife Service. Occasional Paper 93, pp. 18–28.

Prince, P.A. & Morgan, R.A. (1987) Diet and feeding ecology of Procellariiformes. In *Seabirds: Feeding Ecology and Role in Marine Ecosystems* (J.P. Croxall, ed.), pp. 154–71. Cambridge University Press, Cambridge.

Rodriguez-Marin, E., Punzon, A., Paz, J. & Olaso, I. (1994) Feeding of most abundant fish species in Flemish Cap in summer 1993. *NAFO SCR Document*. 94/35. 33 pp.

Tanasichuk, R.W. (1998a) Interannual variations in the population biology and productivity of the euphausiid *Thysanoessa spinifera* in Barkley Sound, Canada, with special reference to the 1992 and 1993 warm ocean years. *Mar. Ecol. Prog. Ser.* **173**, 163–80.

Tanasichuk, R.W. (1998b) Interannual variations in the population biology and productivity of the euphausiid *Euphausia pacifica* in Barkley Sound, Canada, with special reference to the 1992 and 1993 warm ocean years. *Mar. Ecol. Prog. Ser.* **173**, 181–95.

Tanasichuk, R.W. (1999a) Euphausiids in a coastal upwelling ecosystem: Their importance as fish prey, interannual variations in euphausiid population biology and productivity, and some implications of these changes for fish production. PhD. thesis, University of Bergen, Norway. 123 pp.

Tanasichuk, R.W. (1999b) Interannual variations in the availability and utilization of euphausiids as prey for Pacific hake (*Merluccius productus*) along the south-west coast of Vancouver Island. *Fish. Oceanogr.* **8**, 150–56.

Vesin, J.-P., Leggett, W.C. & Able, K.W. (1981) Feeding ecology of capelin (*Mallotus villosus*) in the estuary and western Gulf of St. Lawrence and its multispecies implications. *Can. J. Fish. Aquat. Sci.* **38**, 257–67.

Vinogradov, V.I. (1984) Food of silver hake, red hake and other fishes of Georges Bank and adjacent waters, 1968–74. *Science Council Studies NAFO* **7**, 87–94.

Waddell, B.J., Morris, J.F.T. & Healey, M.C. (1992) The abundance, distribution, and biological characteristics of chinook and coho salmon on the fishing banks off southwest Vancouver Island, May 18–30, 1989 and April 23–May 5, 1990. *Can. Tech. Rep. Fish. Aquat. Sci.* **1891**, 113 pp.

Ware, D.M. & McFarlane, G.A. (1995) Climate-induced changes in Pacific hake (*Merluccius productus*) abundance and pelagic community interactions in the Vancouver Island Upwelling System. In *Climate change and northern fish populations* (R.J. Beamish ed.), pp. 509–521. *Canadian Special Publication of Fisheries and Aquatic Sciences* **121**, 509–21.

7.3 The Southern Ocean

Inigo Everson

The Southern Ocean food web

Generalised descriptions of the Southern Ocean food web have been given by several workers with good examples to be found in Hart (1942), Holdgate (1967) and Murphy *et al* (1988). These, by grouping several related species together and presenting them as single components, highlight what is perceived as the relative simplicity of the system. An oft-quoted example is the food chain which leads from phytoplankton, around a millimetre in size, through Antarctic krill, which grow to around six centimetres, to the blue whale which can grow up to 30 metres in length. Simplicity is only a relative term as even though the system may be simple relative to other oceanic ecosystems, when one examines each component, it is clear that the picture in Fig. 7.3.1 is in reality quite complicated.

The Convention for the Conservation of Antarctic Marine Living Resources (CCAMLR) provides a slightly different focus, concentrating on two categories, the

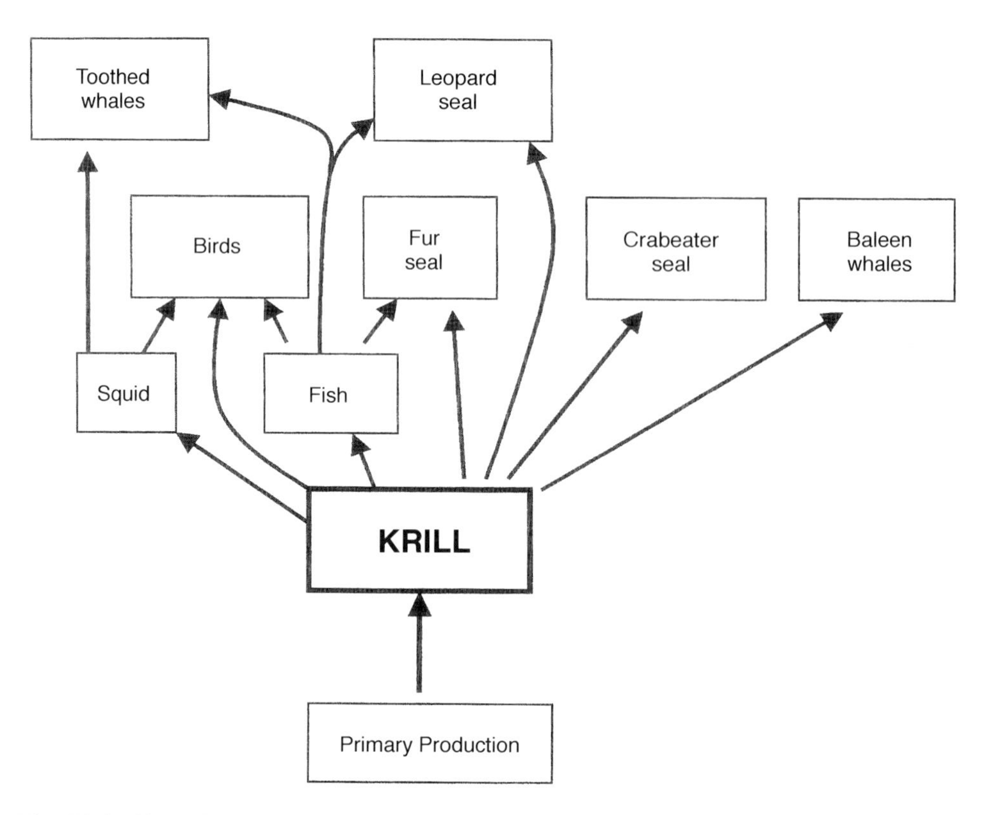

Fig. 7.3.1 Simplified representation of Southern Ocean Food Web linkages that are centred around krill.

harvested and dependent species. If the list is restricted to those species for which there is currently a targeted fishery, then in the Southern Ocean, there are relatively few that fall into the category of harvested species. Apart from the Antarctic krill (*Euphausia superba*), there is the mackerel icefish (*Champsocephalus gunnari*) and Patagonian and Antarctic toothfish *(Dissostichus eleginoides* and *D. mawsoni* respectively). There have been in addition exploratory fisheries for stone crab (*Paralomis* spp) and squid (*Martialia hyadesi*). Other species have been harvested in the past and information on the reported catches can be found in the Statistical Bulletins (CCAMLR, 1999). Of these species the only one which is itself harvested and is at the same time dependent on krill is the mackerel icefish.

Krill fishing, which in the Southern Ocean in the 1960s was very much in embryonic form, was considered by the 1970s to be an opportunity to provide large amounts of protein to satisfy world demands. As the krill fishery expanded there was concern that it might adversely affect dependent species. This heightened awareness was behind the research programme 'Biological Investigations of Marine Antarctic Systems and Stocks' (BIOMASS) (SCAR/SCOR, 1977). This was the research programme which in a short space of time led to the Commission for the Conservation of Antarctic Marine Living Resources (CCAMLR) (Miller & Agnew in Chapter 12). Central to both is a requirement for the identification of the key dependent species and the quantification of interactions between them and krill. The dependent species for which there is greatest information published are the whales, seals, birds and fish.

Impact of major dependent species

Reviewing the diet of Southern Ocean birds, Croxall (1984) identified the following species that are dependent to a great extent on krill. Of the penguins, Adelie (*Pygoscelis adelie*), chinstrap (*P. antarctica*), macaroni (*Eudyptes chrysolophus*) and gentoo (*P. papua*) feed extensively on Antarctic krill (*Euphausia superba*), although in high latitudes, such as in Prydz Bay, Adelies feed on *E. crystallorophias* (Cooper & Woehler, 1994). In the northern part of the Southern Ocean around South Georgia both gentoo and macaroni penguins also feed on *E. frigida* and in the same region Macaroni penguins also feed on *E. triacantha*.

Although black-browed (*Diomedea melanophris*), light-mantled sooty (*Phoebretia palpebrata*) and grey-headed (*D. chrysostoma*) albatrosses feed on krill this constitutes only 40% of the diet for the first two and 16% for grey-heads (Croxall, 1984; Prince & Morgan, 1987). Only about one-fifth of the diet of giant petrels (*Macronectes* spp) contains krill although, as with many of the Procellariidae, their diet appears to be fairly catholic. Some of the smaller petrels, such as the Antarctic petrel (*Thalassoica antarctica*), Cape petrel (*Daption capensis*), snow petrel (*Pagodroma nivea*) and diving petrel (*Pelecanoides* spp) also feed extensively on krill. The smaller storm petrels (*Oceanites spp*) and prions (*Pachyptillia spp*) feed on a wider range of crustacea including krill although the emphasis is much more towards copepods (Prince & Morgan, 1987).

That the birds are significant consumers of krill can be seen from the estimates of the annual amount of krill eaten by the different species. Considering two localities, South Georgia at the northern end of the zone and Prydz Bay close to the Antarctic continent, we can see in Table 7.3.1. the differences in the amounts taken. This also highlights the fact that the bulk of the consumption of krill by birds is attributable to relatively few species, macaroni penguins and prions at South Georgia and Adelie penguins at Prydz Bay.

Table 7.3.1 Estimated annual consumption in thousands of tonnes of krill by major bird species. The figures in parenthesis are the proportion by mass that krill represents of the diet of the species at that site. Data from Croxall & Prince (1987) for South Georgia and Cooper & Woehler (1994) for Prydz Bay.

Species	South Georgia	Prydz Bay
Adelie penguin	–	66 (97)
Chinstrap penguin	2.5 (100)	–
Gentoo penguin	61 (68)	–
Macaroni penguin	3872 (98)	–
Black-browed albatross	8.3 (38)	–
Grey-headed albatross	6.3 (15)	–
Light mantled sooty albatross	1.8 (37)	–
Southern giant petrel	2.7 (82)	0
Northern giant petrel	1.5 (65)	–
Antarctic petrel	–	0.53 (0.77)
Cape petrel	3.1 (86)	0.14 (0.21)
Antarctic fulmar	–	0.98 (1.44)
Snow petrel	0.4 (80)	0.01 (0.02)
Antarctic prion	1345 (58)	–
White-chinned petrel	210 (27)	–
Wilson's petrel	3.8 (40)	0.23 (0.33)
Common diving petrel	52.8 (15)	–
South Georgia diving petrel	126 (76)	–

With the exception of elephant seal (*Mirounga leonina*) all the Antarctic seals feed to some extent on krill (Laws, 1984). In the case of Weddell (*Leptonychotes weddelli*) and Ross (*Ommatophoca rossi*) the proportion of krill in the diet is relatively slight. Krill are the major item in the diet of crabeater seals (*Lobodon carcinophagus*) and make up around half the food eaten by leopard seals (*Hydrurga leptonyx*). Fur seals (*Arctocephalus gazella*) feed extensively on krill but when that is scarce concentrate on fish and squid. This change of diet has been implicated in the large changes in standing stock of mackerel icefish around South Georgia (Everson *et al.*, 1999). The diving depths of fur and crabeater seal have been shown to follow the diurnal pattern of vertical migration of krill (Croxall *et al.*, 1985; Laws, 1984). This indicates that although krill may be present in an area, if they remain deep, they may be out of the diving range of the seals. The latitudinal range of foraging activity of seals is shown in Fig. 7.3.2.

By far the most abundant seal species in the Southern Ocean is the crabeater seal

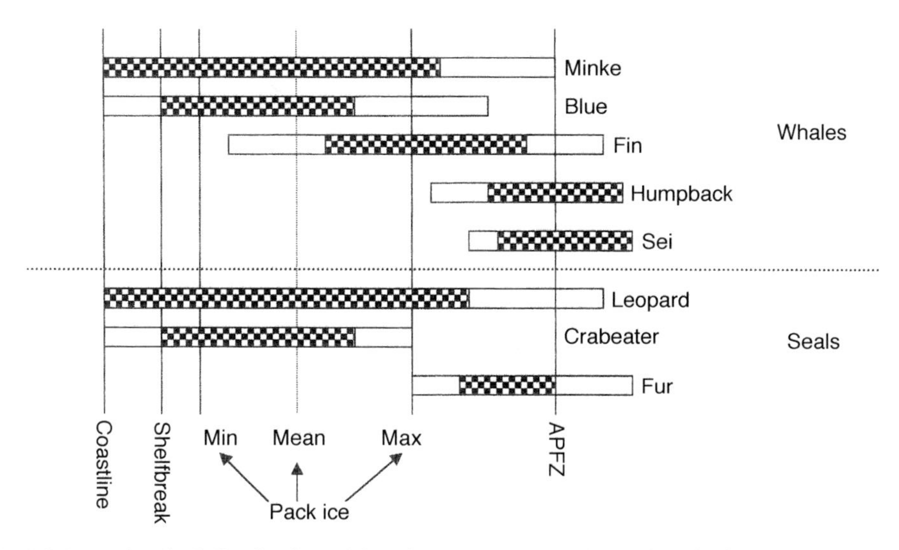

Fig. 7.3.2 Latitudinal distribution of foraging ranges of whales and seals that are dependent on krill in the Southern Ocean. The checked area is the main range of foraging activity. With permission from The Royal Society, London. (Modified from Laws, 1977.)

the total population of which in 1956–57 was thought to number about 15 million individuals (Eklund & Atwood, 1962) and 30 million in 1982 (Laws, 1984). This estimate was further revised by Erickson & Hanson (1990) who gave a population size of 11 to 12 million individuals. The differences between these estimates may represent some underlying changes although they are more likely to reflect the difficulties in counting seals which are concentrated in the pack ice zone. Their distribution within the pack ice zone means that their feeding area will be changing with the seasonal changes in sea-ice distribution; thus in winter they may be present much further to the north than during the summer when the ice cover is at a minimum. Taking the consumption rates given by Laws (1984) this indicates that crabeater seals consume between 50 and 134 million tonnes of krill a year. Although fur seals are important locally around South Georgia where the current population of around 4 million (Boyd, 1993) is estimated to take around 5 or 6 million tonnes of krill per annum (Everson *et al.*, 1999), their impact on a Southern Ocean scale remains quite small.

Seven species of baleen whale are present in the Southern Ocean, all of which have been subject to extensive whaling. The latitudinal range of foraging activity by the five species (Minke *Balaenoptera acutorostrata*, Blue *B. musculus*, Fin *B. physalus*, Sei *B. borealis* and humpback *Megaptera novaeangliae*) which feed predominantly on krill is shown in Fig. 7.3.2. Estimates of stock sizes and their estimated consumption of krill prior to the onset of commercial whaling and from recent whale sighting surveys are set out in Table 7.3.2. Blue, fin and sei whales migrate into the Southern Ocean during the summer to feed (Plate 4, facing p. 182). The southerly limit of distribution on these feeding migrations is the northern limit of the pack ice zone so that their

Table 7.3.2 Stock sizes and estimated krill consumption of initial and present whale stocks present in the Southern Ocean. Stock sizes from Kock & Shimadzu (1994), krill consumption rates derived from Laws (1977).

Whale species	Initial stock size (thousands)	Present stock size (thousands)	Initial whale stock krill consumption (million tonnes)	Present whale stock krill consumption (million tonnes)
Blue and pygmy blue	220	0.7–5.5	80	0.24–1.9
Fin	490	25	100	5
Sei	200	38	15	2.7
Humpback	130	12	14	1.3
Minke	?	760	?	75
TOTAL			>209	~85

foraging region changes as the pack ice retreats. Results of an analysis of whale distributions and pack ice limits over the period 1931–1939 presented by Mackintosh (1970) are shown in Fig. 7.3.3. In a study on Minke whales in the Ross Sea and the area immediately to the west, Ichii *et al.* (1998) found that the amount of body fat was related to the feeding conditions during that season. This was related further to the amount and size of krill and also compared to the extent of sea-ice cover.

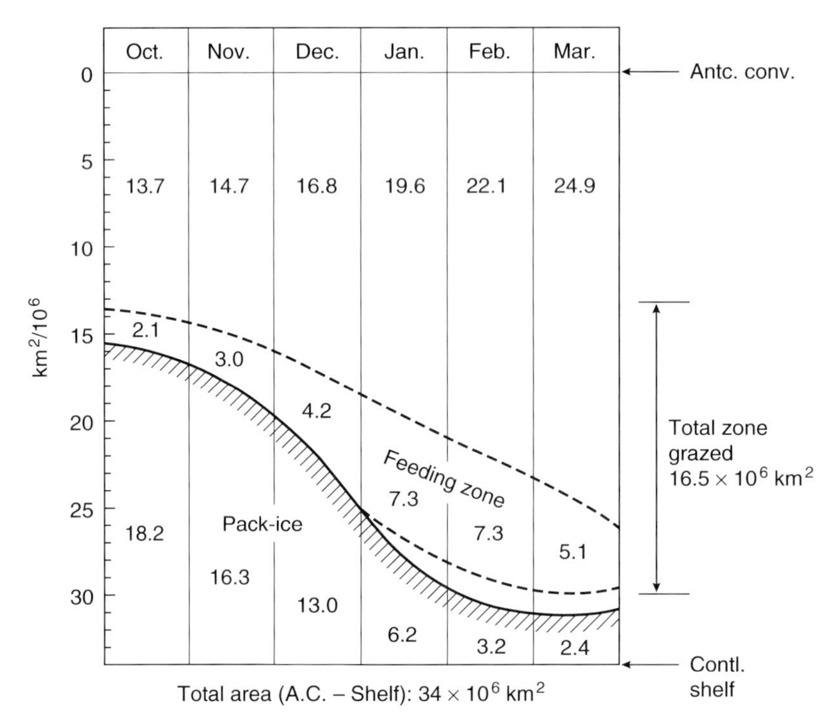

Fig. 7.3.3 Seasonal variation in the ice-free area between the Antarctic Polar Frontal Zone and the Antarctic Continent and indication of the potential grazing area for blue and fin whales. (With permission from Mackintosh, 1970.)

Several species of fish are known to feed on krill (Kock, 1992) and the concentrations of some species, such as mackerel icefish (*Champsocephalus gunnari*) at South Georgia, or Myctophidae in the open sea may be locally significant. There is evidence for local availability affecting the fish populations (Everson *et al.*, 1999) and this topic is developed further in Everson, Chapter 8. In spite of this the overall consumption of krill by fish on a Southern Ocean scale is unlikely to be anything like as great as that of whales, seals or birds. Squid are a group known to be present in the Southern Ocean some of which are known to feed on krill. Kock & Shimadzu (1994) discussed this at some length but concluded that, beyond noting that the amount might be very large and probably at least tens of millions of tonnes, precise figures were impossible to determine.

The foregoing discussion indicates two important points, firstly that relatively few species have a very large impact on krill stocks. Secondly, that if the estimated amount of krill consumed by predators is used as the indicator, the annual production of krill must be of the order of hundreds of millions of tonnes. Even if these consumption figures are overestimated by a substantial margin the total is still well above the reported krill catch for the Southern Ocean (Miller & Agnew in Chapter 12).

Ecosystem monitoring

The static picture of a food web providing qualitative links between the different elements is only the starting point for understanding ecosystem interactions. An obvious development is to monitor key components and assess how they themselves vary and in relation to other parts of the system. Recognising the central role of krill in the Southern Ocean system, CCAMLR established its own Ecosystem Monitoring Programme (CEMP). This arose from a meeting where the objective of ecosystem monitoring was defined as:

> To detect and record significant changes in critical components of the ecosystem, to serve as a basis for the Conservation of Antarctic Marine Living Resources. The monitoring system should be designed to distinguish between changes due to the harvesting of commercial species and changes due to environmental variability, both physical and biological.
>
> (SC-CAMLR 1985)

Discussion, at that meeting and subsequently, centred around programmes that were then in operation and which might, with minor adjustment, be modified to fulfil the aims of the CEMP and additional studies which would need to be set *de novo*. Arising from this a series of key species comprising penguins, flying birds and seals was identified for which information could be collected routinely in a standardised manner so as to include the main components of the life history. These Standard Methods are set out in SC-CAMLR (1991). Analysing these data involves an assessment of the spatial and temporal scales of the interactions, a subject that is considered in more detail in Everson (Chapter 8).

References

Boyd, I.L. (1993) Pup production and distribution of breeding Antarctic fur seals (*Arctocephalus gazella*) at South Georgia. *Antarctic Sci.* **5**, 17–24.

CCAMLR (1999) *Statistical Bulletin*. Commission for the Conservation of Antarctic Marine Living Resources, Hobart, Australia.

Cooper, J. & Woehler, E.J. (1994). Consumption of Antarctic krill (*Euphausia superba*) by seabirds during summer in the Prydz Bay region, Antarctica. In *Southern Ocean Ecology: the BIOMASS Perspective* (S. El-Sayed, ed.), pp. 247–60. Cambridge University Press, Cambridge.

Croxall, J.P. (1984) Seabirds. In *Antarctic Ecology* (R. Laws, ed.), pp. 531–616. Cambridge University Press, Cambridge.

Croxall, J.P. & Prince, P.A. (1987) Seabirds as predators on marine resources, especially krill, at South Georgia. In *Seabirds. Feeding Ecology and Role in Marine Ecosystems* (J.P. Croxall, ed.), pp. 347–68. Cambridge University Press, Cambridge.

Croxall, J.P., Everson, I., Kooyman, G.L., Ricketts, C. & Davis, R.W. (1985) Fur seal diving behaviour in relation to vertical distribution of krill. *J. Anim. Ecol.* **54**, 1–8.

Eklund, C.R. & Atwood, E.L. (1962) A population study of Antarctic Seals. *J. Mammalogy* **43**, 229–38.

Erickson, A.W. & Hanson, M.B. (1990) Continental estimates and populations trends of Antarctic ice seals. In: *Antarctic Ecosystems. Ecological Change and Conversation* (K.R. Kerry & G. Hempel, eds), pp. 253–64. Springer-Verlag, Berlin.

Everson, I., Parkes, G., Kock, K.-H. & Boyd, I. (1999) Variation in the standing stock of the mackerel icefish *Champsocephalus gunnari* at South Georgia. *J. Appl. Ecol.* **36**, 591–603.

Hart T.J. (1942) Phytoplankton periodicity in Antarctic surface waters. *Discovery Rep.* **21**, 261–356.

Holdgate, M.W. (1967) The Antarctic Ecosystem. *Phil. Trans. R. Soc.* **252**, 363–89.

Ichii, T., Shinohara, N., Fujise Y., Nishiwaki, S. & Matsuoka, K. (1998) Interannual changes in body fat condition index of minke whales in the Antarctic. *Mar. Ecol. Prog. Ser.* **175**, 1–12.

Kock, K.-H. (1992) *Antarctic Fish and Fisheries*. Cambridge University Press, Cambridge.

Kock, K.-H. & Shimadzu, Y. (1994) Trophic relationships and trends in population size and reproductive parameters in Antarctic high-level predators. In *Southern Ocean Ecology: the BIOMASS perspective* (S. El-Sayed, ed.), pp. 287–312. Cambridge University Press, Cambridge.

Laws, R.M. (1977) Seals and whales in the Southern Ocean. In *Scientific Research in Antarctica*. Discussion meeting organised by V.E. Fuchs & R.M. Laws. *Phil. Trans. Roy. Soc. London* B279 81–96.

Laws, R.M. (1984) Seals. In: *Antarctic Ecology* (R. Laws, ed.), pp. 531–616. Cambridge University Press, Cambridge.

Mackintosh, N.A. (1970) Whales and krill in the twentieth century. In *Antarctic Ecology*, Vol. 1 (M.W. Holdgate, ed.), pp. 195–212. Academic Press, London.

Murphy, E.J., Morris,D.J., Watkins, J.L. & Priddle, J.L. (1988) Scales of interaction between Antarctic krill and the environment. In *Antarctic Ocean and Resources Variability* (D. Sahrhage, ed.), pp. 120–30. Springer-Verlag, Berlin.

Prince P.A. & Morgan, R.A. (1987) Diet and feeding ecology of Procellariiformes. In *Seabirds. Feeding Ecology and Role in Marine Ecosystems* (J.P. Croxall, ed.), pp. 154–71. Cambridge University Press, Cambridge.

SCAR/SCOR (1977) *Biological Investigations of Marine Antarctic Systems and Stocks (BIOMASS)*. Vol 1: Research Proposals, 79 pp. University Library, Cambridge.

SC-CAMLR (1985) Report of the ad hoc Working Group on Ecosystem Monitoring, Report of the Fourth Meeting of the Scientific Committee for the Conservation of Antarctic Marine Living Resources. CCAMLR, Hobart Australia.

SC-CAMLR (1991) *CCAMLR Ecosystem Monitoring Program (CEMP). Standard Methods for Monitoring Studies*, 131 pp. CCAMLR, Hobart, Australia.

Chapter 8
Ecosystem Dynamics Involving Krill

Inigo Everson

8.1 Introduction

The picture of krill that has been developed in the preceding chapters will have demonstrated the central role of krill in oceanic food webs. Recognition of this, and the potential effects that overfishing on krill might have on dependent species, was behind the requirement for an ecosystem approach to management enshrined in the Convention for the Conservation of Antarctic Marine Living Resources (CCAMLR). This has in turn focused attention on the provision of scientific advice and for the ecosystem approach this is much more complicated than the traditional 'single-species' assessments.

Having accepted the ecosystem approach, the supporting scientific advice needs an increased level of complexity. This is because we need to determine not only the direct effects of harvesting on the resource but also whether a given level of fishing will adversely impact dependent species. This is further complicated by the need to take account of natural variation in the system. Understanding these sources of variation, whether they be associated with the krill, the fishery, the environment, dependent species or interactions between them, is a topic which requires a careful examination of the available information leading to the development of models which seek to provide explanations.

Traditional single-species fisheries models use the population as the basic unit for which a total allowable catch (TAC) is determined. This is useful in providing an upper limit beyond which the target fishery should not expand although it should not be used in isolation. When this first requirement has been met, further measures can be put in place to take account of the spatial and temporal scales of the interactions between the ecosystem components. This immediately raises a further level of complexity beyond single species population models because there is an enormous number of possible interactions within the system which are themselves operating at different time and space scales. In this chapter I follow a line starting with the simplest approach at the large scale and develop this through the time and space scales appropriate for the interactions leading to a consideration of ecosystem assessment and how this can be used to provide advice for management of krill fisheries.

8.2 Large-scale interactions

At the large scale we are looking for models which will provide an umbrella value with which to define an absolute maximum for the fishery. One of the simplest models of this type is that defined by Gulland (1971) who was looking for a simple approximation for maximum sustainable yield (MSY) in situations where data are sparse. His basic formulation:

$$Y = 0.5MB_0$$

where M is the coefficient of natural mortality and B_0 is the unexploited biomass, was further developed by Beddington and Cooke (1983). They demonstrated that the factor 0.5 was generally too high due to uncertainties over the estimation of M and recruitment (R) and that this would result in many instances in MSY being overestimated.

Their formulation, $Y = \lambda MB_0$, was used by Butterworth *et al.* (1992, 1994) to provide a method for calculating the potential yield of krill. This Krill Yield Model (KYM) was then developed further as the Generalised Yield Model (GYM) by Constable & de la Mare (1996) and which incorporated the terms λ and M into a single constant γ to give a basic formulation: $Y = \gamma B_0$. The derivation and characteristics of γ are described by Miller & Agnew in Chapter 12. Thus with an M of 0.6 the original Gulland formula would have given an MSY of 0.3 times the unexploited biomass; the value of γ used for the recent estimates, as explained by Miller & Agnew, Chapter 12, is 0.116. The CCAMLR approach is therefore erring very much towards protecting the stock and taking account of uncertainty.

Within the CCAMLR area the precautionary catch limits derived in this way apply to very large tracts of ocean, several million square kilometres. These umbrella values are deemed 'precautionary' because they allow for dependent species. If all the catch were taken within a limited region, and if that same region was critical for a dependent species, then, even though the catch limit allows for the requirements of the dependent species *in toto* they may, even so, be adversely affected. To cater for this eventuality we need further models at smaller scales.

8.3 Smaller-scale interactions

Development of models at smaller scales is a tricky business requiring knowledge of those dependent species which should be considered and their spatial and temporal foraging behaviour. The various sections of Chapter 7 have identified the species or groups that should be considered, so here I generalise to highlight the scales of interest. Spatially we need to consider both horizontal and vertical planes. Horizontally we have the geographical scale while vertically we have the location in the water column. Temporally the most important scales are probably season and time of day although, depending on the topic of interest, months and long-term trends are also important.

Whales are present in all the main oceans and a large proportion of the balaenopterids feed on krill. Their large size and high mobility has meant that they have often been considered to have the ability to move between feeding areas. Thus if food is scarce in one region they might simply move to a new feeding area. Consideration of the actual feeding grounds would indicate that the continental shelf or shelf break, regions of krill aggregation, would define the geographical scale while vertically they might have a diving depth of perhaps 500 m. Seasonally they tend to migrate to the feeding grounds during the summer. Thus although able to feed over a wide area their main impact is probably local to the shelf and shelf break.

Pinnipeds are thought to be of major importance in the Southern Ocean and are typified by two contrasting species. The crabeater seal (*Lobodon carcinophagus*) is widespread throughout the pack ice zone throughout the year. Diurnally, the number of seals visible on the ice, the inverse of those diving at any time, increases when the krill are at their deepest (Laws, 1984). This optimises the diving depth over the top 60 metres of the water column. In contrast adult fur seals (*Arctocephalus gazella*) during the breeding season are restricted to the coastal region, probably within 150 kilometres of their breeding site (Boyd, pers. comm. reported in Everson & de la Mare, 1996). During the winter they are able to travel far although, if they remain in the Southern Ocean feeding on krill, they may be concentrated on the shelf or at the shelf break. Vertically most dives are restricted to little more than 60 m (Boyd & Croxall, 1992; Boyd *et al.*, 1997).

Birds fall into two broad categories, those that feed relatively close to the surface and those that are able to dive. Black-browed albatross (*Diomedea melanophris*), in the Southern Ocean, and petrels in all oceans have feeding ranges which cover large areas, even during the breeding season. Their ability to catch krill is constrained by the depths to which they can dive, a limitation of around 5 or at maximum 10 m. The other functional group of flying birds is the diving petrels and alcids, which can cover moderate distances rapidly, are constrained by range from breeding sites, and vertically restricted to around the top 30 m of the water column (Bradstreet & Brown, 1986). Penguins, a functional group of birds that cannot fly, have a restricted foraging range during the breeding season and dive to depths over 50 m (Croxall & Lishman, 1987), somewhat deeper than the alcids.

Fish are a group which are known to feed extensively on krill in North Pacific, North Atlantic and to a lesser extent in Southern Ocean waters. The main interactions involving fish and krill identified in Chapter 7 appear to occur on the continental shelf and shelf break. Variation in feeding intensity is probably a function of the distribution of both the krill and the fish, as for example krill being carried onto the shelf by the coastal current off the Japanese coast (Endo, Chapter 7). In the vertical plane, fish are likely to be able to feed over the whole water column on the continental shelf, some species, such as the mackerel icefish (*Champsocephalus gunnari*) around South Georgia, migrating off the bottom at night to feed (Kock, 1992).

The importance of understanding the time and space scales of the predator foraging activity was highlighted by Everson (1983) who considered the foraging activity of two penguin species during the breeding season. The example that was used there was of macaroni and gentoo penguins, breeding on an isolated island at the critical time of year when the chicks are being fed. Their maximum foraging ranges at that time are known to be quite different although both species are to a great extent dependent on krill. The aggregated nature of the krill distribution means that krill may not be available to both species at the same time; this is indicated in Fig. 8.1.

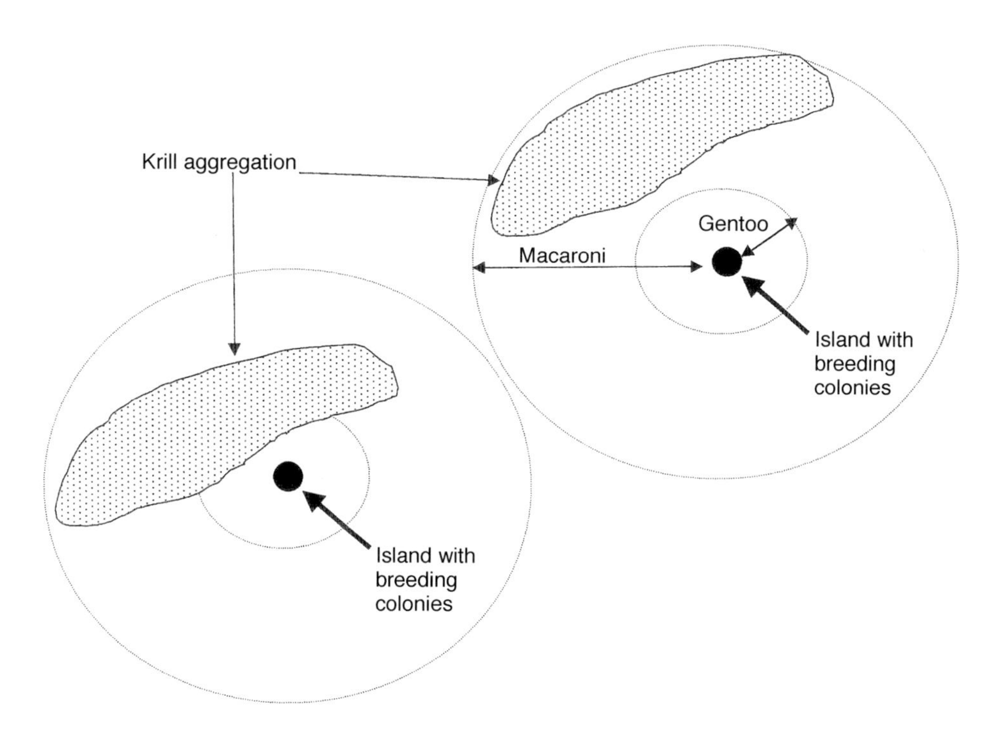

Fig. 8.1 Diagrammatic representation of a krill patch relative to the foraging range of gentoo and macaroni penguins at the height of the breeding season. The left-hand map shows an aggregation within the range of both species whilst the aggregation in the right-hand map would only be available to macaroni. (Redrawn from Everson, 1983.)

The vertical distribution of krill is complicated, as described in Watkins (Chapter 4) with no simple pattern emerging to describe all situations. The general pattern is that krill tend to be deep by day while after dark they tend to be at or near the surface. They are also present at the surface by day in some instances (see Endo, Chapter 3). These rhythms mean that krill may only be available to surface feeding predators for a brief period as demonstrated in Fig. 8.2.

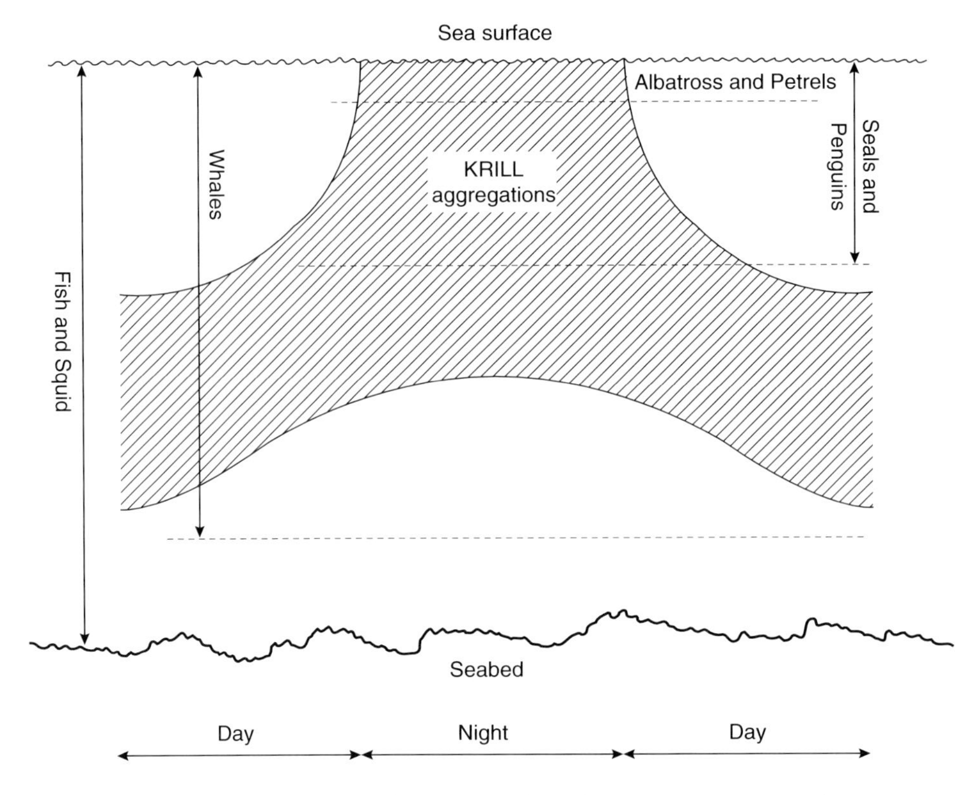

Fig. 8.2 Representation of diurnal krill availability to different types of predator. (Redrawn from Everson, 1983.)

8.4 Dependent species indices

In order to develop interaction models it is necessary to obtain time-series of data on the key characteristics of the dependent species. Unfortunately this does not appear to have been undertaken as a routine in many areas. An exception is the Southern Ocean where the CCAMLR Ecosystem Monitoring Programme (CEMP) has set up regular monitoring of what are thought to be key parameters associated with several krill eating species. A full description of the programme is beyond the scope of the present chapter so what follows is a very brief summary. The dependent species which have been identified for the CEMP are: Adelie (*Pygoscelis adelie*), chinstrap (*P. antarctica*) and gentoo (*P. papua*) penguins, black-browed albatross (*Diomedea melanophris*), Antarctic (*Thalassoica antarctica*) and Cape petrels (*Daption capensis*), fur (*Arctocephalus gazella)* and crabeater seals (*Lobodon carcinophagus*). In addition krill as a harvested species and major prey item and certain key environmental variables are also included. Most of these parameters have been monitored as part of the programme since 1989 or 1990 although in some instances national programmes were in place over a decade prior to that time. The CEMP database consequently forms a very powerful archive with which to study ecosystem interactions.

Initially these indices were considered separately and in many instances provided indications of what have become known as good and poor krill seasons. Two major difficulties have emerged with interpreting these data. The first is that the magnitude and diversity of the data is so great that it is impossible to view them all in a single table and gain a reasonable overview. The second, probably more serious problem is that the response of each index to the krill status has to be assessed carefully with respect to the other indices. For example when looking at penguin breeding success (the number of chicks fledged) this indicator may go down in a poor krill season whereas the fledging weight of the chicks may be normal; this might be because only the well-fed chicks survive. A fuller consideration of these interactions and their attendant problems can be found in the reports of the SC-CAMLR Working Group on Ecosystem Monitoring and Management.

Refining these indices into a form whereby they can be interpreted speedily has proved quite a difficult task. Currently groups of indices which integrate krill availability over approximately similar time and space scales are being amalgamated into combined standardised indices (CSI). This process involves estimating the standardised normal deviate for each variable, for each species at each site and for each season. Currently the amalgamation is in two forms using parameters in the first instance that integrate over the summer and secondly those indices appropriate to the winter months. These are then combined to form a time-series for each site, an example of which is shown in Fig. 8.3.

While having considerable merit in providing a pictorial view of the situation and highlighting the 'poor krill season' the approach fails to take account of interactions between parameters or provide any scaling factors which might take account of their relative influence. The other problem is that it does not take account of the time taken for any change in krill availability being noted in the CEMP index. Therefore to gain a better understanding of the way in which the system operates we need to take into account the extent and timing of the responses. In essence this means recognising the onset of a poor krill period and knowing the delay before this effect is recognisable in the predator index. This requires an understanding of the functional relationship between the predator and its prey.

8.5 Functional relationships

In the ecosystem context we are looking to understand the functional relationships between predator performance and krill availability, the latter expressed in terms of abundance and distribution. In order to achieve this we need to understand the time and space scales of the interactions, the underlying theme of this chapter. This was considered in some detail by CCAMLR during a meeting in Vina del Mar, Chile

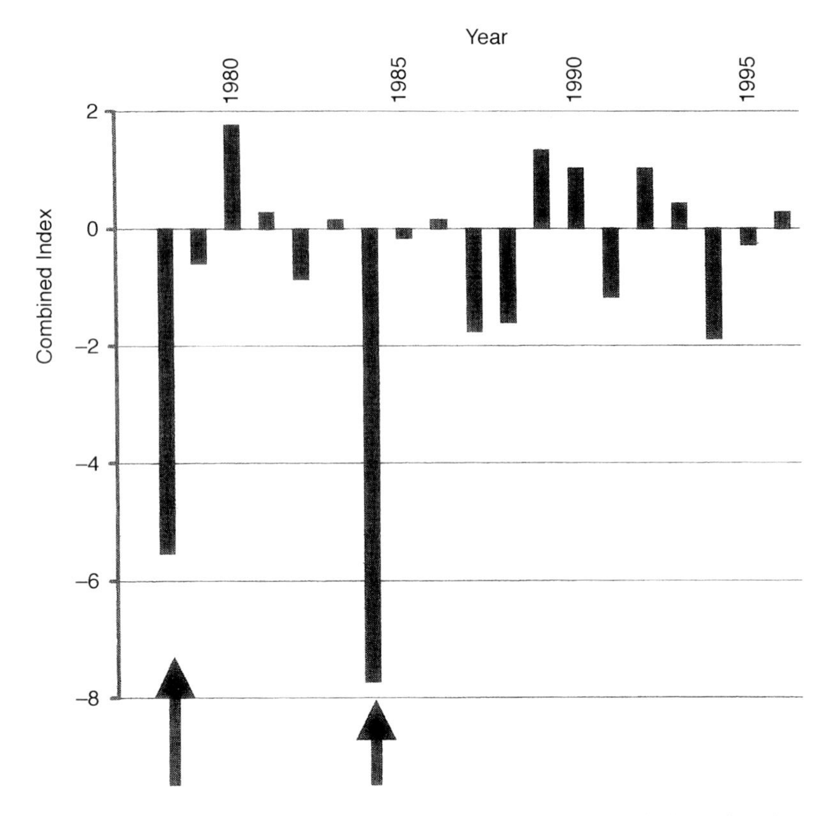

Fig. 8.3 Combined Standardised Index for CEMP parameters at South Georgia. The two arrows indicate seasons of very low krill availability to the predators. (Data from SC-CAMLR XVI, p. 227).

(SC-CAMLR, 1992) and during that meeting questions of functional relationships, krill availability and natural effects were discussed.

At the largest scale, Butterworth & Thomson (1995) considered the possible effects of different levels of krill harvesting on several dependent species. This work is discussed by Miller & Agnew in Chapter 12 and will not be referenced further here beyond noting that considerable work will be required before the method can be taken as the basis for scientific advice for management of the krill fishery.

Viewed on the local scale of the continental shelf, fish provide a convenient example for study as they are present in all the major krill zones considered in this book and in many instances have been harvested. Here I take the example of the interactions between the mackerel icefish (*Champsocephalus gunnari*) and its main food krill at South Georgia. Biological information, length, mass, sex, maturity state and stomach contents, are available from a series of trawl surveys in the region. Similar data have also been obtained from the commercial fisheries that took place in the 1970s and 1980s. Using these data several indices have been calculated.

Analysis of the stomach contents of the fish from different locations on the South Georgia shelf is quite instructive. Results from the January–February period in three seasons are presented in Kock *et al.* (1994) which indicate that there is a large degree of variation both within and between seasons in terms of stomach fullness and stomach contents. In 1985 and 1992 no clear pattern emerges but in 1991 there was a much larger proportion of empty stomachs and furthermore krill were only present in the stomachs of fish caught on the eastern and northeastern portion of the shelf. Evidence from other sources, referenced in Everson *et al.* (1999) indicates that 1991 and 1994 were seasons when krill were scarce and that over much of the area mackerel icefish were feeding on the hyperiid amphipod *Themisto gaudichaudii*. These observations are consistent with the known patchy distribution of krill in the region.

Condition indices, the ratio of measured mass to the estimated mass from a generalised length to mass relationship, varies around unity. When the index is greater than one the fish are assumed to have been feeding well and increasing their overall body mass. When food is in short supply body reserves will be mobilised leading to a reduction in mass and thus a reduction in condition index. This relationship is demonstrated in Fig. 8.4 where the krill density estimates are plotted against the total mass condition index for mature fish from research surveys undertaken during the same month. During this study two condition indices have been used, one using total mass and the other using gutted mass. These show essentially the same story although the total mass index varies more widely due to feeding and reproductive status. This difference is more accentuated when the condition index is greater than one as there will be variation due to stomach fullness and gonad size. When it is less than one the stomach is more likely to be empty and the fish are less likely to come into spawning condition.

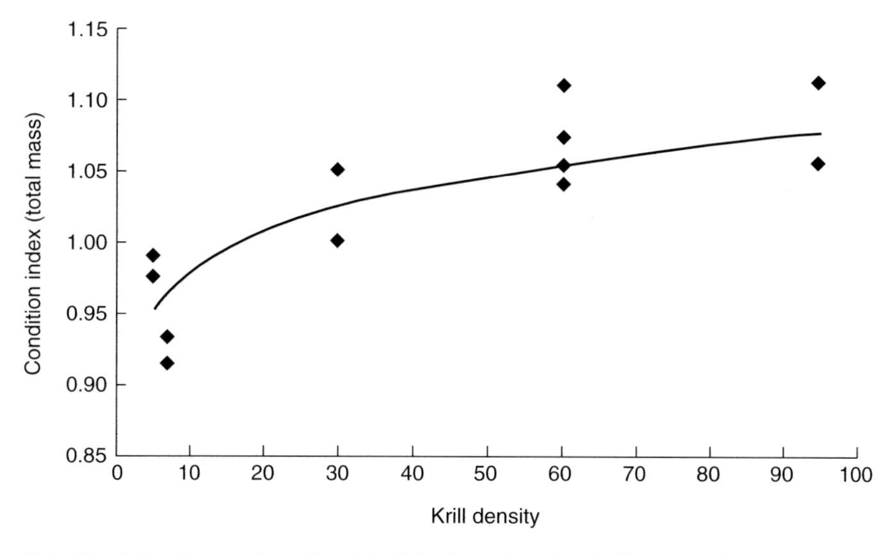

Fig. 8.4 Condition index of mackerel icefish plotted against krill density from acoustic surveys undertaken during the same month.

Although the fit to the data is good it could be improved by two refinements. The first improvement would be to ensure that the acoustic survey to estimate the abundance of krill is over the same area as the fish are present and likely to be feeding. The second is a refinement which relates to the response time of the fish to the availability of their food. A fish in poor condition is likely to feed actively when krill become available. This will lead to an immediate and rapid response in stomach fullness, which might be 5% of the total mass when the fish is replete, thus having an immediate effect on the condition index estimated from total mass. After a brief period, food will be converted to muscle so that the gutted mass condition index is likely to follow the total mass condition index but with a short time lag. The opposite sequence would be expected when the situation changed from one of high to low food availability. A diagrammatic representation of the time difference between the date of observation and the time period over which the index is integrating is indicated in Fig. 8.5. In the absence of direct observations on condition indices and feeding status these effects have been inferred from the work of Love (1988) who studied the effects of feeding condition and starvation on fish.

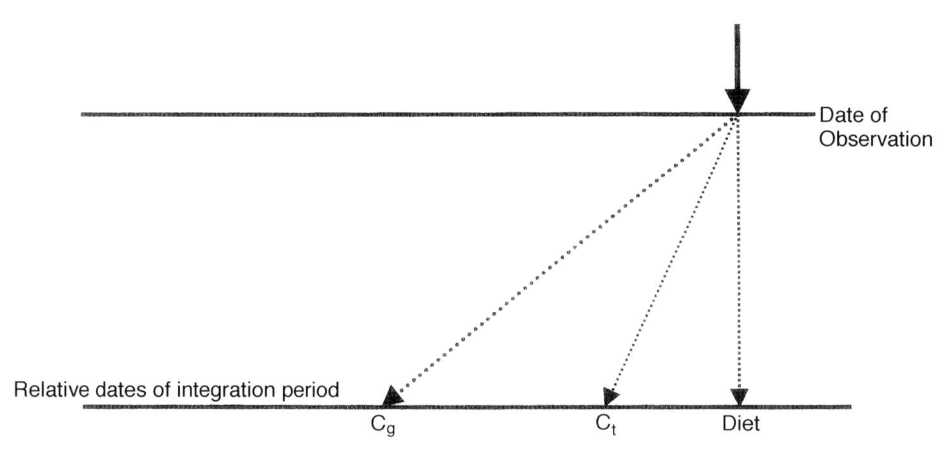

Fig. 8.5 Conceptual framework indicating the possible response times of three mackerel icefish indices. Diet = stomach contents or stomach fullness; C_t = condition index using total mass; C_g = condition index using gutted mass.

Feeding status is thought also to affect the reproductive cycle of the icefish. Periods of krill scarcity prior to the onset of the gonad maturation process have been shown to delay the start of the process (Everson *et al.*, 2000). In extreme cases this delay may be for three or four months but, except in extreme years, it does not appear to prevent the fish spawning completely. Diet, condition and spawning status all provide some indication of the feeding conditions encountered by the icefish. The diet, stomach fullness and composition of the contents, relate to conditions in the immediate vicinity of the fish at the time of capture. Condition takes a little time to work through the system although since this index correlates with the acoustic estimates of krill the response is recognisable within a month (Fig. 8.4). The reproduction index appears to be influenced by the krill status such that when krill are scarce the maturation process may be delayed until conditions improve.

The conceptual framework that has been developed for mackerel icefish can, with some modification be applied to other dependent species, a process which would have more utility than the simple combination of indices into CSI in the CEMP. To achieve this we need to consider the proximity in time of the observation to local krill availability, the integration period of the index and its response time to changes in krill availability.

In the case of penguins, at the shortest time and smallest space scales, we have, as with the fish, stomach contents and fullness during the breeding season. The analogue from the CEMP penguin indices is diet. We also have from the CEMP foraging trip duration (CEMP Index A5), an index that uses the time taken for the adult penguin to obtain sufficient food for itself and its chick; the longer the trip, the lower the availability of krill. These indices have a very short response time and all provide spot indications of the availability of krill. They are not sufficient to provide integration over a period of anything other than, at the most, a few days and over an area local to the breeding colony.

Krill availability over periods of a few weeks to a few months is tracked, as we have seen above, by the condition index of the fish. This index is available for any time of year during which sampling takes place, be it commercial fishing or research survey. The more that this sampling is spread out over the geographical range of the fish and throughout the year the clearer the picture that will emerge of the availability of krill.

The analogues for the penguins are much more closely specified to time of year. Thus the weight of the adult birds on arrival at the breeding colony (CEMP Index A1) is integrating over around four or five months prior to a set calender date. This arrival weight may be set by a ramp function of 'increasing importance' as that date approaches which, in the case of penguins, means that it is probably related to local krill availability, although at this stage local is not defined. At South Georgia, where the krill fishery is concentrated in the winter months, this may be the period of greatest predator–fishing overlap, a topic considered later. The duration of the first incubation shift (CEMP Index A2) may be influenced not only by the arrival weight (CEMP Index A1) but also by the availability of krill to the female while she is replenishing reserves post-egg-laying and provisioning herself for the second incubation shift. The second shift, during which the male seeks to replenish his reserves is likewise affected by krill availability but this time is several weeks later in the year. On a longer time-scale we have breeding success (CEMP Index A6), monitored throughout the season, and fledging weight (CEMP Index A7). Together these indicate how successful the adults have been in finding krill and bringing it to the colony to provision the chick over the same period.

Using the icefish analogue when considering a still longer time-scale, the reproductive status of the fish is integrating krill availability over much of the year, although the precise form of that relationship is still not fully described. In addition fish survey data can provide age structured population information with which to describe demographic changes in the icefish population. The analogous CEMP indices for penguins are population size and demography (CEMP Indices A3 and A4 respectively).

Using these indices it should be possible to construct a series of models which integrate over progressively longer time periods. The starting point would be the short and medium term CEMP indices. Using the mass at first arrival, incubation shift, diet and foraging trip indices a picture could be developed to infer local krill status from May or June through to the end of the breeding season. The conceptual framework for such a model is shown in Fig. 8.6. This model could then be incorporated into an age-structured model in order to determine population trajectories. Such an approach has obvious advantages over the simple amalgamation of indices into CSI because it is taking into account the ecological changes in the system.

Thus far in this chapter the consideration has been largely at the level of predator–prey interactions. However one of the main reasons for undertaking this type of study is to understand how commercial fishing might indirectly impact the predators by affecting the abundance of their prey.

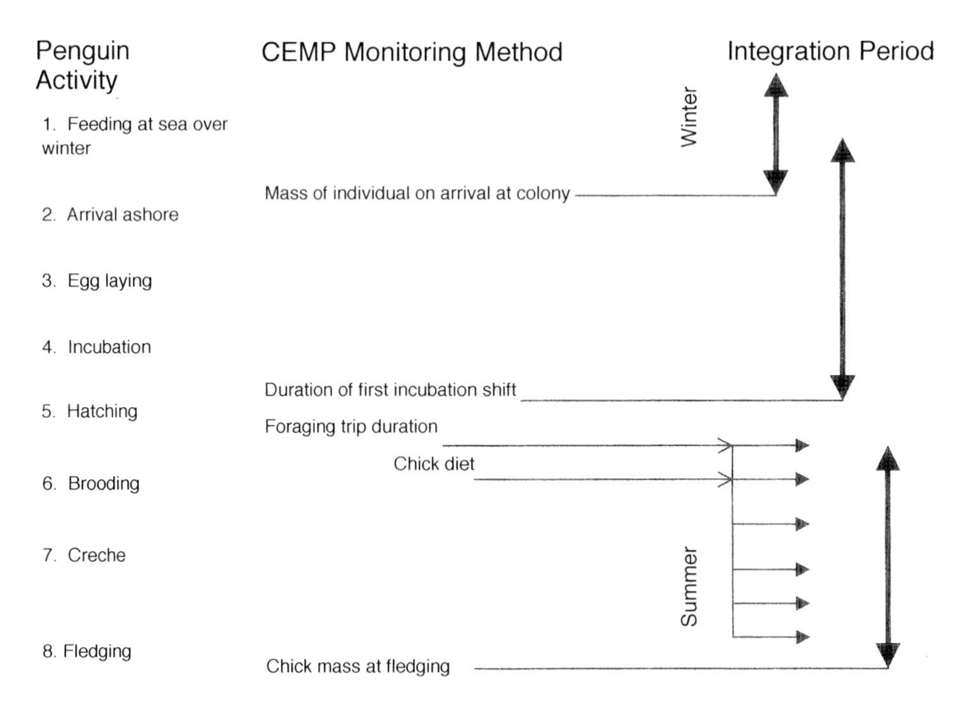

Fig. 8.6 Conceptual framework indicating the integration period for CEMP penguin indices. The foraging trip duration and chick diet are both multiple observations during the course of the season. The other indices are from single observations but at different times during the season.

8.6 Predator fisheries overlap

The introduction of reporting of Antarctic krill catches by fine-scale rectangles, half a degree of longitude by one degree of latitude, demonstrated that the bulk of the catch was being taken on the shelf or at the shelf break (Everson & Goss, 1992). These localities were the same as those where penguins forage during the chick rearing period. This then raised questions regarding the degree of overlap and impact between fishing and dependent species foraging.

No simple procedure has been proposed to describe adequately this interaction so, in order to understand the process we need to break it down into manageable components. For simplicity I consider the problem from two perspectives, qualitatively in terms of time and space, and quantitatively in terms of the amounts of krill taken by the fishery in relation to the requirements of dependent species.

Penguins during the breeding season provide a good example of a 'central place forager' since they are constrained by the range to which they can venture in search of food and return in good time to feed a chick. Thus for any given colony the foraging area can be described by the segment of a circle whose radius is the maximum foraging range. This has been used to define a critical period distance (CPD) for dependent species during the breeding season. Foraging activity can be determined by direct observations from research vessels at sea leading to the identification of particular foraging localities as shown in Fig. 8.7. It is also possible, due to the miniaturisation of instrument packages, to monitor individual animals at sea. This has been done for fur seal and to a lesser extent, due to the smaller size of the animals, for penguins. Outside of the breeding season, it is assumed that the birds forage over a much greater area; this may not be true necessarily but since they are then not constrained by the need to return to a breeding site, they are able to search for food in other localities.

The situation for fish is somewhat different. The mackerel icefish are known to be restricted to the continental shelf region. It is not known to what extent feeding migrations take place so spatially it has to be assumed that the whole shelf region is a reasonable descriptor of their feeding range at all times of year. In Canadian waters, hake are found on the shelf and slope regions (Simard & Mackas, 1989). Off the coast of Japan there are pollock feeding on krill that are carried onto the shelf (Endo, Chapter 7).

Several approaches to the problem of determining the relationship between predator foraging and commercial fishing have been considered by CCAMLR. These were defined in terms of the type of overlap.

- *Precautionary overlap*, which in considering the largest spatial scale is intended to cover the whole distribution of krill and all krill predators. It is covered by the krill yield model described earlier in this chapter and by Miller & Agnew, in Chapter 12).
- *Potential overlap*, is on a very broad scale such that local overlaps or separations

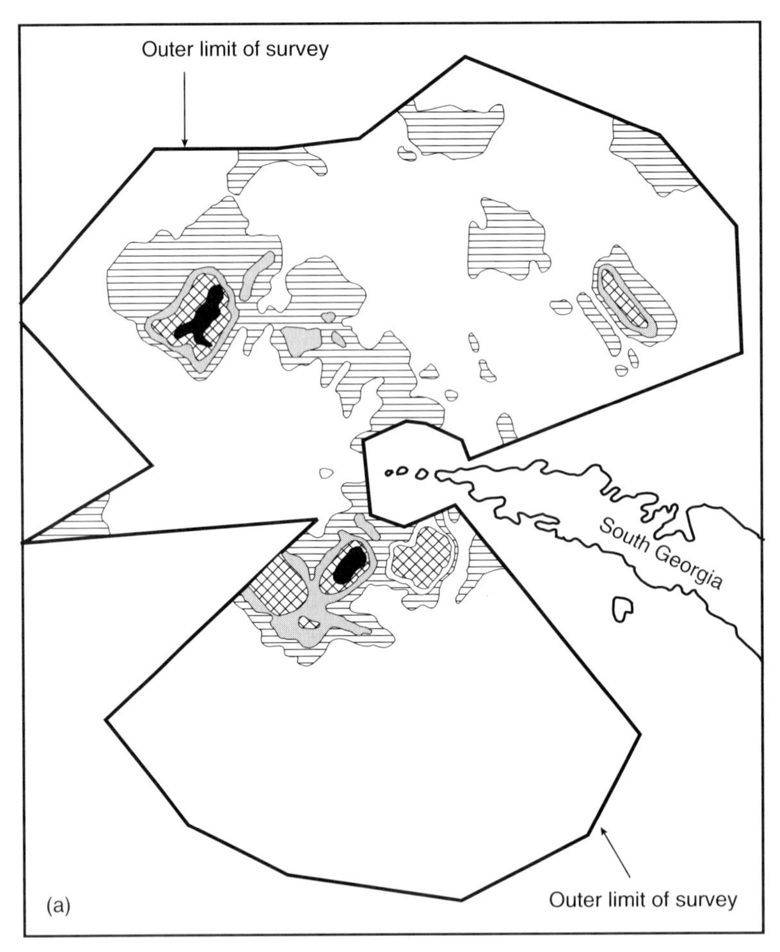

Outer limit of survey
South Georgia
Outer limit of survey
(a)

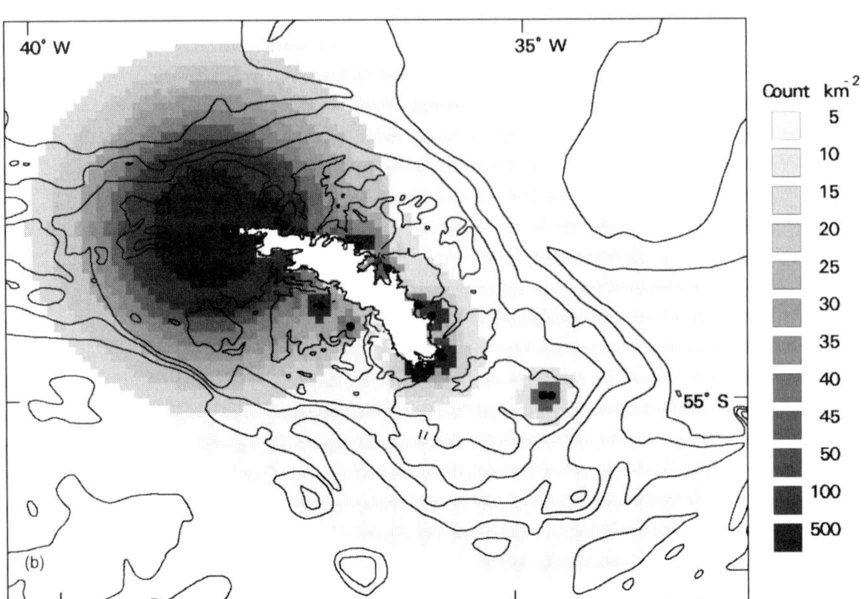

40° W
35° W
55° S
(b)
Count km^{-2}
5
10
15
20
25
30
35
40
45
50
100
500

between predators and the fishery may be missed or misrepresented. This has been considered by CCAMLR as the critical period distance (CPD). The index is currently calculated as the krill catch within 100 km of predator colonies during the period December to March. It is not a measure of competition between predators and the fishery, but is a simple expression of potential niche overlap. Ichii *et al.* (1994a, b) considered the spatial distribution of colonies and catches in the Antarctic Peninsula region. They showed that although a high proportion of the catch may be taken within the foraging distance of land-based colonies, those colonies containing the largest numbers of penguins were not adjacent to the main fishing grounds. In order to make further progress with this type of work a more refined approach is needed.

- *Realised overlap*, in which fine-scale overlap is measured but without taking account of any movement of krill through the region. To describe this a simple standardised index, such as Schroeder's index, has been used (SC-CAMLR, 1997). This has the form:

$$I_t = 1 - 0.5 \sum | p_{i,t} - q_{i,t} |$$

Where $p_{i,t}$ is the proportion of krill consumed by a predator in grid square I during time period t and $q_{i,t}$ is the proportion of krill taken by the fishery in grid square I during time period t. The index ranges from $I_t = 0$, indicating no spatial overlap during period t to $I_t = 1$, complete overlap. Currently this index is used by CCAMLR where the grid square is defined as the fine-scale rectangles already mentioned above.

This approach has considerable merit in defining the localities and times during which the interaction is likely to be most significant. However, since the two proportions are determined independently and without reference to the other component, care needs to be exercised in interpreting the results. Thus, to take an extreme example, if the fishery were taking less than 1% by mass of the krill required by penguins and all those krill were caught from a high index foraging locality, the index would have a high value and yet no problem might exist. Some mechanism is therefore needed to incorporate the total available krill into the index.

Another approach that attempts to incorporate fishing impact, and in fact predates the use of the Schroeder's Index by CCAMLR, was described by Agnew & Phegan (1995). For this index the whole area was divided into small units of 10 × 10 nautical miles and the proportion P of the total number of penguins that were likely

Fig. 8.7 *(Opposite.)* (a) Foraging distribution of Antarctic fur seals in the vicinity of South Georgia determined from direct visual observations at sea (Madureira, 1992). (b) Foraging footprint for the macaroni penguin at South Georgia. The figure links at-sea observations made during the same study as that for Fig. 8.7(a) but integrates that information with data on breeding colony location and size. (Part (b) with permission from Trathan *et al.*, 1999.)

to feed in each unit estimated. This is multiplied by the total krill requirement of the penguins to give a penguin krill requirement for the location K_p. The other component is the krill catch K_c over the same period in the same area. The index is the product ($K_p . K_c$) and increases where either predation pressure or fishing increases. Even so this still does not get around the problem of how much krill is available within the small units.

Dynamic overlap, whereby the interaction would be described by the functional link between predators and the fishery, would take account of fine-scale vertical and horizontal distributions of predators and the fishery and the availability of prey to both resource users. Undoubtedly this is the most difficult and complicated scenario but does provide the greatest insight into the functioning of the system. A model that makes comparisons of penguin reproductive success and adult survival in the absence or presence of a krill fishery was developed by Mangel & Switzer (1998). Over the range of catch used in the study, the reductions in reproductive success are essentially linear functions of krill catch. The reductions in offspring and parent survival as determined by the model were mainly determined by how long the fishing season lasts and the capacity for harvest, rather than when, during the season, the fishing commences. In their paper the authors note further refinements which might be incorporated into the model by refining the prey field, including a component for chick mortality related to krill biomass and finally allowing for plasticity in the fledging period. Clearly other approaches are also possible which take account of other indices of the dependent species as well as incorporating components which describe more closely the krill environment.

In this chapter so far I have been considering interactions in time and space from the perspective of a system that was more or less enclosed and which might change from year to year and between summer and winter. Such a view may be appropriate for an enclosed sea or lake but misses an essential variable in the ecosystems in which krill are key components. This additional variable is the movement of krill through a region and is often referred to as krill flux.

8.7 Krill flux

In Chapter 7 several instances are noted whereby krill are known to be carried into, through, and out of key regions. The amount of krill taken out of a region by dependent species during the breeding season can be viewed as one estimator of krill production in that region. Viewed from the perspective of krill production, that amount of krill arose as a result of the balance of influx to and efflux from the region allied to production by those krill present within the region. Consideration of the flux of krill is therefore a further complication that must be included in the consideration of models from which management advice is developed. Krill are able to move actively about in the water column; Watkins (Chapter 4) has highlighted their swarming and vertical migratory capabilities. In spite of this their ability to move over significant distances against major ocean currents is probably not great. Taking a simple approach to addressing this problem, which might appear as something of a contradiction in terms, we might assume that krill are passive particles that can migrate vertically.

Everson & Murphy (1987) followed krill patches along the north side of the Bransfield Strait and noted that they were moving at about the same rate as the current in that region. The movements of the commercial fleets off the north coast of the South Shetland Islands indicate that the fishing vessels are following patches along the coast (Endo & Ichii, 1989). These movements appear to be at about the same rate as the water movement in that region.

As an extension of this type of study, Murphy (1995) developed a model of the South Georgia system in which the local prey abundance was regulated by a continuous, hydrodynamically mediated, supply rate and the concentrating effects of abiotic–biotic interactions. Using estimates of annual predator demand and krill concentration in the area the relationship between flow rate and depletion in prey concentration as a function of range from the colony was estimated. The model indicated that concentrating factors need to be large to produce the build up of krill densities of the order estimated to occur in the South Georgia area. As might be expected those predators which were restricted to the inshore region encountered the greatest changes in prey abundance. This was further exacerbated when random fluctuations in the inter-annual prey availability were introduced.

This approach was extended further by Murphy *et al.* (1998) by an examination of the effects of inter-annual variation in year class strength on the standing stock and also teleconnections between major areas. Taking information from the Fine Resolution Antarctic Model (FRAM) to define the deep ocean current field, particles were seeded into the circumpolar current and their drift modelled across the Scotia Sea. The rates of movement, whereby particles would take 140 to 160 days to travel from the Antarctic Peninsula to South Georgia were in line with the timing of arrival of different life history stages of krill across the region (Hofmann *et al.*, 1998). These results are consistent with the results from drogued buoys that took 110 to 120 days to drift over the same area.

Arising from this we can see that krill flux is operating in two major ways. On the large scale, in deep oceanic water, the flux appear's to be consistent with estimates of flow rate. Locally on the shelf the picture is much more complicated with krill being concentrated in eddy systems and remaining there for extended periods, in some instances of the order of months.

Much of this section has been concerned with highlighting differences due to movement through and between regions. In spite of this large-scale movement there is an underlying consistency over the locations where krill concentrations are found. This can be seen in the distribution of whale catches during the 1920s (Everson, 1984). More recently it has been possible to identify what are becoming traditional fishing grounds. For example at South Georgia the fleets regularly begin fishing at the shelf break to the north-east of Cumberland East Bay (Murphy *et al.*, 1997;

Trathan *et al.*, 1998), while in the South Shetlands the bulk of the fishing is along the north side of King George Island up to Elephant Island (Ichii & Naganobu, 1996). This level of consistency allied to the distribution of dependent species highlights the need for an ecosystem approach to the management of krill, the starting point of this chapter.

8.9 Selectivity

The preceding discussion has taken no account of the size or age of the krill being taken by dependent species. For those krill species whose lifespan is less than a year this is probably of little consequence for assessments. However, in situations where several age or size classes are present this presents a further dimension for consideration.

Recent developments in sampling and analysis have meant that it is possible to obtain information on the size distribution of the krill eaten, either from direct measurements from stomach contents or scat samples (Reid *et al.*, 1999). These provide a good indication of the size, sex and in some cases maturity stage of the krill eaten. This is important information to analyse in developing energy budgets for the dependent species. The samples provide good indications of what the predators actually consumed although they are not necessarily representative of the size composition of the krill that were present or available within the foraging range. The question then arises whether the predators are concentrating on taking a particular type of krill. Such a question does not have a simple answer because it will be dependent on the type and amount of krill that are available within range of the predator. Thus when krill are scarce the predator is likely to 'take what it can find' whereas when krill are plentiful the predator has some degree of choice.

It is possible to obtain samples from land-based predators with reasonable ease and thus build up a short time-scale temporal picture of change through a season. This needs to be checked against some direct and independent estimates of the krill size distribution at intervals to determine the extent of selectivity. A study of this type was undertaken by Reid *et al.* (1999) who was able to follow size classes through a summer season (Fig. 8.8). As well as showing the growth of the dominant size class through the season, part of the study compared the size distribution of krill taken in net hauls from a short part of the season. This study shows two important points. The first is that a single indication of the length–frequency distribution of the krill being taken can be a very poor indication of the situation at other times during the breeding season. The second point raises the question of whether the fur seals were feeding on krill that were essentially passing the island as if on a conveyer, or else whether the predation effect was influencing the size distribution of the krill that were concentrated for a period in the region. This last point is important because it raises the question of the direct influences of predation on krill mortality rates.

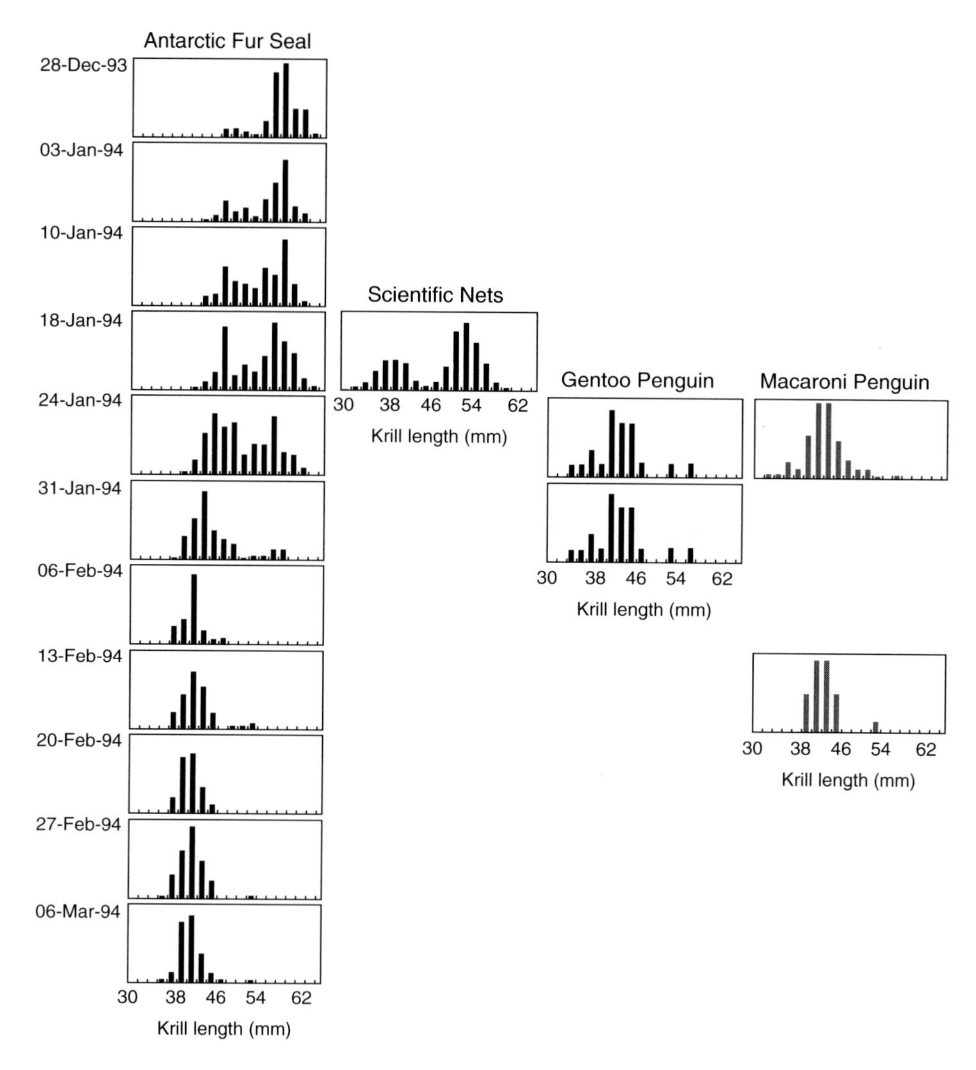

Fig. 8.8 Sequence of krill length frequency distributions from Antarctic fur seals, scientific nets, gentoo penguins and macaroni penguins during the period from December 1993 to March 1994. (With permission from Reid *et al.*, 1999.)

Providing the information on krill length distribution can be linked to predator numbers and feeding intensity it can be compared to the amounts of krill present from direct estimates. This then opens the way to developing more sophisticated age structured models of the type outlined by Everson (1977).

8.10 Ecosystem monitoring and management

The concept of an ecosystem approach to the management of resources is on the one hand intuitively obvious and on the other hand, due to the complexity of the system, extremely difficult to define. Even though the concept has been around for many years – the BIOMASS programme (El-Sayed, 1977) from which CCAMLR developed was the earliest formal fisheries approach – there is still some way to go before it can be considered to be properly implemented. Within CCAMLR the Working Group on Ecosystem Monitoring and Management (WG-EMM) (SC-CAMLR, 1995) has been developing the concept of an ecosystem assessment. In this context an ecosystem assessment would consist of two parts:

(1) an analysis of the status of key biotic components of the ecosystem, and
(2) a prediction of the likely consequences of alternative management actions for the future status of these components.

In terms of ecosystem assessment, WG-EMM in 1995 agreed that:

> 'status would need to include not only the points necessary for a single species assessment, which are:
>
> current abundance and productivity of the harvested species, with abundance related to some level prior to the onset of exploitation; and
>
> if possible, the relationships (links) between these quantities and the state of the environment.
>
> but also points related to dependent species which may be summarised as:
>
> current abundance of dependent species (usually expressed as breeding population size or an index thereof) in relation to previous values, where possible in conjunction with data on current and recent adult survival and recruitment rates.'

These components can be expressed slightly differently as the following series of questions which address problems of krill availability and dependent species demography:

(1) Is the availability of krill changing?
(2) Are populations of dependent species in decline?
(3) How much krill is required by the dependent species?
(4) What is the extent of overlap between krill fishing and the foraging by dependent species?

Addressing each of these questions in turn should provide a structure within which a management regime can operate. Again I take as an example the Southern Ocean and in particular CEMP. The CEMP parameters can be broadly divided into two categories, those that indicate the availability of krill to the predator and those that help interpret changes in predator demography. Such a distinction, however, should not be considered as absolute, it is convenient in providing a breakdown of the information on the dependent species so that it can be interpreted in a manner which will lead to the development of conservation measures.

Moving through the breeding season of Macaroni penguin, one of the CEMP species, we can establish a chronology whereby krill availability can be inferred. At this stage, as indicated in Fig. 8.6, the consideration is purely temporal.

At the start of the breeding season the penguins arrive ashore. The total mass of each individual on arrival will indicate how well they have been able to feed in the weeks prior to arrival at the breeding site. Natural mortality will have had some effect on the population but providing the recruitment balances deaths the numbers arriving will be stable. Although mortality and recruitment are likely to be affected most strongly by predation due to higher trophic levels, these may be mediated by feeding conditions experienced by the penguins. The arrival mass of individuals and the number of birds arriving at the site can therefore be taken as providing a long-term and medium-term index of krill availability over the winter months; the temporal scale at this stage is coarse, of the order of months. The establishment of the nest, egg laying and departure date of the female to feed, are all dependent upon how the bird has fared in finding food over the recent past prior to its arrival back at the colony. The temporal scale is shorter here when using the parameters to provide an index of krill availability. The return of the female to end the first incubation shift is critical, and her condition and the date at which she returns will be dependent on how well she has been able to provision over the period. The temporal scale of the interactions becomes progressively shorter as we move through the remainder of the incubation phase and into chick rearing. By this time the temporal scale is down to days. Difficulties in obtaining krill over this period will result in chick mortality but if the situation improves then those chicks that survive are likely to achieve a normal fledging mass.

This overview demonstrates that from the dependent species perspective the temporal scale over which krill availability is assessed needs to be varied. Accepting the need for this flexibility on the temporal scale necessitates that, because the dependent species has a finite maximum swimming speed, the spatial scale is also varying. However, providing we can accept that the dependent species is free to forage at will within the temporal framework imposed by its breeding cycle this does not matter because it is the temporal scale that is of primary importance. The spatial scale is important in determining a 'foraging footprint' and is discussed later.

Using this process we can develop rolling series of composite indices, the composition and temporal scale of which changes through the season. Examination of the patterns of these indices for different years should provide indications of patterns within season and also, when there were serious breeding failures, the pattern which led up to the failure. It is important to note that the time-scale during which these indices are combined together will be species dependent, a feature which although adding complication does at the same time provide an extra dimension by allowing inter-species comparisons to be made. An analysis of these rolling indices for the different species over a series of years will highlight those parts of the season when krill availability was affecting the dependent species. Combining this information with that obtained from studies on foraging range, analysed in a similar manner to that shown in Fig. 8.6, provides a picture of the krill availability in time and space through the breeding season. This information can be used to extend knowledge on the distribution of krill obtained by directed surveys in the vicinity of the breeding site.

The combined indices while providing a good indication of the amount of krill available to the predators do not in this form provide indications of demographic changes. This leads into the second question 'Are populations of dependent species in decline?'; what is needed here is the traditional population demographic information whose key components are population size, growth and mortality. These can be determined from regular estimates of population size and a consideration of how the CEMP-type parameters might influence the values. It is, however, important to define the 'population' in terms of practical management units. Thus although there may be significant mixing between breeding colonies of dependent species at, for example, adjacent island groups it may be desirable to separate them due to the differing availability of krill experienced at the two localities. An example might be the islands of the South Shetlands. Such a division might be determined by the extent of the foraging footprint of key predators.

Regular and routine monitoring of population size is clearly indicated as the best approach to this question. However this requirement needs to be set against the cost in undertaking such a task and although it would be ideal to undertake such a task annually that approach may not be cost effective. Reducing the frequency to a three- or five-year interval means that there will be a greater period of time before a downward trend can be positively identified. That is clearly a management risk that falls outside of this chapter. Direct estimation of population size is not the only approach because the same CEMP parameters can be used to investigate variation in, for example, survivorship of adults, survivorship of chicks through to first breeding. Essentially this is taking the basic components traditionally used in fish stock assessment, recruitment, mortality and production and providing estimators derived from direct observations.

Answers to the second question when considered in conjunction with the energetic requirements and dietary information lead to estimation of the amount of krill required by the dependent species, the third question. Croxall *et al.* (1984) provided estimates of food consumption at regular time intervals throughout the year. Refinement of such estimates and examining them in conjunction with foraging activity provides the 'average' krill consumption term in the Agnew & Phegan (1995) model previously mentioned. Studies on foraging behaviour have also been developed to indicate the area over which the predators move in search of food, the 'foraging footprint' for the dependent species (Trathan *et al.*, 1999). This foraging footprint (Fig. 8.7b) will vary during the year due to the constraints imposed by the breeding cycle. This spatio–temporal component leads naturally into the fourth question, the extent of overlap between krill fishing and dependent species foraging.

The Schroeder index and the Agnew & Phegan model provide indices of overlap and predation pressure on the krill. There is a third component which would assist in determining whether fishing is likely to be having an effect on krill locally and that is a simple ratio of the amount of krill taken by commercial fishing compared to the amount required by dependent species. When this fishing to predation index, (K_c/K_p) increases it indicates that fishing is taking a larger proportion of the available krill and consequently fishing is more likely to have an impact on the dependent species. It is important to note that this index does not necessarily indicate that any management action is required but rather provides a warning that an effect might be noticeable in some of the predator demographic parameters.

Two further points need to be made on this theme. Firstly that because the krill sampling techniques differ widely it is important that the estimates of krill standing stock and production are properly estimated. Commercial fishing estimates krill by applying a conversion factor to the amount of product, research surveys provide estimates based on scaling acoustic signals while figures for dependent species are derived from a combination of population size, energy requirement and diet; all of these are subject to different degrees of imprecision. The second point relates to prey switching by predators. In times of krill scarcity predators are likely to take alternative prey. A good example is the Antarctic fur seal at South Georgia, which feeds on fish during periods of krill scarcity. This is thought to have a major impact on the mackerel icefish population (Everson *et al.*, 1999) but since this goes beyond the direct effects of krill in the ecosystem it is not considered further here.

With the practical framework to providing an ecosystem assessment outlined above we are now in a position to consider how this can be taken forward as management advice.

8.11 An ecosystem approach to krill management

The problem of krill management in an ecosystem context can be reduced to two key questions. The over-riding and first question is: 'Taking account of the requirements of dependent species, is the krill fishery sustainable in the long term?' This question is adequately addressed by the KYM and providing there is a maximum limit for the krill fishery in place no further comment is needed.

The second question is: 'Are dependent species being adversely affected by krill fishing?'. This in turn breaks down into a series of questions and action points as shown in Fig. 8.9. The primary consideration is to determine whether there is any evidence for a decline in the populations of dependent species. This may come from the direct estimation of population size but might also appear as an increase in mortality rate before a significant trend was recognisable in population size.

A decline in population size does not itself indicate that the cause was com-

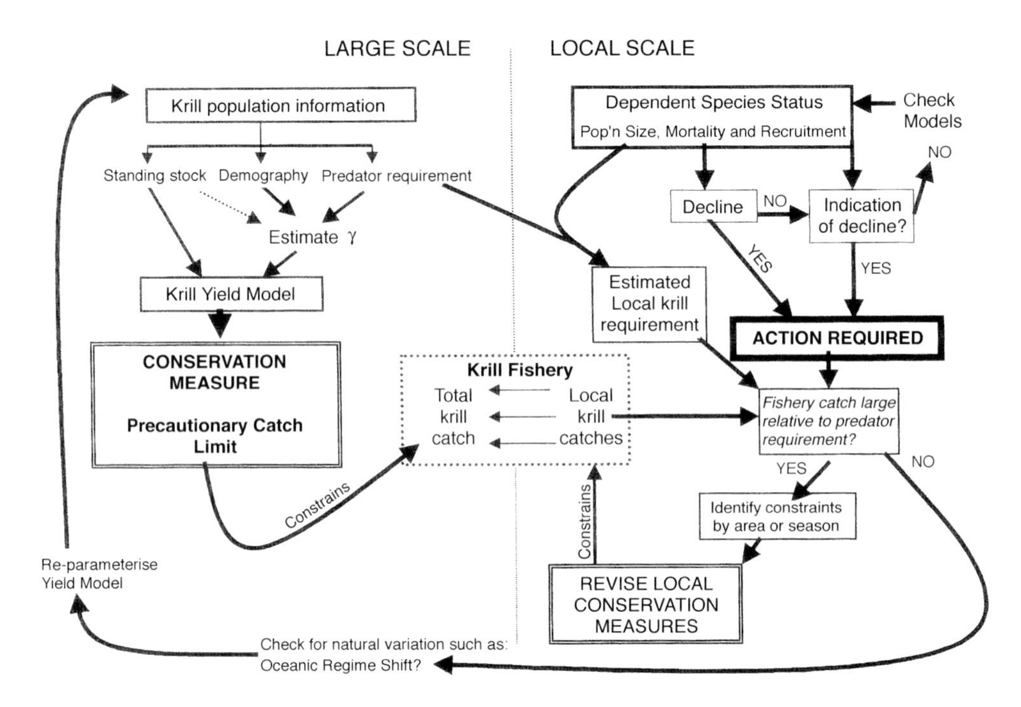

Fig. 8.9 Decision processes incorporating information from dependent species into a mechanism to provide advice for management of a krill fishery.

mercial fishing so the next level of enquiry should be towards investigating the various possibilities. Analysis of indices which relate to krill availability will indicate whether the cause lies with the ability of the predators to find krill. Such an investigation needs to take account of the oceanography of the region to determine whether underlying cause might be due to variation in oceanic circulation or production. The investigation should also take account of the degree of overlap between fishing and predation as indicated by the Schroeder, Agnew & Phegan and fishing–predation indices. With this information it is then possible to provide advice on locations and times of year when fishing pressure might need to be reduced. Such information would translate directly to conservation measures for the krill fishery in question. There is also the possibility that the underlying cause for the decline in the population of a dependent species might not be due to fishing but environmental change, either natural or man-induced, such as through global warming. Considerations of that type fall outside the remit for this chapter but are nonetheless vitally important topics in a global management context.

Using the decision processes envisaged in Fig. 8.9 we have a structure within which a science plan can be formulated that can provide advice for management of krill fisheries in an ecosystem context. Further refinements are clearly possible but establishing this first link would be a major step in fisheries management.

References

Agnew, D.J. & Phegan, G. (1995) A fine scale model of the overlap between penguin foraging demands and the krill fishery in the South Shetland Islands and Antarctic Peninsula. *CCAMLR Sci.* **2**, 99–110.

Beddington, J.R. & Cooke, J.G. (1983) The potential yield of fish stocks. *FAO Fisheries Technical Paper* **242** FIRM/T242, 53 pp.

Boyd, I.L. & Croxall, J.P. (1992) Diving behaviour of lactating Antarctic fur seals. *Can. J. Zoology* **70**, 919–28.

Boyd, I.L., McCafferty, D.J. & Walker, T.R. (1997) Variation in foraging effort by lactating Antarctic fur seals: response to simulated increased foraging costs. *Behav. Ecol. Sociobiology* **40**, 135–44.

Bradstreet, M.S.W. & Brown, R.G.B. (1986) Feeding ecology of the Atlantic Alcidae. In *The Atlantic Alcidae* (D.N. Nettleship & T.R. Birkhead, eds). Academic Press, London.

Butterworth, D.S. & Thomson, R.B. (1995) Possible effects of different levels of krill fishing on predators – some initial modelling attempts. *CCAMLR Sci.* **2**, 79–98.

Butterworth, D.S., Punt, A.E. & Basson, M. (1992) A simple approach for calculating the potential yield of krill from biomass survey results. *SC-CAMLR Selected Scientific Papers* **8**, 207–217.

Butterworth, D.S., Gluckman, G.R., Thomson, R.B., Chalis, S., Hiramatsu, K. & Agnew, D.J. (1994) Further computations of the consequences of setting the annual krill catch limit to a fixed fraction of the estimate of krill biomass from a survey. *CCAMLR Sci.* **1**, 81–106.

Constable, A.J. & de la Mare, W.K. (1996) A generalised model for evaluating yield and the long-term status of fish stocks under conditions of uncertainty. *CCAMLR Sci.* **3**, 31–54.

Croxall, J.P. & Lishman, G. (1987) The food and feeding ecology of penguins. In *Seabirds. Feeding Ecology and Role in Marine Ecosystems* (J.P. Croxall, ed.), Cambridge University Press, Cambridge.

Croxall J.P., Prince, P.A. & Ricketts, C.R. (1984) Impact of seabirds on marine resources especially krill, of South Georgia waters. In *Seabird Energetics* (G. Causey Whittow & H. Rahm, eds), pp. 285–317. Plenum, New York.

El-Sayed, S.Z. (ed.) (1977). *Biological Investigations of Marine Antarctic Systems and Stocks*, Vol. 1, 79 pp. Scott Polar Research Institute, Cambridge, England.

Endo, Y. & Ichii, T. (1989) CPUEs body length and greenness of Antarctic krill during 1987/88 season in the fishing ground north of Livingston Island. Scientific Committee for the Conservation of *Antarctic Marine Living Resources*, Selected Scientific Papers. SC-CAMLR-SSP/6, pp. 323–45.

Everson, I. (1977) *The Living Resources of the Southern Ocean.* Food and Agriculture Organisation of the United Nations, Southern Ocean Fisheries Survey Programme. GLO/SO/77/1, 156 pp.

Everson, I. (1983) Estimation of Krill abundance. In *On the biology of Krill Euphausia superba*. In Proceedings of the Seminar and Report of the Krill Ecology Group, Bremerhaven, 12–16 May 1983. (S.B. Schnack, ed.), *Berichte zur Polarforschung*, Sonderheft 4, December 1983, pp. 156–68.

Everson, I. (1984) Marine interactions. In *Antarctic Ecology*, Vol. 2., Chapter 14 (R.M. Laws, ed.), pp. 785–819). Academic Press, London.

Everson, I. Klock, K.-H. & Ellison, J. (2000) Inter-annual variation in the gonad cycle of the mackerel icefish. *J. Fish Biol.*

Everson, I. & de la Mare, W.K. (1996) Some thoughts on precautionary measures for the krill fishery. *CCAMLR Sci.* **3**, 1–11.

Everson, I. & Goss, C. (1992) Krill fishing activity in the southwest Atlantic. *Antarctic Sci.* **3**, 351–58.

Everson, I & Murphy, E.J. (1987) Mesoscale variability in the distribution of krill *Euphausia superba*. *Mar. Ecol. Prog. Ser.* **40**, 53–60.

Everson, I., Parkes, G., Kock, K.-H. & Boyd, I.L. (1999) Variation in standing stock of the mackerel icefish *Champscephalus gunnari* at South Georgia. *J. Appl. Ecol.* **36**, 591–603.

Gulland, J.A. (1971) *The Fish Resources of the Ocean*. Fishing News (Books), West Byfleet, Surrey, for FAO, 255 pp.

Hofmann, E.E., Klinck, J.M., Locarnini, R.A., Fach, B., & Murphy E. (1998) Krill transport in the Scotia Sea and environs. *Antarctic Sci.* **10**, 406–415.

Ichii,T. & Naganobu, M. (1996) Surface water circulation in krill fishing areas near the South Shetland Islands. *CCAMLR Sci* **3**, 125–36.

Ichii, T., Naganobu, M. & Ogishima, T. (1994a) An assessment of the impact of the krill fishery on penguins in the South Shetland Islands. *CCAMLR Sci* **1**, 107–113.

Ichii, T., Naganobu, M. & Ogishima, T. (1994b) A revised assessment of the impact of krill fishery on penguins in the south Shetland Islands. Document WG-Joint-94/17, 20 pp. CCAMLR, Hobart.

Kock, K.-H. (1992) *Antarctic Fish and Fisheries.* Cambridge University Press, Cambridge.

Kock, K.-H. , Wilhelms, S., Everson, I. & Gröger, J. (1994) Variations in diet composition and feeding intensity of mackerel icefish *Champsocephalus gunnari* at South Georgia (Antarctic). *Mar. Ecol. Prog. Ser.* **108**, 43–57.

Laws, R.M. (1984). Seals. In *Antarctic Ecology* (R. Laws, ed.), pp. 531–616. Cambridge University Press, Cambridge.

Love, R.M. (1988) *The Food Fishes their intrinsic variation and practical implications.* Van Nostrand Reinhold, New York.

Madureira, L. St.P. (1992) Application of a dual frequency echosounder to the identification and quantification of Antarctic krill *Euphausia superba* distribution around South Georgia in relation to its principal predators. PhD dissertation, University of Cambridge, July 1992. 167 pp.

Mangel, M. & Switzer, P.V. (1998) A model at the level of the foraging trip for the indirect effects of krill (*Euphausia superba*) fisheries on krill predators. *Ecol. Model.* **105**, 235–56.

Murphy, E.J. (1995) Spatial structure of the Southern Ocean ecosystem: predator-prey linkages in Southern Ocean food webs. *J. Anim. Ecol.* **64**, 333–47.

Murphy, E.J., Trathan, P.N., Everson, I., Parkes, G. & Daunt, F. (1997). Krill fishing in the Scotia Sea in relation to bathymetry, including the detailed distribution around South Georgia. *CCAMLR Sci.* **4**, 1–18.

Murphy, E.J., Watkins, J.L., Reid, K. *et al.* (1998) Interannual variability of the South Georgia marine ecosystem: biological and physical sources of variation in the abundance of krill. *Fish. Oceanogr.* **7**, 381–90.

Reid, K., Watkins, J.L., Croxall, J.P. & Murphy, E.J. (1999) Krill population dynamics at South Georgia 1991–1997, based on data from predators and nets. *Mar. Ecol. Prog. Ser.* **177**, 103–114.

SC-CAMLR (1992) *Report of the Eleventh meeting of the Scientific Committee for the Con-*

servation of Antarctic Marine Living Resources, pp. 426–27. SC-CAMLR, Hobart, Australia.

SC-CAMLR (1995) *Report of the Fifteenth meeting of the Scientific Committee for the Conservation of Antarctic Marine Living Resources*, SC-CAMLR, Hobart, Australia.

SC-CAMLR (1997) *Report of the Sixteenth meeting of the Scientific Committee for the Conservation of Antarctic Marine Living Resources*. SC-CAMLR, Hobart, Australia.

Simard, Y. & Mackas, D.L. (1989) Mesoscale aggregations of Euphausiid sound scattering layers on the continental shelf of Vancouver Island. *Can. J. Fish. Aquat. Sci.* **46**, 1238–49.

Trathan, P.N., Everson, I., Murphy, E.J. & Parkes, G. (1998) Analysis of haul data from the South Georgia krill fishery. *CCAMLR Sci.* **5**, 9–30.

Trathan, P.N., Murphy, E.J., Croxall, J.P. & Everson, I. (1999) Use of at-sea distribution data to derive potential foraging ranges of macaroni penguins during the breeding season. *Mar. Ecol. Prog. Ser.* **169**, 263–75.

Chapter 9
Krill Harvesting

Taro Ichii

9.1 Introduction

A krill fishery was first conducted for *Meganyctiphanes norvegica* in the Mediterranean in the 19th century where the catch was used as bait (Fisher *et al.*, 1953). In recent years, large-scale krill fishing has targeted *Euphausia superba* in the Antarctic and *E. pacifica* in coastal waters off northeastern Japan, with smaller scale harvesting of *E. pacifica* off western Canada (Eddie, 1977; Everson, 1978; Nicol & Endo, 1997). Minor, intermittent fisheries for *Thysanoessa inermis* and for *E. nana* off Japan, were reviewed by Nicol & Endo (1997). This chapter presents information on krill harvesting in the former three fisheries, including the history of each fishery, fishing strategy, fleet organisation, characteristics of fishing grounds, detection and catching of krill swarms and the primary concerns of fishermen. Finally, comparisons of the three krill fisheries highlight the characteristics of each.

9.2 Antarctic waters (*Euphausia superba*)

History of the fishery

Ever since a high abundance of *E. superba* became apparent (Marr, 1962; Miller & Agnew, Chapter 12) speculation existed that the species might be harvested commercially. During the 1960s, it was estimated that a total catch of upwards of 150 million tonnes of krill might be taken. This amount, often referred to as a 'krill surplus' created by the great reduction in baleen whale stocks (Laws, 1977) caused heightened interest in an Antarctic krill fishery. However, the principle reason for the initiation of harvesting was probably the displacement of fishing fleets from Japan, the former USSR, and other Eastern Bloc countries from 200 n. miles Exclusive Economic Zones, as a result of developments in the Law of the Sea negotiations in the 1960s and early 1970s (Kock, 1992). In Japan, krill from the stomachs of baleen whales (Plate 4, facing p. 182) caught by whaling were initially used as food for aquaculture, a practice which also promoted the development of the Antarctic krill fishery (Kaneda, personal communication).

The commercial krill fishery started in the early 1970s (Fig. 9.1). Pelagic fishing

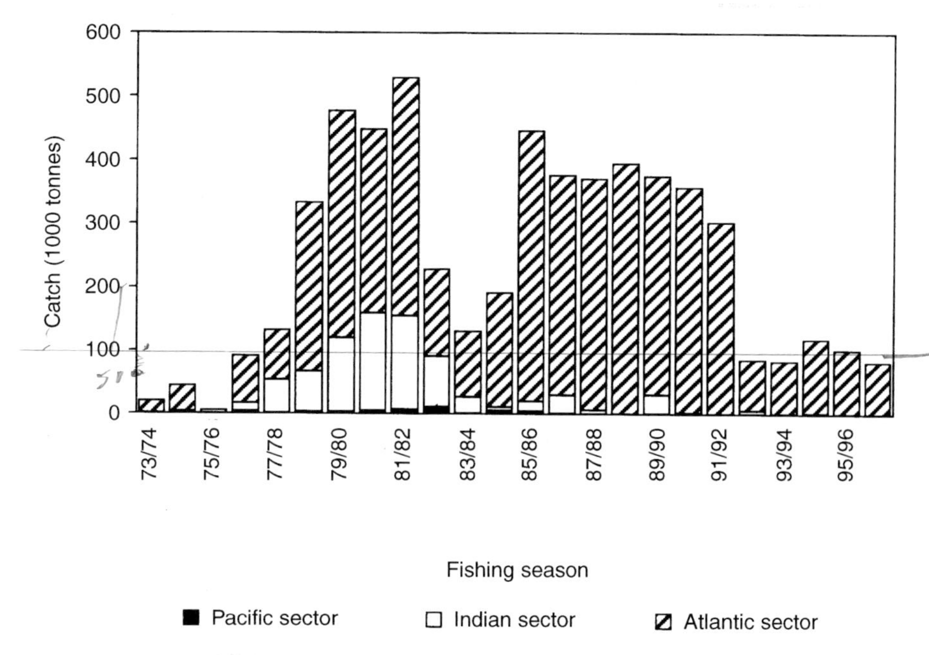

Fig. 9.1 Annual catches of Antarctic krill in the Atlantic (Statistical Area 88), Indian (Area 58) and Pacific (Area 88) sectors of the Southern Ocean (CCAMLR, 1990a; 1990b; 1997; SC-CAMLR, 1997). Statistical Areas shown in Fig. 9.2.

activities from former Eastern Bloc countries were heavily subsidised which facilitated fishing operations in waters where vessels from countries with market economies, and operating without a significant subsidy, had great difficulties fishing economically (Kock, 1994). The fishery reached a peak in the early 1980s when over half a million tonnes were caught. However, catches dropped substantially thereafter due to problems in krill processing and increased interest in finfish (Kock, 1992). From 1986–87 to 1990–91, annual catches became stabilised at around 350 000 to 400 000 t, the fishery accounting for about 13% of the world catch of crustaceans at that time (Nicol, 1991). Catches declined again in the 1991–92 season and even more dramatically to approximately 87 000 t in the 1992–93 season, when economic considerations caused the states of the former USSR to discontinue krill fishing. Thus, most of the catch until 1991–92 was taken by the USSR, the dominant nation thereafter being Japan (Table 9.1). Other nations, South Korea, Poland, Chile, Bulgaria, Latvia, Panama, Spain, East Germany and UK, have also been involved in the krill fishery, but have always been 'minor players' compared to the former USSR and Japan.

The krill fishery in the Southern Ocean is regulated under the auspices of the Commission for the Conservation of Antarctic Marine Living Resources (CCAMLR), which came into force in 1982 (Fig. 9.2) The annual fishing period for the Antarctic is referred to as split-year, which extends from 1 July to 30 June. However, in the South Georgia area (Subarea 48.3), as the fishing season falls in

Table 9.1 Reported catches of Antarctic krill by different nations fishing in the CCAMLR area 1973–1998.

Year	Bulgaria	Chile	France	GDR	Japan	Korea	Latvia	Panama	Poland	Russia	S. Africa	Spain	Ukraine	USSR	UK
1973/74	–	–	–	–	646	–	–	–	–	–	–	–	–	21 700	–
1974/75	–	–	–	–	2676	–	–	–	–	–	–	–	–	38 900	–
1975/76	–	276	–	–	4750	–	–	–	21	–	–	–	–	500	–
1976/77	–	92	–	–	12 802	–	–	–	6966	–	–	–	–	105 049	–
1977/78	94	–	–	8	25 219	–	–	–	37	–	–	–	–	116 601	–
1978/79	46	–	–	102	36 961	511	–	–	–	–	–	–	–	295 508	–
1979/80	–	–	6	–	36 276	–	–	–	226	–	–	–	–	440 516	–
1980/81	–	–	–	–	27 698	–	–	–	–	–	–	–	–	420 434	–
1981/82	–	–	–	–	35 755	1429	–	–	–	–	–	–	–	491 656	–
1982/83	–	3752	–	–	42 326	1959	–	–	360	–	–	–	–	180 290	–
1983/84	–	1649	–	–	49 572	2657	–	–	–	–	–	–	–	74 381	–
1984/85	–	2598	–	–	38 274	–	–	–	–	–	–	–	–	150 538	–
1985/86	–	3264	–	–	61 074	–	–	–	2065	–	–	–	–	379 270	–
1986/87	–	4063	–	–	78 389	1527	–	–	1726	–	–	450	–	290 401	–
1987/88	–	5938	–	–	73 112	1525	–	–	5215	–	–	–	–	284 873	–
1988/89	–	5394	–	–	78 928	1779	–	–	7871	–	–	–	–	301 498	–
1989/90	–	4500	–	396	62 187	4039	–	–	1275	–	–	–	–	302 376	–
1990/91	–	3679	–	–	67 582	1210	–	–	9571	–	–	–	–	275 495	–
1991/92	–	6065	–	–	74 325	519	–	–	8607	151725	–	–	61 719	–	–
1992/93	–	3261	–	–	59 272	–	–	–	15 909	4249	–	–	6083	–	–
1993/94	–	3834	–	–	62 322	–	71	–	7915	965	2	–	8852	–	–
1994/95	–	–	–	–	60 303	–	–	141	9384	–	–	–	48 884	–	–
1995/96	–	–	–	–	60 546	–	–	495	20 610	–	–	–	20 056	–	–
1996/97	–	–	–	–	58 798	–	–	–	19 156	–	–	–	4246	–	308
1997/98	–	–	–	–	63 233	1623	–	–	15 312	–	–	–	–	–	634

Data from CCAMLR (1990a; 1990b; 1997) and SC-CAMLR (1997)

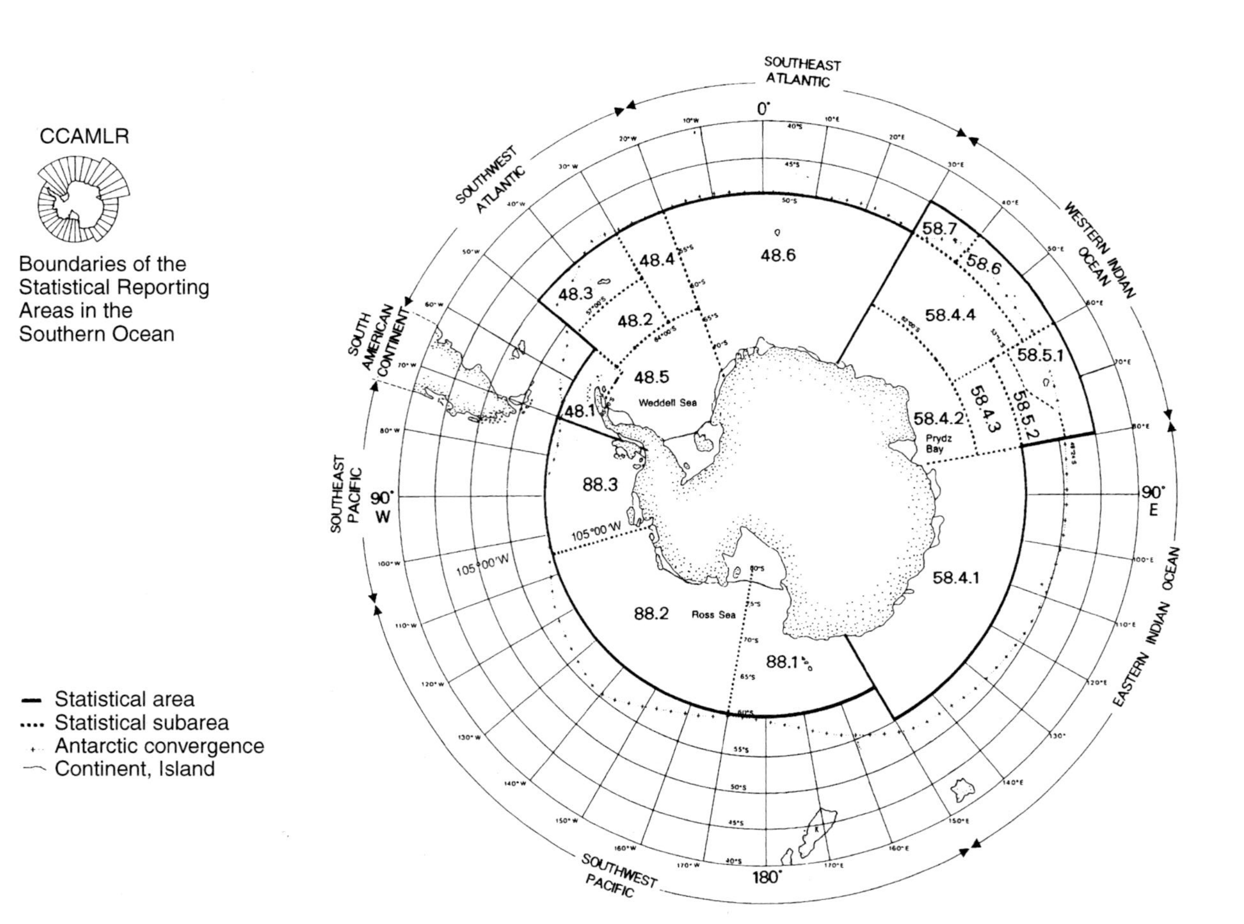

Fig. 9.2 CCAMLR Statistical Areas, Subareas, and Divisions. With permission from the CCAMLR.

June and July, thus encompassing the end of one split-year and the beginning of the next, the annual period is expressed as calendar years.

Fishing strategy

A number of difficulties have been associated with the *E. superba* fishery. Besides the limited market for krill products, the greatest problem is distance; the fishery being centred far from major ports and being undertaken in very inhospitable seas. The geographical remoteness is reflected in the areas that have been fished for krill. Although krill have a circumpolar distribution, as shown in Figs 9.3 and Table 3.3.2, the fishery has tended to concentrate in areas close to other fisheries off South America, South Africa and New Zealand. Thus, when krill fishing is less profitable, the vessels engage in other fishing activities, such as for squid and finfish off South America for the reminder of the year.

Krill quality is also a major concern. Important from the commercial view point, *E. superba* products are graded by 'greenness', body size and body colour. 'Green' krill are those which have been feeding intensively on phytoplankton, that accumulates in the frontal part, specifically in the hepatopancreas inside the carapace. This is unimportant if peeled krill meat or meal is being produced, but 'green' krill are actively avoided if fresh-frozen or boiled-frozen products are required, owing to the dirty appearance, unfavourable smell and inferior taste of products from the former. 'Green' krill tend to occur in the early austral summer (December–January) (Kawaguchi *et al.*, 1998).

In the Japanese fishery the krill are graded by the fishermen according to their size: LL (larger than 45 mm), L (between 35 and 45 mm), and M (below 35 mm). Generally, the LL size category, especially if it contains gravid females, is preferred for both human consumption and sport fishing. It is also easier to peel larger krill. However, because the smaller classes (L and M sizes) are in some demand for sport fishing and aquaculture, it is advantageous to operate in fishing grounds where krill show size segregation, allowing size-specific targeting to be employed (Kigami, personal communication).

Krill are also graded according to body colour. Transparent krill, termed 'white' are firm with a pleasing appearance, whereas brick red/pink krill, termed 'pink', are flaccid and easy to crush. The former, more valuable than the latter, are more frequently present from late in the austral summer (Ichii, 1987). Because of the substantial variability in quality of *E. superba,* the prime fishing strategy is to harvest the most marketable krill, rather than to concentrate solely on maximising the catch (Kigami & Kaneda, personal communication).

Fishing fleet organisation

Principally two types of operational pattern occur, which for convenience can be termed individual ship operation and closely cooperating fleet operation. The

Japanese fishery has essentially been structured around the first type with stern trawlers mostly of around 3000 to 4000 t GRT, although communications are maintained between vessels belonging to the same company (Ichii, 1987). In contrast, the former USSR vessels, which were mostly 3800 t stern trawlers, used to work in close cooperation, their fishing fleets usually operating in concert with scouting vessels (Everson, 1978, 1988). The latter continually searched for new krill concentrations and surveyed their properties (location, horizontal extension, catch rate, krill size, 'greenness') so as to inform the catching vessels (Dolzhenkov *et al.*, 1988)

Detailed information is available for the 'mother-ship' operation conducted by Japan in the early years as a subsidised venture from 1977–78 to 1981–82 (Shimadzu, 1984). It consisted of a mothership (8000 t class) with freezing and processing capabilities, and 7 to 10 stern trawlers (349 t class) with boiling and freezing facilities. One or two trawlers operated as 'scouts' outside of the fleet's fishing ground. Even though the trawlers were able to process krill, they frequently transported catches to the mothership to make full use of the latter's more effective processing capabilities. A small transportation boat, associated with the mothership, received the cod-end from the trawler and then transferred it to the mothership, a transporting procedure conducted so smoothly that little difference in quality occurred between krill products made by the mothership and the trawlers. Economically, however, the mothership operation did not compare favourably with individual trawler performances since the fleet typically caught 200–300 t/day in total compared to the 50–70 t average for a 3000 t class vessel in those days, and accordingly was discontinued (Shimadzu, 1984).

The remote operation of this fishery is supported by cargo vessels, which frequently, twice a month during high-season, visit the trawlers to replenish food and fuel supplies, and pick up krill products. When sea conditions are rough, cargo transfer is carried out close to the ice edge or in the lee of islands so as to take advantage of the calmer conditions there. These days, many of the trawlers are based in foreign ports such as in Chile, Argentina and South Africa, so that both the krill and other fisheries have become a much less complicated logistic operation.

Characteristics of the fishing grounds

Antarctic krill characteristically aggregate into concentrations, comprising many swarms, and these are targeted by fishing trawlers (Butterworth & Miller, 1987; Ichii, 1987). Since the concentrations are formed in limited areas, vessels make use of a range of information to locate harvestable concentrations. Most important is the historical distribution of good harvesting areas (Ichii, 1987). During searching activity, surface water temperatures are monitored in order to detect oceanographic fronts, which tend to hold krill aggregations for prolonged periods. Icebergs with many resting seabirds or covered by bird droppings can also be indications of the existence of nearby krill aggregations. Once such 'indicators' have been found, vessels intensify their searching activities by increasing the frequency of directional changes so as to locate the high density part of the krill concentration. Communication with other fishing vessels within the same fishing company enhances the likelihood of finding high krill concentrations. The fishing vessels avoid operations near the ice edge, because of the problems encountered when trawling in icy waters. Historical fishing grounds thus exploited are divided geographically into three regions: off Kemp Land, off Wilkes Land and the south-west Atlantic (Fig. 9.3). The krill fisheries in these three regions have been combined with demersal fishing off Africa, squid fishing off New Zealand and fishing off South America for squid and finfish. The greatest krill catches have been taken between 50°E–70°E in Kemp Land, 110°E–150°E in Wilkes Land, and 30°W–65°W, South Shetland Islands, South Orkney Islands and South Georgia, in the south-west Atlantic. Some effort has been extended over wider ranges than these limits indicate (Fig. 9.3). Good fishing grounds are closely associated with ice-free continental and insular shelf break–slope regions (Ichii, 1990; Everson & Goss, 1991). In the most recent years, almost all vessels have operated in the south-west Atlantic.

Krill fishing off Kemp Land and Wilkes Land used to be conducted in the austral summer, with fishing activity shifting from the offshore to the continental shelf break as the ice edge retreated southwards (Fig. 9.4). In austral mid-summer, krill trawling was concentrated along the continental shelf break. In the south-west Atlantic, on the other hand, the krill fishery has recently operated from the austral summer through to the winter (Fig. 9.5). The fishery begins in the South Shetland Islands area in mid austral summer, moving north as the sea-ice advances through the South Orkney Islands area to the northernmost fishing area around South Georgia by austral mid-winter (Everson & Goss, 1991). The most popular fishing grounds are formed near the shelf break of the respective islands.

The daily catch rate has tended to be highest, around 50–150 t/vessel, in the south-west Atlantic sector, an area that includes the South Shetland Islands, South Orkney Islands and South Georgia. Lower catch rates, of around 30–90 t/vessel, were made off Wilkes Land (Fig. 9.6), this suggests that krill are more abundant and the fishing grounds more stable in the Atlantic sector than off Wilkes Land. The fishing grounds are most likely influenced by complex interactions between krill behaviour and environmental factors. Regarding the latter, the continental slope front is considered to be responsible for the formation of fishing grounds off Wilkes Land (Ichii, 1990). In the South Shetland Islands area, which has been intensively studied, the coupling of oceanographic features with favourable feeding conditions is important; the shear zones and sluggish currents caused by the insular shelf may provide stable sites that retain krill. Since these sites also have abundant food sources, krill may actively move to them, resulting in dense quasi-stable concentrations (Ichii *et al.*, 1998).

Krill in the South Shetland Islands area show offshore–onshore segregation in both size and maturity stages (e.g. Siegel, 1988; Siegel *et al.*, 1997; Ichii *et al.*, 1998). Hence fishing vessels can select larger krill by operating in the offshore and slope

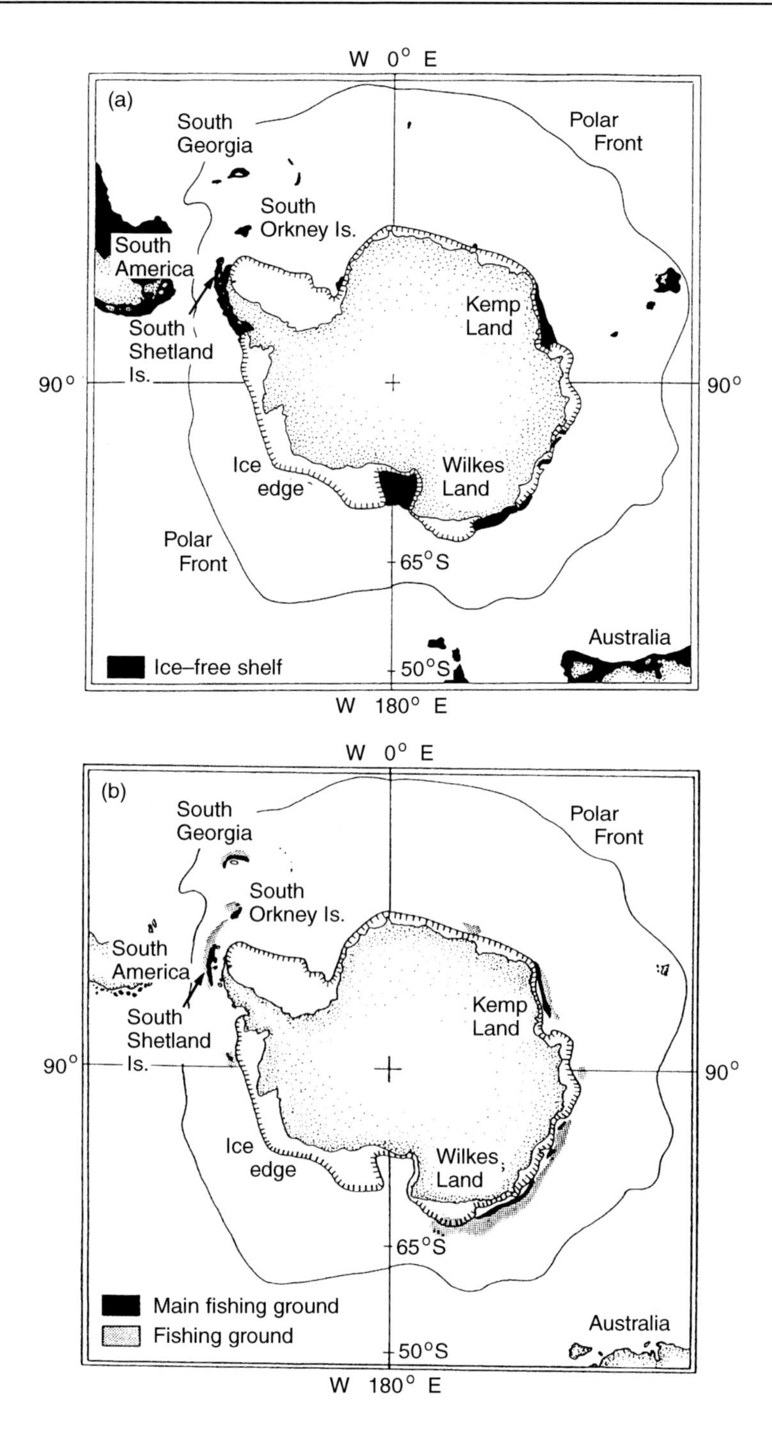

Fig. 9.3 Distribution of Antarctic krill fishing grounds in relation to continental and insular shelves and pack-ice edge during austral summer in the Antarctic. (a) Mean pack-ice edge in February, ice-free continental shelf and Polar Front. (b) Fishing ground distribution.

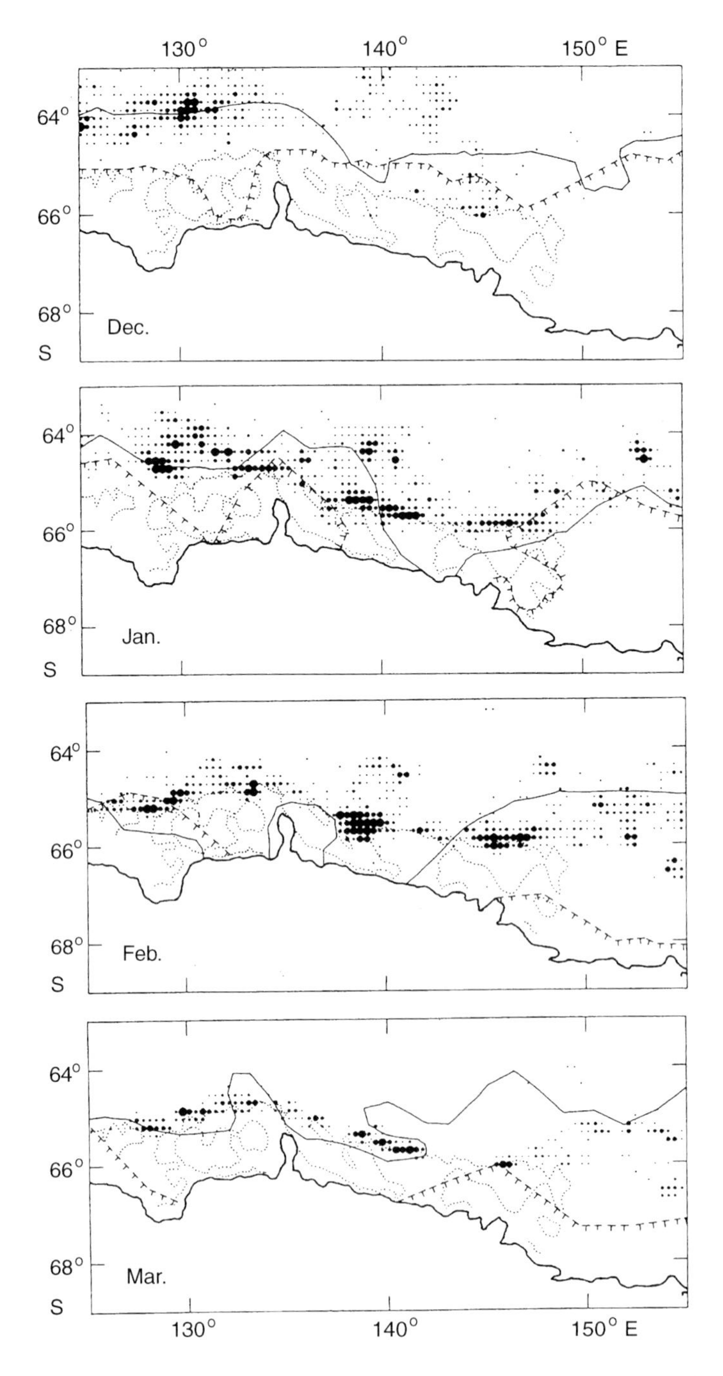

Fig. 9.4 Distribution of accumulated catches of Antarctic krill (tonnes/10–10 n. miles) from 1976/77 to 1993/94 for each month off Wilkes Land. Maximum (—) and minimum (⊤) – pack-ice edge in each month during 1976/77–1993/94 are shown. Dotted lines indicate 500 m depth contours. Catch data from Japanese fishery. Size of closed circles proportional to catch size, e.g. closed circles of maximum diameter correspond to >500 tonnes.

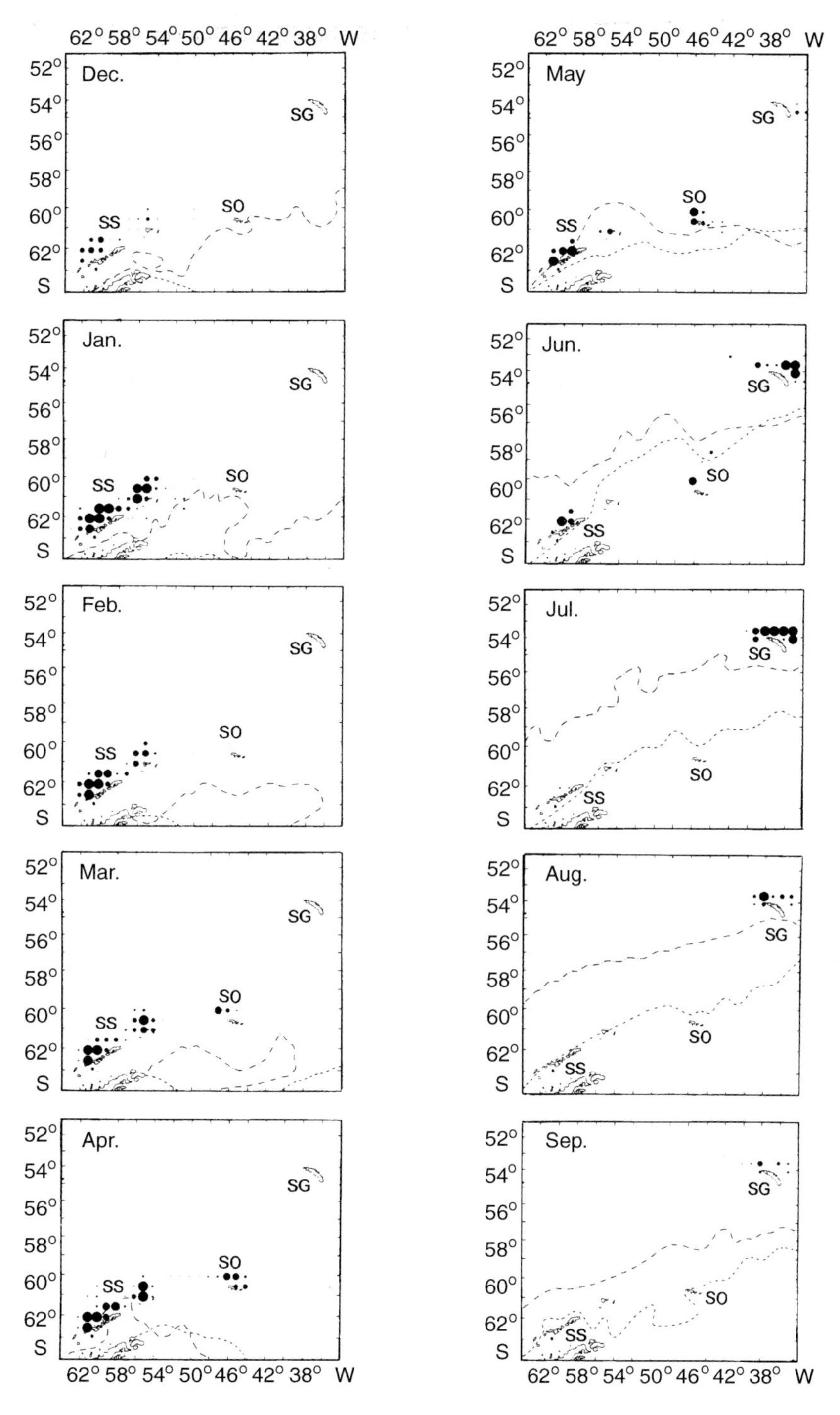

Fig. 9.5 Distribution of recent (1989/90–1996/97) accumulated catches of Antarctic krill (tonnes/30–30 n. miles) for each month in the Southwest Atlantic. Maximum (–·–·–) and minimum (––––) – pack-ice edge in each month during 1989/90–1996/97 are shown. Catch data from Japanese fishery. Size of closed circles proportional to catch size, e.g. closed circles of maximum diameter correspond to > 1000 tonnes. SS = South Shetland Islands, SO = South Orkney Islands, SG = South Georgia.

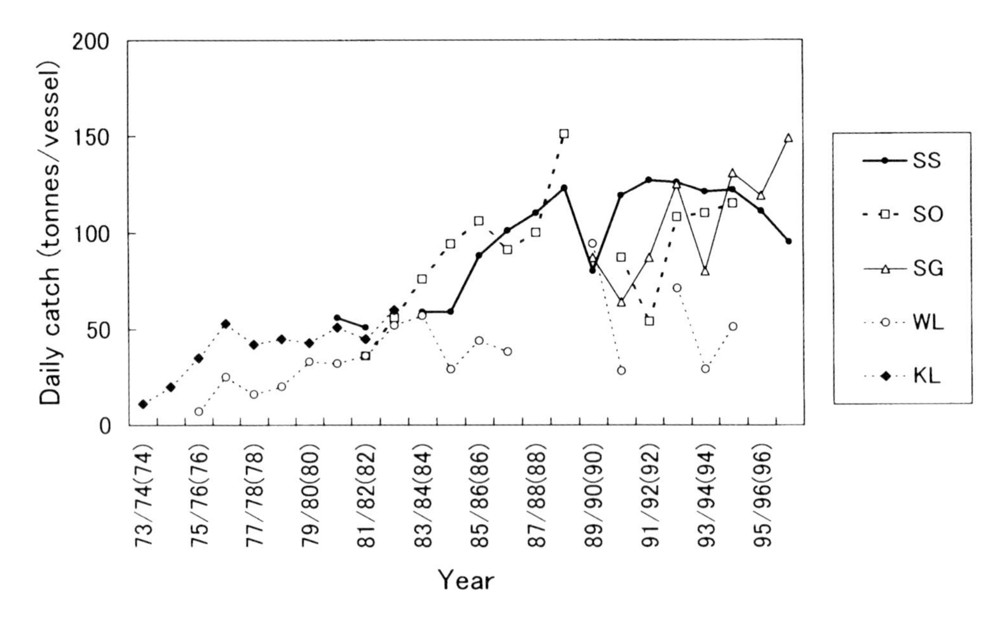

Fig. 9.6 Interannual variations in average daily catch (tonnes/vessel) in each fishing area for Antarctic krill. Data from Japanese fishery. SS: South Shetland Islands; SO: South Orkney Islands; SG: South Georgia; WL: Wilkes Land; KL: Kemp Land. Calendar years in parentheses are for the South Georgia area.

regions. Such selection is difficult in other areas since segregation is not apparent. The size distribution of the krill caught during the austral winter in South Georgia is distinctly different to that from other fishing grounds in that larger krill (>50 mm) are usually absent. This may be because krill may shrink due to food shortages during winter as Ikeda & Dixon (1982) suggested although it is more likely due to the advection of small krill into the region.

Advantages in operating off Wilkes Land lay in the quality of krill caught, which tended to be better than in the south-west Atlantic during the austral summer such that there were fewer 'green' krill, leading to a greater proportion of white firm products rather than red–pink and soft ones (Ichii, 1987). In fact processing of the fresh-frozen krill needed to be completed within three hours of capture, compared with two hours in the Southwest Atlantic. These advantages did not compensate however for the lower catch rates already noted above. By-catches of salps may also be a problem in the Southwest Atlantic (Ichii, 1987).

Detection and catching of krill swarms

On the fishing grounds, the detection by echo-sounder and sonar leading to capture of individual aggregations are the basic activities of the harvesting operation. Echo-sounders can detect swarms at most depths immediately below the vessels, but have very limited horizontal range, no more than vessel width. Sonars, on the other hand, have an effective horizontal detection range of as much as 300 to 500 m when swarms are dense and well below the surface, but only a limited vertical range that is constrained by the transducer beam width, which could well miss swarms even directly below the vessel.

Echo-sounders are the most important swarm detection device although in early summer, when swarms generally are small, sonar is indispensable. In areas where the number of swarms per unit area is so high that detection by sonar is unnecessary, the main purpose of the latter is for determining the direction of the swarm extension, so as to judge the optimal direction for towing (Ichii, 1987). In areas where krill form dispersed aggregations greater than 3–4 km in chord length, their densities are usually too low for sonar detection, location by echo-sounder being the only method. Surface swarms, which occur regularly only in the half-light period after sunset off Kemp and Wilkes Land during the early austral summer, can be detected by the naked eye, but usually only by experienced observers (Ichii, 1987).

Fishing vessels usually concentrate on the larger and denser swarms for harvesting (Ichii, 1987). When a fishable swarm is detected, the vessel is manoeuvred on to a course that will result in the swarm being intercepted by the trawl. When the trawl is shot, final adjustments to the course are made by reference to sonar. Adjustments to the trawl depth are made by tracking the swarm depth as seen on the echo-sounder and net sounder. The latter, which relies on vertically scanning transducers attached to the headline of the trawl net, has two principal functions: to indicate net depth in relation to the surface and the quantity of krill that have actually entered the net. For krill harvesting, a midwater trawl net with a mouth opening of 500 to 700 m^2 is used (Plate 5, facing p. 182). Since krill appear to take no effective avoidance action to commercial nets, towing speeds are generally around two knots. If wind speeds are more than 2–3 on the Beaufort scale (about 5–10 knots), the vessels trawl with the wind from behind.

Haul duration is generally kept to a time which will ensure a maximum catch of between 7 and 10 t. This is for two reasons, firstly because with larger hauls, since the krill are thin and easily crushed, product quality suffers, and secondly because operations need to be linked to the vessel's processing capabilities. To maintain such a catch level, fishermen use the net sounder to determine the optimal number of swarms fished or else adjust the towing time (Fig. 9.7). When larger cohesive aggregations of several hundred metres in horizontal length are trawled, a single swarm is usually sufficient for the optimal catch. Even so, when widely dispersed low-density aggregations extending over many kilometres are fished, it may take at least an hour to reach the desired catch level.

Towing is conducted continuously throughout the day and night, enabling processing to be conducted in an unbroken sequence. Analyses of catch per unit of effort (CPUE), where effort is determined by trawling time, and trawling depth show different diurnal patterns between fishing grounds in the south-west Atlantic (Fig. 9.8). In the South Shetland Islands area during the austral summer (January–April), only slight diurnal changes in CPUE occurred, being larger in the daytime

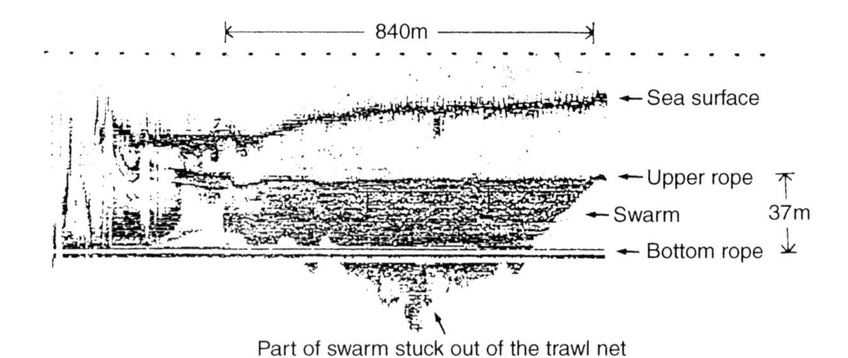

Fig. 9.7 Net recorder chart when a dense swarm was fished. Height of net was 37 m. 20 tonnes (too much!) of Antarctic krill was caught in 14 min. of towing. Absolute density was estimated as 40 g/ m^3. The fishing was conducted at 66°03′S, 145°53′E on 2 March, 1986.

(160–300 kg/min) than at night (100–240 kg/min). There was also a small and slight diurnal change in trawl depth with fishing being deeper in the daytime (60–80 m depth) than at night (40–50 m). At South Georgia during the austral winter months of June through to August, on the other hand, a marked distinct diurnal change in CPUE was apparent with much higher values in the daytime (220–700 kg/min) than at night (80–380 kg/min). Trawling depths were distinctly deeper than those in the South Shetland Islands area and showed substantial inter-annual variations, especially in the daytime where they varied from 50 to 250 m. It should be noted that daytime trawling depths were significantly correlated with krill size (Fig. 9.9) because the larger krill tend to descend deeper in daytime (also see Kawaguchi *et al.*, 1996). In the South Orkney Islands during the austral autumn months of April to June, both CPUE and trawling depth show an intermediate pattern between those in the South Shetland Islands and South Georgia. These differences in CPUE and trawling depth among the three fishing areas may reflect both regional and seasonal changes in distributional behaviour of krill. Thus, because of the generally higher CPUE, winter krill fishing operations are not disadvantageous by comparison with those in the austral summer, despite the rougher sea conditions encountered during winter.

Primary concerns of the fishermen

On the south-west Atlantic fishing grounds fishermen are most concerned with the quality of krill, since catching large quantities is still fairly straightforward. Lower quality krill 'green' and 'red-coloured' can be avoided by operating during the late austral summer, autumn and winter months. Recently, for example, Japan has extended the fishing season into the austral winter season, so as to avoid catching early season 'green' krill and thereby increase the catch of colourless (white) krill, which become available later in the season (SC-CAMLR, 1992; Kawaguchi *et al.*,

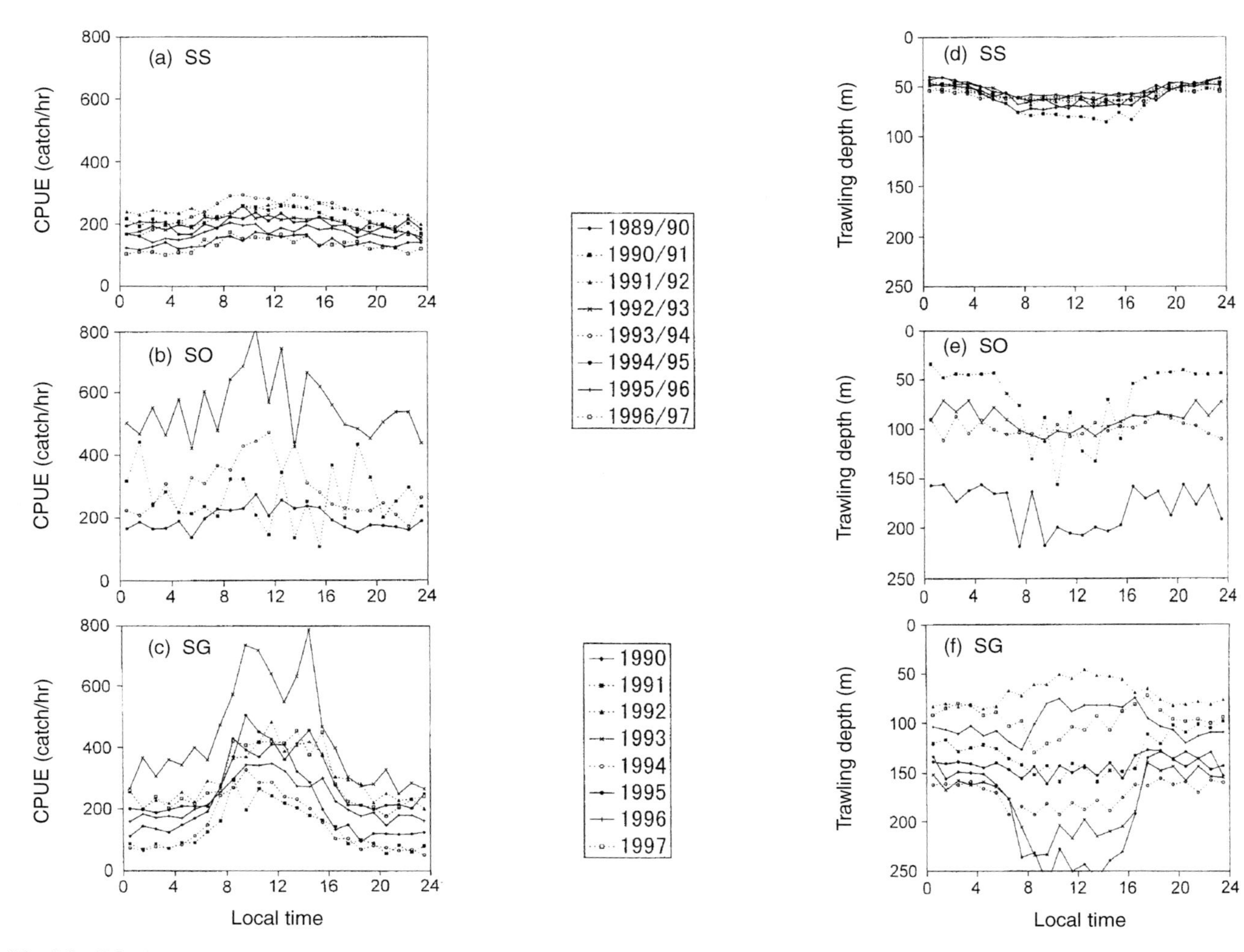

Fig. 9.8 Diurnal variations in average catch per towing time (a)–(c) and trawling depth (d)–(f) from 1989/90 to 1996/97 in each fishing area for Antarctic krill in the Southwest Atlantic sector. SS: South Shetland Islands (January to April); SO: South Orkney Islands (April to June); SG: South Georgia (June to August). Data from Japanese fishery.

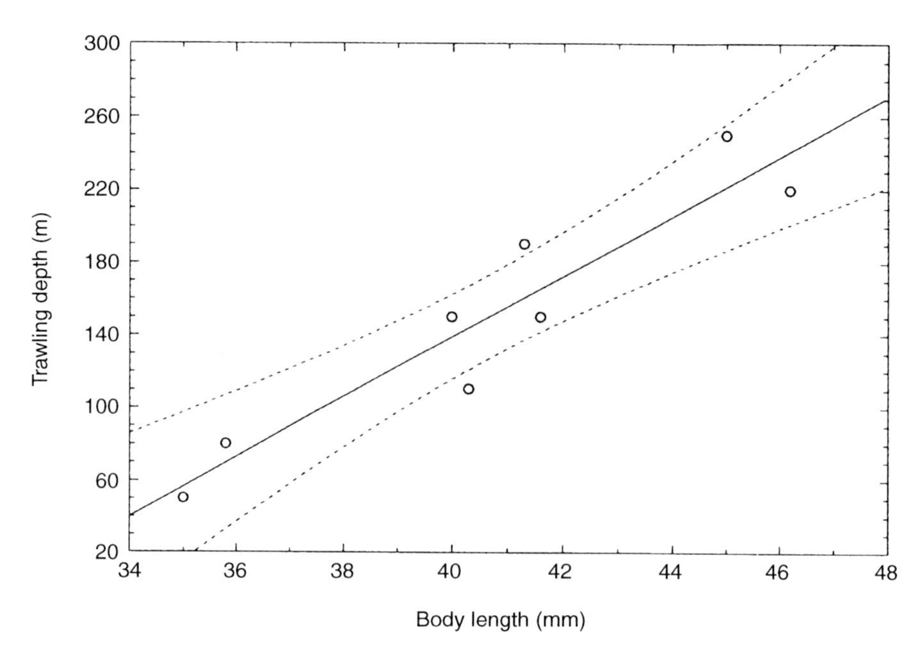

Fig. 9.9 Relationship between average trawling depth at noon and average total body length of Antarctic krill caught during the austral winter in the South Georgia area. Linear regression (solid line) with 95% confidence limits (broken lines) are shown. $y = -521.4 + 16.5x$, where y is trawling depth (m) and x is body length (mm). ($r = 0.94$, $p < 0.0005$, $n = 8$.)

1998; Plate 2, facing p. 182). However, 'body size', varying between years rather than following a consistent seasonal pattern, is often a major problem. This is caused by recruitment of *E. superba* showing great inter-annual variability (Siegel & Loeb, 1995), with subsequent growth to the largest size class (>50 mm) requiring a period of 4–5 years (Siegel, 1987). Figure 9.10 shows substantial inter-annual variability in the size composition of krill in catches for each fishing area, the patterns appearing to be synchronous to some extent across the South Shetland Islands, South Orkney Islands and South Georgia areas. This suggests a similarity in year-class strength of krill across the Scotia Sea. Hence fishing companies face the problem of an overall shortage of larger krill when krill are predominantly small-sized throughout the south-west Atlantic.

There is some suggestion of long-term declines in krill abundance and recruitment in the South Shetland Islands area due to regional warming (Loeb *et al.*, 1997). CPUE of the Japanese fishery expressed in terms of catch per towing volume also shows a long-term decreasing trend in this area (Kawaguchi *et al.*, 1997). A possible reason for the decreasing CPUE, is the use of different fishing strategies in response to a demand for a higher product quality as well as a general decrease in krill density in the area. Furthermore, in 1997–98, gravid females were caught in May (an unusually late season for spawning) in the South Shetland Islands area (Kigami, personal communication). The prevailing feeling among fishers, therefore, is that

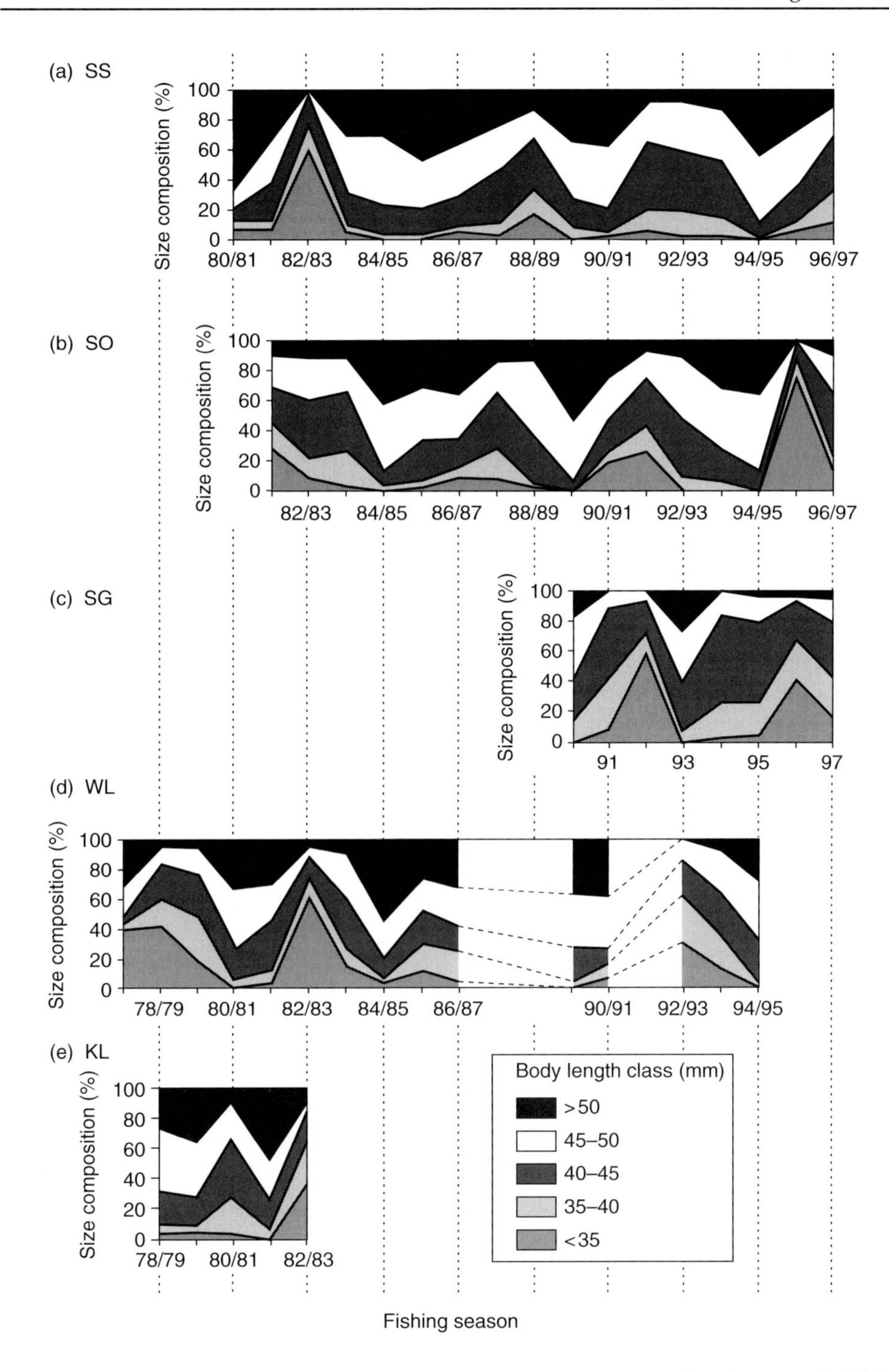

Fig. 9.10 Interannual variations in body length composition of Antarctic krill for each fishing area. SS: South Shetland Islands; SO: South Orkney Islands; SG: South Georgia; WL: Wilkes Land; KL: Kemp Land.

regional warming may well be affecting krill abundance and their biological characteristics in the south-west Atlantic.

9.3 Japanese northeastern coastal waters (*Euphausia pacifica*)

History of the fishery

About 100 years ago, dip net fishing, which later developed into the use of a bow-mounted trawl, operated in Sanriku (Iwate and Miyagi Prefectures) coastal waters for sand lance (*Ammodytes personatus*). The same method was applied to *E. pacifica* in the mid-1940s by fishermen off Oshika Peninsula, Miyagi Prefecture (Odate, 1991; Fig. 9.11). Increasing bait requirements for sport fishing in the late 1960s caused the *E. pacifica* fishery to expand along the northern and southern coasts of Miyagi Prefecture. A fishery was subsequently established off Ibaraki Prefecture in 1972, although large amounts of krill were not landed until 1977 (Fig. 9.12), when a coastal branch of the Oyashio Current extended far to the south. Shortly thereafter in 1974, Fukushima Prefecture began a krill fishery to be followed by the Iwate Prefecture in 1975 (Odate, 1991). In the 1980s, increasing food demands for sea bream culture, and the introduction of high-speed fishing boats and effective fishing gear, such as fish pumps, echo-sounders, sonar and larger net haulers, resulted in an increased catch rate (Kuroda, 1994). With recent poor catches in other fisheries, such as that for sand lance, due to overfishing, the krill fishery has developed into an important coastal water fishery, with an average annual catch of around 60 000 t over the last 14 years (Fig. 9.12).

Fishing strategy

Fishermen engaged in coastal fisheries attempt to maintain economic stability by participating in diverse fishery operations throughout the year such as boat seines for sand lance and juvenile sardine, squid jigging, gill nets for salmon and aquaculture for sea mustard. Coastal krill fishing was initially conducted during the off-season for the other species (Kuroda & Kotani, 1994), the importance of the fishery being much greater in the Sanriku coastal area than the Joban (Fukushima and Ibaraki Prefectures) coastal area, since catches were more stable in the former locality.

There has been little market competition between Antarctic krill (*E. superba*) and *E. pacifica* products, because they are sold for different purposes. For example, in the case of sport fishing, *E. pacifica* is smaller and cheaper and used as chumbait, whereas *E. superba* are mostly used as individual lures (Kigami, personal communication). Compared with *E. superba*, *E. pacifica* show less variability in body size and feeding condition, resulting in the product of this species being graded only by

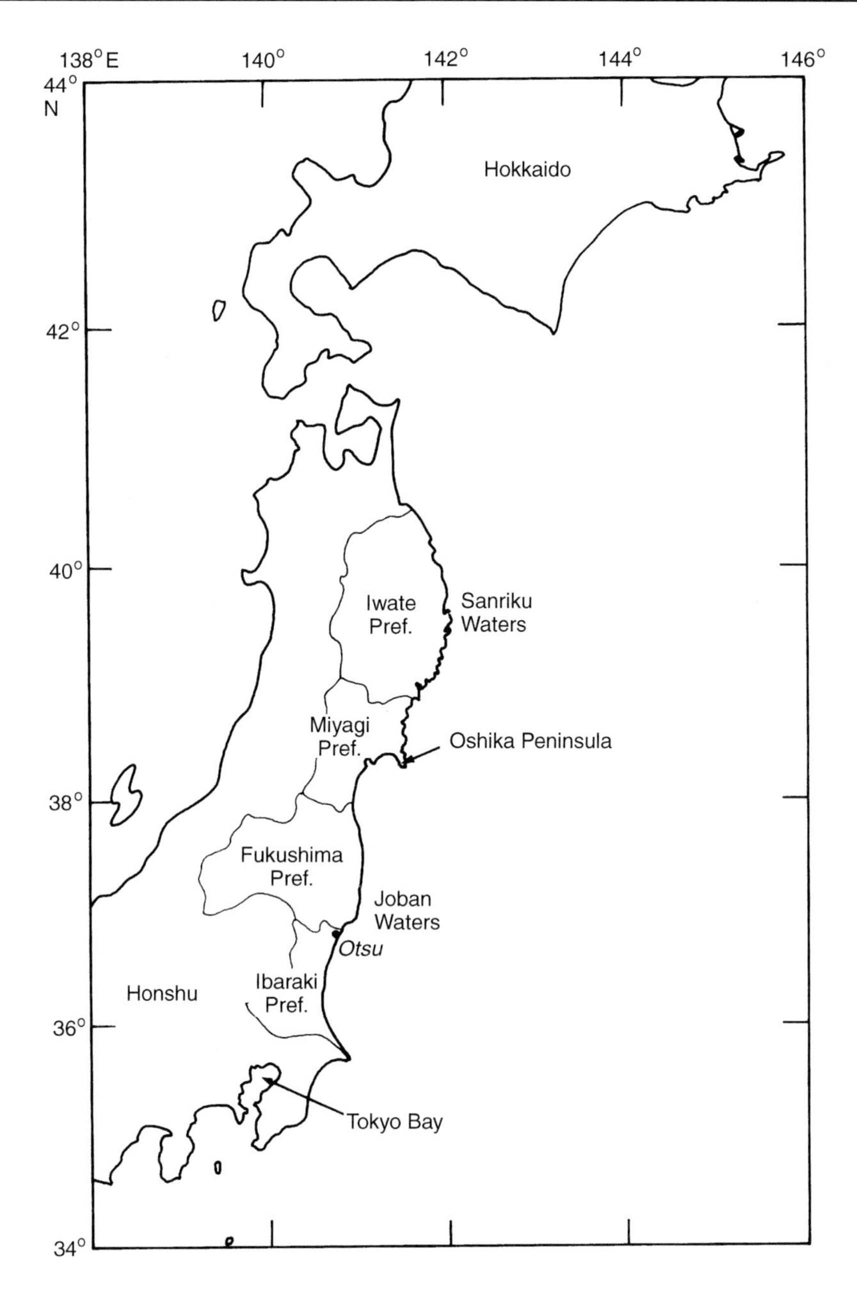

Fig. 9.11 Map of northeastern Japan showing *Euphausia pacifica* fishing prefectures.

body colour. In contrast to *E. superba,* white-coloured *E. pacifica* are cheaper than red ones by 10–15%. Fishermen, therefore, generally have less concern over krill quality needing only to regulate catches in collaboration with their counterparts in adjacent prefectures so as to keep the price high.

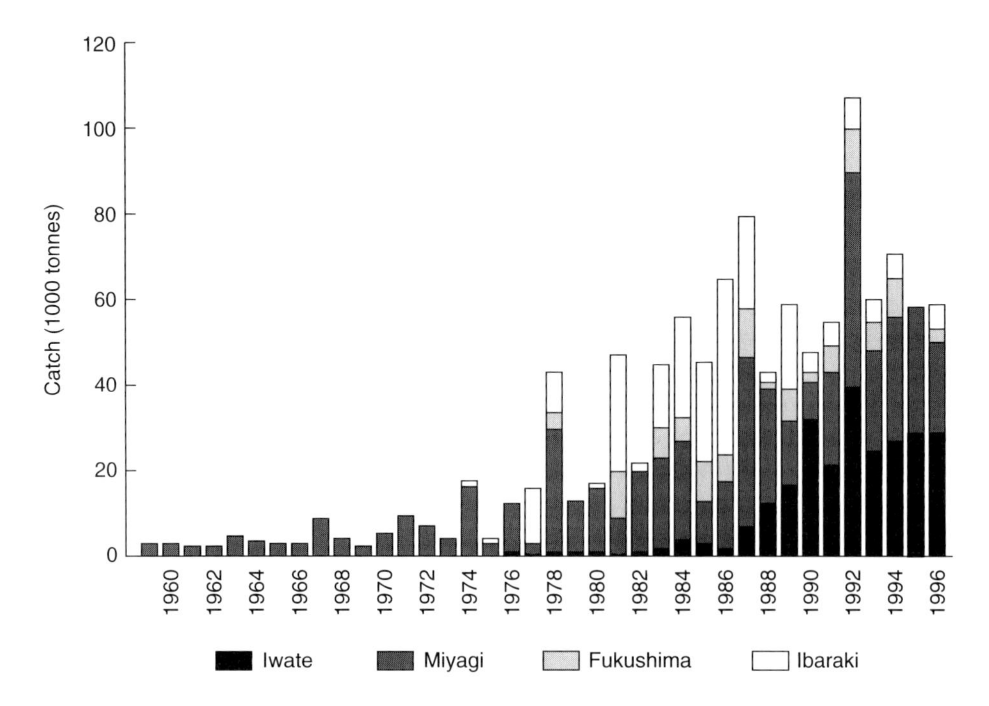

Fig. 9.12 Annual catches of *Euphausia pacifica* in northeastern Japan. Data from Anon. (1998a).

Fishing fleet organisation

Only small boats of less than 20 t GRT are engaged in the fishery. Boat seines are currently used in all prefectures, including Miyagi Prefecture, where they have been used alongside bow-mounted trawls since 1991. The number of boats licensed for the krill fishery varies from 300 to 500 in each prefecture, although not all operate currently. The average number operating per day was 40–100 in each prefecture from 1986 to 1993 (Kuroda & Kotani, 1994) with each fishing vessel operating more or less individually.

Fishing operations are restricted to daytime for two reasons. Firstly, because ports are only open during the daytime (i.e. from dawn to *c.* 15:00 h in the Sanriku coastal area and from dawn to dusk in the Joban coastal area); and secondly, because fishing vessels are not equipped with freezing facilities, and so they must return within a day for onshore processing and freezing of their catch (Odate, 1991; Sawadate, 1993).

Characteristics of the fishing grounds

The use of small boats and single day operations has restricted the fishing areas to nearshore waters. Fishing conditions are highly dependent upon the strength of the cold Oyashio Current, the fishing grounds being formed near the front between the coastal branch of the current (<5°C) and coastal waters with an optimal surface water temperature of 7° to 9°C (Odate, 1991). The onset of the fishing season is, therefore, determined by the nearshore occurrence of optimal surface waters; such information is available from the Japan Fisheries Information Service Center, which provides sea-surface temperature data recorded by NOAA-9 AVHRR (Sawadate, 1993). Fishers also use local determinants, such as occurrence of certain seabirds and marine mammals in coastal waters and the presence of *E. pacifica* in the stomachs of demersal fish landed at the ports (Kotani, 1992). Individual vessels maintain communication with each other and spread out to search for harvestable concentrations (Sawadate, 1993). The fishing grounds thus exploited correspond to the continental shelf (<200 m), within 10–20 nautical miles from the shore (Kuroda & Kotani, 1994). It is yet to be determined whether or not fishable aggregations of krill exist in the offshore regions (Taki, personal communication).

In the Sanriku coastal area, the krill fishing grounds tend to be stable since the optimal water temperature occurs regularly, whereas in the Joban coastal area, it is unstable since the optimal water temperature does not occur in years when the Oyashio current is weak (Fig. 9.13). For example, the lack of fishing grounds in the Joban coastal area in 1997 was related to a northward anomaly of the southern limit of the first Oyashio current branch from the mean inter-annual value. By contrast, fishing grounds established in the Joban coastal area in 1993 corresponded to the southward extension of the first Oyashio current branch. As shown in Fig. 9.14, the average daily catch rate expressed as tonnes/vessel for boat seines was high (> 5 t) every year in the Sanriku coastal area, compared with substantial inter-annual variability (0–4 t) in the Joban coastal area. Poor catches usually occur in years characterised by the northward anomaly of the southern limit of the Oyashio Current, such as occurred in 1997 (Fig. 9.13).

The fishing season also varies from area to area and from year to year, usually occurring from February to April/May in the Sanriku coastal area, compared with considerable inter-annual variation in the Joban coastal area. In the latter area, the fishing season usually commenced in February in the mid 1980s, but often became as late as May in the 1990s (Kuroda & Kotani, 1994).

Detection and catching of krill swarms

Swarms are detected using echo-sounders, sonar, and direct visual sightings of surface aggregations and feeding seabirds. In the early days, when fishing boats were not equipped with either echo-sounder or sonar, only surface swarms could be fished. These were detected by the reddish colour of the sea surface or else the presence of flocks of feeding seabirds. The latter provided an especially useful cue for locating surface aggregations, owing to the greater distance from which the observation could be made (Odate, 1991). However, with the introduction of echo-sounders, aggregations in mid- and deep-water were able to be detected (Suzuki, 1986). This opened the way for mid-water and bentho–pelagic aggregations to be fished (Nakamura, 1992). Decreasing use of sighting cues in the 1990s is attributed

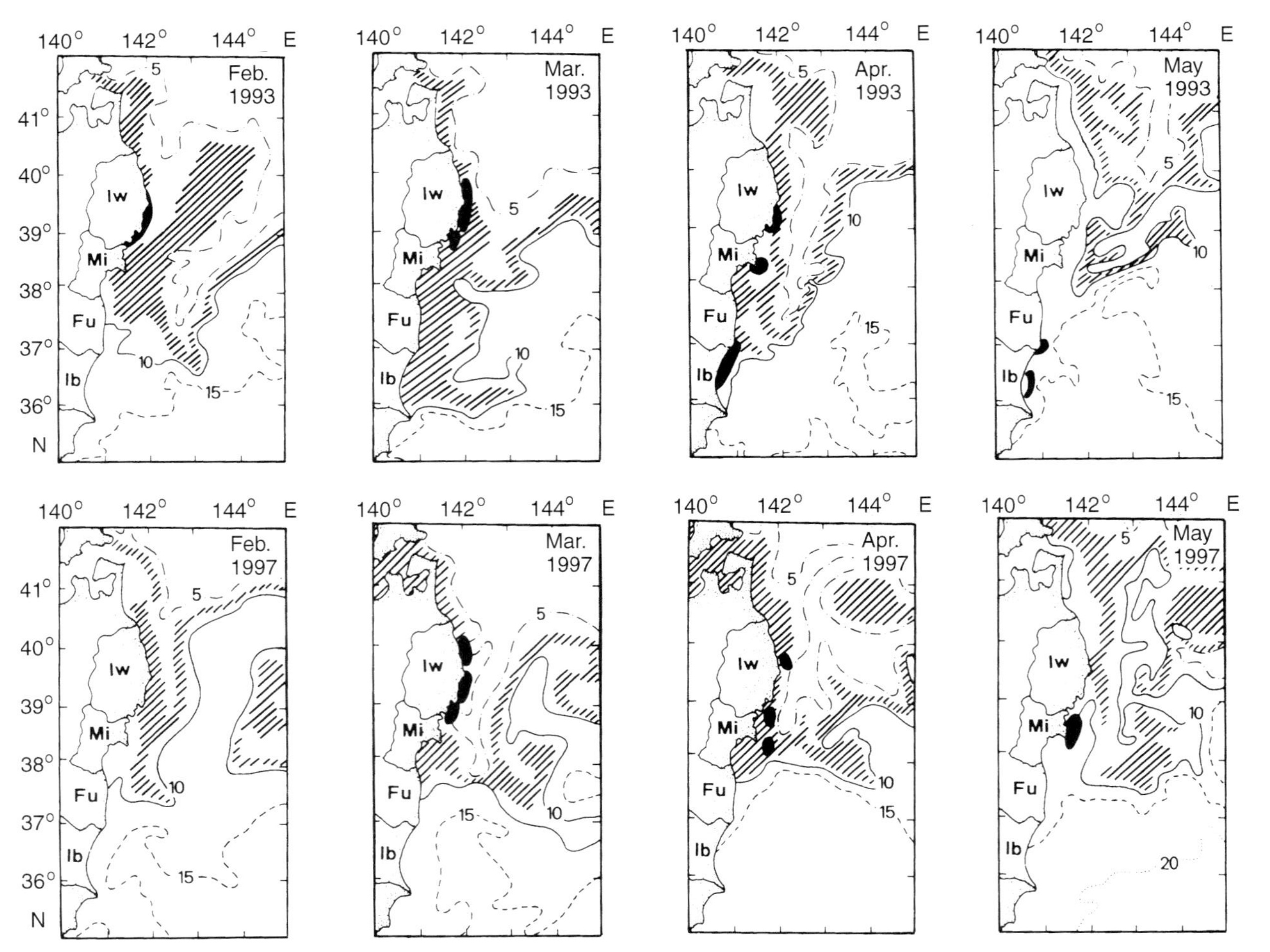

Fig. 9.13 Locations of *Euphausia pacifica* fishing grounds (black area) in relation to sea-surface temperatures in 1993 (cooler year) and 1997 (warmer year) off northeastern Japan. Iw: Iwate; Mi: Miyagi; Fu: Fukushima; Ib: Ibaraki. Areas of sea temperature from 7–9°C are shaded. Modified from 'Quick reports on fishing and oceanographic conditions' from Japan Fisheries Information Service Center.

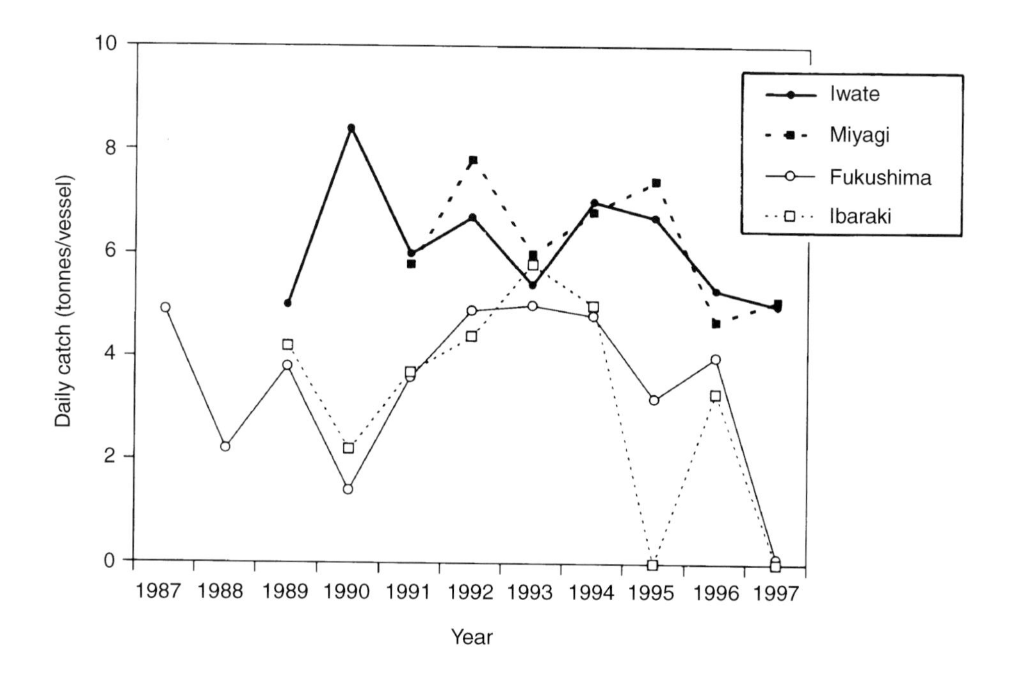

Fig. 9.14 Interannual variations in average daily catch of *Euphausia pacifica* (tonnes/vessel) by boat seine in each prefecture off northeastern Japan. Data from Anon. (1992; 1993; 1994; 1995a; 1996; 1998a.)

also to the decreasing occurrence of surface swarms due to increasingly warm surface water temperatures (Odate, 1992).

In Japanese coastal waters, not only *E. pacifica* but also finfish such as sand lance can be predominant. Acoustic identification of species is thought to be possible using characteristics in echo traces. At a speed of 10–12 knots, echo traces of krill are either round or layered, while those of sand lances are vertical and needle-like (Ebisawa, 1995). Discrimination between krill and walleye pollock (*Theregra chalcogramma*) is also possible using characteristic differences in SV (volume backscattering strength) between two frequencies (38 kHz and 120 kHz) (Miyashita *et al.*, 1997). The SV difference is large for krill, but small for walleye pollock.

Of the two krill fishing methods employed, bow-mounted trawls, which have a net opening of approximately 75 m^2, only catch aggregations within 8 m of the surface. In this method, when an aggregation is sighted, the boat approaches and two poles slide forward like probing antennae (Fig. 9.15a). The booms tilt and plunge, causing the net to spread open beneath the bow. The boat pushes the net slowly through the aggregation, engulfing the krill (Kodama, 1995a). The average catch per haul is approximately 300 kg (Odate, personal communication).

Boat seines, the other fishing method, can catch aggregations as deep as 150 m. When an aggregation is detected, a buoy fixed to the free end of a rope is thrown

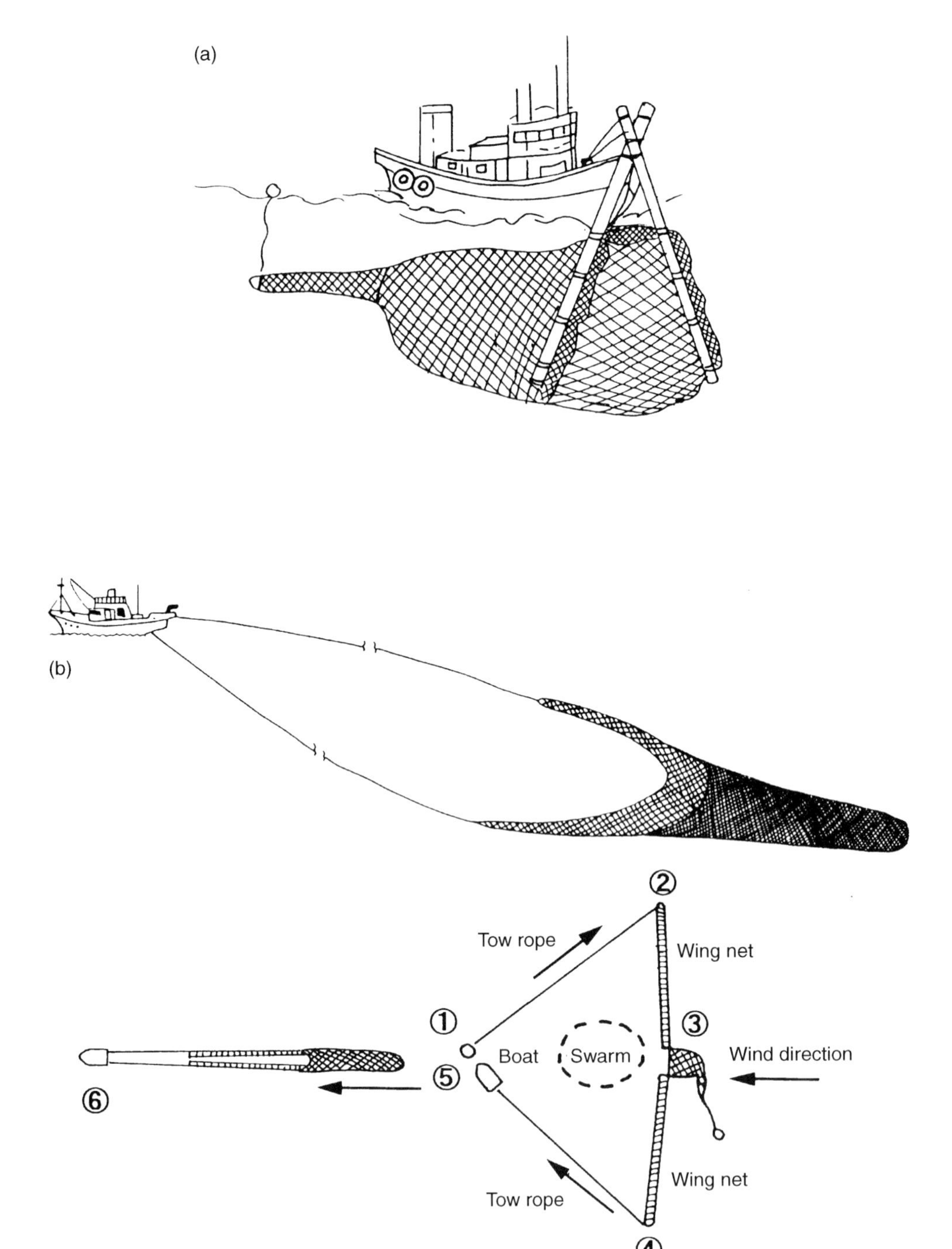

Fig. 9.15 Methods for catching *Euphausia pacifica*: by bow-mounted trawl (a), and boat seine (b) off northeastern Japan. Redrawn from Kodama (1995b).

overboard, and the wing net, 100 m long by 50 m high, pocket net and second wing net are streamed out to surround the krill aggregation (Fig. 9.15b). After the buoy has been retrieved, both ropes are hauled whilst ensuring that the vessel orientates downwind and down current (Kodama, 1995a). The average catch per haul is greater than 1000 kg (Sawadate, 1993).

Bow-mounted trawls, which are now only used in the Miyagi Prefecture, are in the process of being replaced by boat seines, due to the former's limitations in trawling depth and low CPUE. The proportion of catches by bow-mounted trawls relative to the total catches in this Prefecture has been negligible for the past two years, 1996 and 1997.

In the Sanriku coastal area, krill are not distributed in deeper water due to the strong upwelling of very cool water of less than 4°C, well below their optimal temperature. Arising from this, the fishing depths tend to be shallower than 50 m. In the Joban coastal area, on the other hand, as the surface water temperature increases towards the end of the fishing season, surface aggregations of *E. pacifica* descend to cooler mid- and deep-water layers and fishing depths can be deeper than 100 m (Odate, 1992). Nakamura (1992) speculated that in the Joban coastal area, when the water is cold (5°C) and unstratified, krill aggregations occur throughout the water column, whereas mid-water and benthopelagic aggregations occur when the water column is stratified, with water warmer than 10°C at the surface and an extensive, cold deep layer where the temperature varies from 6° to 9°C.

In both types of fishing operation, boat seine and bow-mounted trawl, fish pumps are used to transfer krill from the cod end to plastic containers, each of which can hold about 30 kg of krill. The catch is regulated by controlling the number of these containers allowed on board each vessel.

Primary concerns of the fishers

For fishers, especially in the Joban coastal area, besides the limited market for krill products, the most important concerns related to the krill fishery are the ability to predict fishing conditions, the length of the fishing season and the area of occurrence of the fishery.

Kotani (1992) undertook a correlation analysis between krill catches in various coastal areas and the southernmost latitude of the coastal branch of the Oyashio Current. A significant correlation was found between the annual catch in the Ibaraki Prefecture, where fishing conditions were the least stable, and the southernmost latitude of the coastal branch of the Oyashio current during the fishing season ($r = -0.822$, $p < 0.05$, $n = 8$). In other words, the greater the southward extension of the coastal branch of the Oyashio Current, the higher the catch in this prefecture. Thus, the southernmost latitude of the coastal branch of the Oyashio Current during the fishing season is crucial to predictions of fishing conditions. Kotani (1992) also found a close correlation between the southernmost latitude of the Oyashio Current during the fishing season and that in the preceding January ($r = 0.71$, $p < 0.01$, $n =$

13). Based on this, a prediction of krill catch for a fishing season, may be possible a few months prior to the beginning of that season.

The first fishing day of the year is thought to be best predicted according to the distribution of water masses. Assuming that the optimal temperature for *E. pacifica* aggregations is 5° to 10°C, Kotani (1992) found a significant correlation between the distance from the Otsu fishing ground in Ibaraki Prefecture to the water mass with a temperature of 10°C at 100 m depth and the number of days from 25 January to the date when fishing commenced in that year for each prefecture ($r = 0.917$, $p < 0.01$, $n = 7$). In other words, the closer the water mass with a temperature of 10°C at a depth of 100 m to the Otsu fishing ground, the greater the likelihood of fishing commencing in late January.

The southernmost latitude of the coastal branch of the Oyashio Current showed long-term inter-annual changes (Kodama, 1995b). From 1948 to 1973, a warm water period, the current tended to shift northward, but from 1974 to 1987, a cool water period, it tended to shift southward. From 1988 to the present, another warm water period, it has frequently shifted northward again. Historical catches of *E. pacifica* seem to be related to these long-term changes in the southern limit of the Oyashio Current. In the early- to mid-1980s, Ibaraki Prefecture expanded its catch to around 50% of the total catch, but after 1988, Iwate and Miyagi Prefectures became the main fishing prefectures (Fig. 9.12). Since the krill fishery is licensed by prefectural governors and hence fishing activity permitted only within prefectural boundaries, drastic changes in the location of fishing grounds across boundaries impact greatly on the overall management of these coastal fisheries overall. Therefore, long-term prediction of fishing conditions is an important consideration for management.

To date, catch and CPUE data are used to evaluate fishing conditions, but they are unlikely to be of much use as indicators of krill abundance because both catch and fishing effort are regulated by the industry. For more reliable predictions of fishing conditions (including location of fishery and length of the fishing season), fishery-independent monitoring such as periodic surveys of abundance and distribution of krill in relation to oceanographic conditions, is necessary.

9.4 Canadian west coast waters (*Euphausia pacifica*)

History of the fishery

A trial commercial fishery for *E. pacifica* began in about 1970 in the Strait of Georgia and east of Vancouver Island (Fig. 9.16). Over the ensuing years a market for frozen and freeze-dried euphausiids developed, with the products being used for aquaculture and the aquarium pet food industry (Haig-Brown, 1994). Until 1985, annual catches were less than 200 t, with fishing concentrated initially in Saanich Inlet. Subsequent developments extended the fishery into Howe Sound and most recently in Jervis Inlet (Anon, 1995b). In 1990, the largest catch, totalling 530 t, was

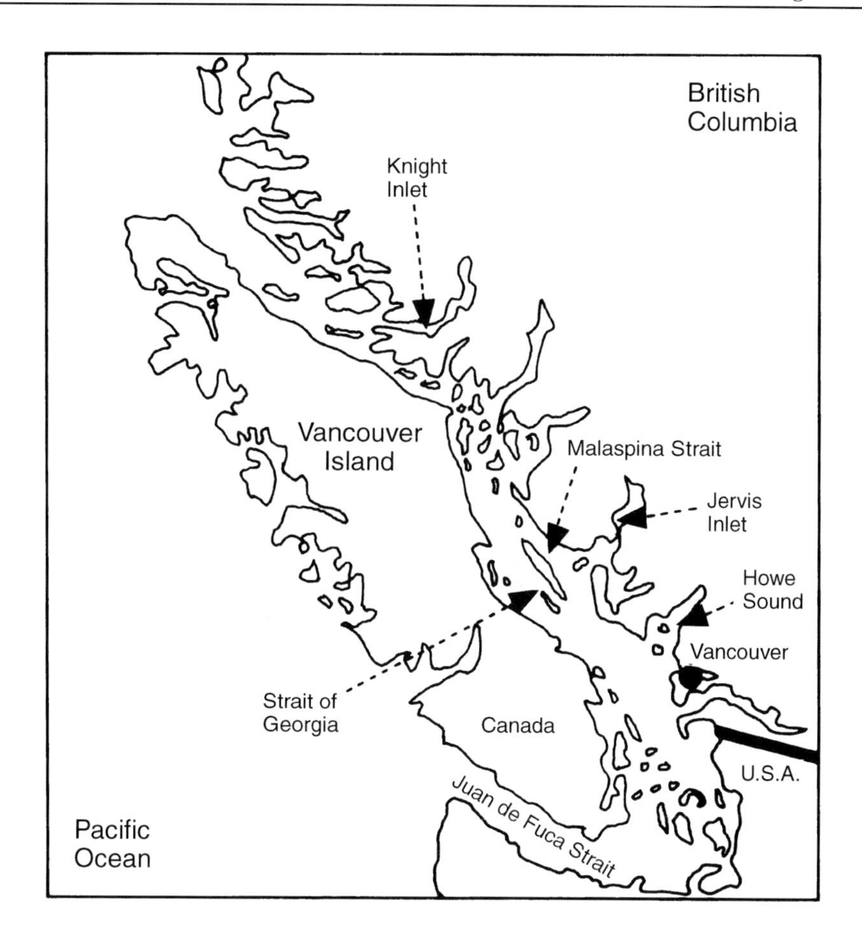

Fig. 9.16 Location of *Euphausia pacifica* fishery off the west coast of Canada. Redrawn from Nicol & Endo (1997). With permission from the Food and Agriculture Organisation of the United Nations.

reported (Fig. 9.17a). In 1992, catches totalling 381 t were reported, with catch rates during November being exceptionally high as fishers found extremely dense aggregations of krill. As a result of this glut a large quantity of krill was landed over a short period of time which exceeded the processing capacity. Consequently, some poor quality and spoiled krill reached the market, discouraging buyers. Subsequent buyer resistance in 1993 resulted in a very much decreased catch that year, totalling only 53 t. However, krill harvests during 1994 recovered to the previous high levels due to renewed market interest (Nicol & Endo, 1997).

Fishing strategy

This fishery is conducted during the out-of-season period for the salmon fishery. The market is limited, with the majority of the products being frozen for export to

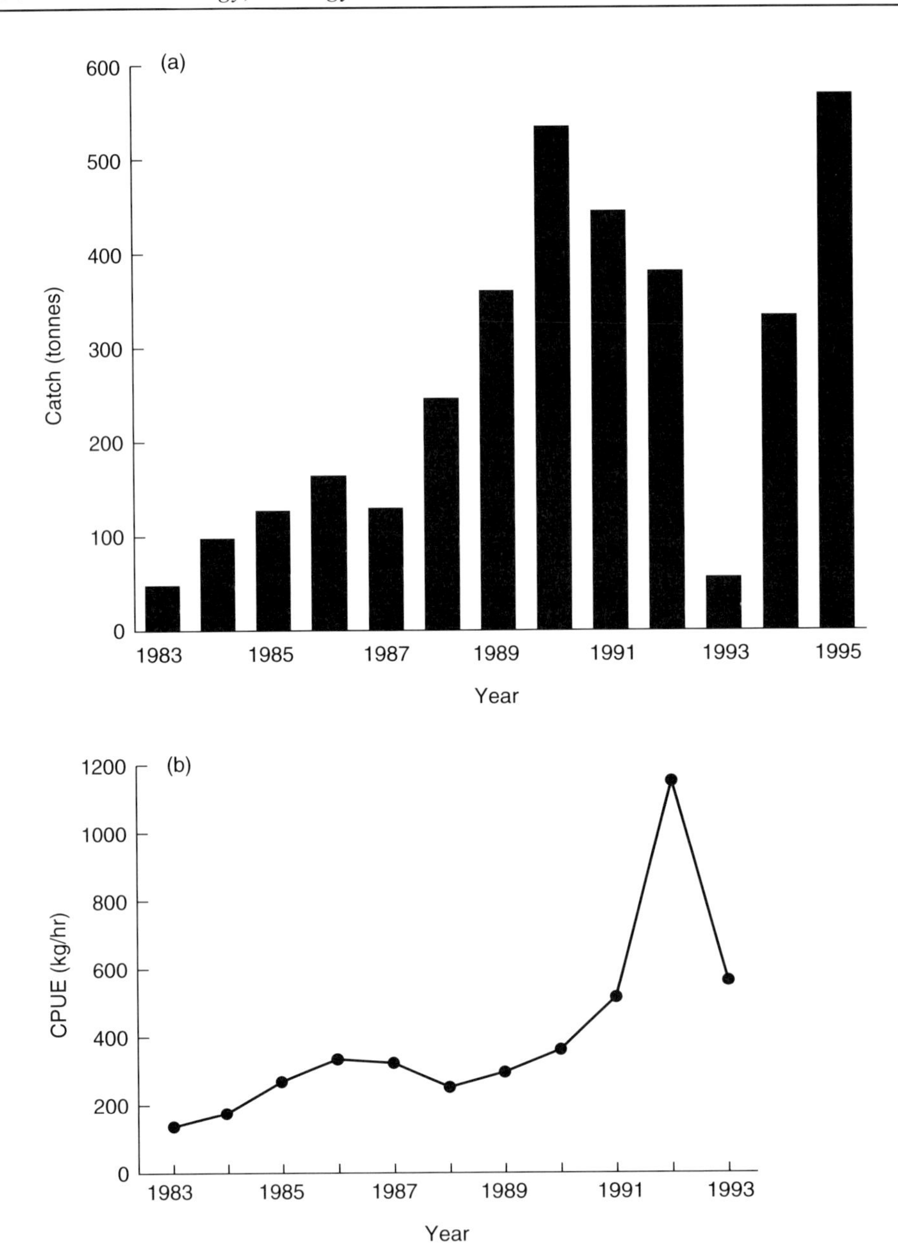

Fig. 9.17 (a) Annual catches of *Euphausia pacifica* off the west coast of Canada. From Nicol & Endo (1997). (b) Interannual variations in average catch per towing time for *Euphausia pacifica* fishery off the west coast of Canada. Data from Nicol & Endo (1997).

the US where it is used in the aquaculture and pet food industries. As in the case of the krill fishery in Japanese northeastern coastal waters, fishermen regulate catches in order to maintain a high price.

Fleet organisation

Two types of vessels participate in the fishery; smaller freezer vessels, whose catches are limited by their low freezing capacity (5–6 t/day) and larger vessels which land greater quantities of krill for onshore processing and freezing (Haig-Brown, 1994). The catch must be frozen within 24 h so as to avoid significant deterioration in product quality. The fishing season can be as short as 20 days (actual fishing days), with individual vessels landing as little as 32 t in a season. The larger vessels, which use seine nets and have no onboard freezing capacity, are usually out-of-season salmon fishing boats, their presence in the krill fishery being usually dependent upon the success of the salmon fishery. If there has been a bad salmon catch, krill are fished to boost revenue (Nicol & Endo, 1997).

The number of licences issued for this fishery increased annually from 7 in 1983 to 56 in 1990, but decreased to 45 in 1991 (Table 9.2). In 1992 licences were limited to 25 vessels following industry advice and because the annual quota was being landed by the vessels already operating. Despite the number of licences issued, the number of vessels reporting landings has been varied from only one in 1993, to a maximum of 17 in 1990. Seven vessels reported euphausiid landings in 1995 and ten in 1996 (Anon, 1998b).

Table 9.2 Numbers of licences issued and fishing vessels and reported landings for the *Euphausia pacifica* fishery off the west coast of Canada.

Year	Number of licences issued	Number of vessels fishing	Fishing days	Quota (t)
1983	7	2	50	500
1984	8	4	67	500
1985	5	2	65	500
1986	11	2	67	525
1987	18	3	36	525
1988	24	4	11	785
1989	45	15	225	500
1990	56	17	255	500
1991	45	14	152	500
1992	25	10	87	500
1993	18	1	29	500
1994	18	6	122	500
1995	18	7	163	500

Data from Nicol & Endo (1997)

Characteristics of the fishing grounds

Currently, most of the commercial krill harvest has come from either the mouth of Jervis Inlet or the adjacent Malaspina Strait within the Strait of Georgia. Because Jervis Inlet (*c.* 316 km) is mostly enclosed, short-term changes in krill abundance due to physical transport are unlikely (Romaine *et al.*, 1995). The commercial fishery is limited to the winter months, from November to March each year, to minimise the incidental catch of larval and juvenile fish, and shellfish. Not only does this ensure the least impact on fish and shellfish resources, but also by minimising by-catch, it improves the quality of krill products (Nicol & Endo, 1997).

Detection and catching of krill swarms

As in other fisheries, krill are located by echo-sounders. The nets that are used have a mouth area of around 80 m^2 the trawl mouth being kept open by a beam and being buoyed so as to keep it from flipping over when the ship turns, while weights on the footrope maintain the net shape. Fishing is carried out, at speeds of less than one knot, close to the surface – often less than 20 m deep at night and on moonless nights when the krill rise to the surface forming layers less than 10 m in vertical extent. Fishing depth is determined largely by the speed of the net through the water (Haig-Brown, 1994).

Catch per unit effort (CPUE in kg/h) reported on harvest logs has remained relatively constant since 1986 at approximately 300 to 400 kg/h (Fig. 9.17b). During the year of greatest effort (1988) CPUE was only 255 kg/h. The highest CPUE (1153 kg/h) was reported in 1992, when fishers reported fishing extremely dense concentrations of krill in Malaspina Strait during November. The CPUE for 1993 was 565 kg/h (Fig. 9.17b). (Nicol & Endo, 1997).

9.5 Comparisons of the three krill fisheries

In Table 9.3 the main features of the three current krill fisheries are summarised. Fishing grounds in the Antarctic and off western Canada are associated with topographical features such as islands, continental shelves, inlets and straits, while those in Japanese northeastern coastal waters are associated with the frontal area between the Oyashio Current and coastal waters. In the latter case, the fishing grounds are also restricted to the continental shelf within 10–20 n.miles from the shore, which means that they tend to form in a similar area each year. Thus, any krill operations tend to be concentrated within certain restricted areas. This should be noted for any assessment of the impact of fishing on local krill abundance and subsequent effect on dependent species.

In the Antarctic, the USSR formerly harvested krill throughout the year, but the Japanese Antarctic krill fishery, which takes most of the current catch, operates only

Table 9.3 Comparative information on current krill fisheries.

	Antarctic Ocean	Off northeastern Japan	Off western Canada
Target species	*Euphausia superba*	*E. pacifica*	*E. pacifica*
Fishing ground	The vicinity of islands and continental slope	Front between the Oyashio current and coastal waters	Inlet and strait
Fishing season	December–August	February–May	November–March
Type of fishing boat	Stern trawlers (3000–4000 t)	Small boats (<20 t)	Small boats
No. of fishing boats	<10	>300	<20
Fishing method	Mid-water trawling	Boat seine	Surface trawling
Fishing time	Day and night	Daytime	Moonless night
Annual catch	*c.* 100 000 t	*c.* 60 000 t	*c.* 300–400 t
CPUE	*c.* 100 tonnes/vessel day *c.* 10 tonnes/haul	*c.* 5 tonnes/vessel/ day *c.* 1 tonne/haul	*c.* 3 tonnes/vessel/ day *c.* 0.5 tonne/haul
Favourite krill quality	Larger white krill are preferred	There is no quality aspect except body colour	There is no quality aspect
Fisherman's concern	Harvesting of preferred krill	Prediction of fishing condition, length of fishing season and location of fishing grounds	
Availability of haul-by-haul data	Yes	No	Yes
Current human consumption	23% of total catch	6–7%	Negligible

from December to August. This allows alternative fisheries such as for squid and fin fish to operate, and at the same time allows vacation time for crews. The fishing season for *E. pacifica* in northeastern Japan is from February to May, when that species is available in the coastal waters. In western Canada, however, fishing is limited to the period from November to March, when incidental catches of larval fish and shellfish are likely to be at a minimum. More than 300 fishing vessels are involved in the krill fishery off northeastern Japan, which makes the fishery very important to the local coastal region. Accordingly, the North Pacific Krill Resources Research Group was established in 1992 to conduct demographic studies, undertake biomass estimates and develop methods for predicting fishing conditions.

The time of day when fishing operations take place for *E. pacifica* is completely different in northeastern Japan when compared with western Canada. In the former fishing takes place during the daytime when ports are open for onshore processing and freezing, whereas in the latter fishing is only at night when krill aggregations are near the surface. Annual catches are restricted by market limitations in all the fisheries. Since boats fishing in the Antarctic are larger in size compared with those in the other two fisheries, CPUEs (tonnes/vessel/day and tonnes/haul) in the former are at least one order of magnitude greater than elsewhere.

E. superba is a larger species, growing to a maximum size of 65 mm (Baker *et al.*, 1990) and is thus subject to substantial size variability at harvest. Furthermore, other quality aspects, degree of 'greenness' and body colour show great variability (Plate 2, facing p. 182), with a resulting emphasis on making catches which will result in 'high quality' products being at a premium. *E. pacifica,* on the other hand, reaches a length of 25 mm (Baker *et al.*, 1990), and has less variability in body size when harvested compared with *E. superba.* Variability in feeding conditions is also less distinct. Hence, body colour of *E. pacifica* is the only quality of concern to the fishers.

In the Antarctic krill fishery, detailed fishing data, i.e. haul-by-haul catch and effort data, have been collected on a continuing basis, thus providing much useful information on spatial and temporal changes in catches, CPUE and size of krill (Endo & Ichii, 1989; Ichii, 1990; Marin *et al.*, 1991; Fedulov *et al.*, 1996; Kawaguchi *et al.*, 1996; 1997; Murphy *et al.*, 1997; Trathan *et al.*, 1998). For example, Japanese haul-by-haul data include the following:

(1) Trawling information – date, position trawling started, time trawling started, trawling time, trawling depth, direction and speed of trawling.
(2) Environment – weather, wind direction and force, atmospheric pressure, air and sea surface temperatures, and pack-ice conditions.
(3) Catch record – weight of each commercial size category, weight of discard, and total catch.

Biological data include the body length of 50 individuals of krill randomly sampled from one haul per day. Currently, measurements of 'greenness' and salp by-catch information have been included in the haul-by-haul data (see Kawaguchi *et al.*, 1998). Comparable haul-by-haul data are collected in the Canadian krill fishery (Anon, 1998b). Analyses of these should provide more details on the fishing activities for and biological aspects of *E. pacifica.*

Acknowledgements

I wish to thank K. Taki and T. Ogishima, Tohoku National Fisheries Research Institute, K.Odate, and Y. Kotani, Nansei National Fisheries Research Institute, for their valuable information on the *E. pacifica* fishery off northeastern Japan. I am also grateful to Gregar Saxby, Biozyme Systems Inc., for information on the *E. pacifica* fishery in Canadian west coast waters. I also thank M. Kigami, Maruha Co., Ltd., and S. Kaneda, Nissui Co., Ltd. for information on current *E. superba* fishery.

References

Anon. (1992) Report of the Research Meeting on North Pacific Krill Resources 1, Tohoku National Fisheries Research Institute, Fisheries Agency (in Japanese).

Anon. (1993) Report of the Research Meeting on North Pacific Krill Resources 2, Tohoku National Fisheries Research Institute, Fisheries Agency (in Japanese).

Anon. (1994) Report of the Research Meeting on North Pacific Krill Resources 3, Tohoku National Fisheries Research Institute, Fisheries Agency (in Japanese).

Anon. (1995a) Report of the Research Meeting on North Pacific Krill Resources 4, Tohoku National Fisheries Research Institute, Fisheries Agency (in Japanese).

Anon. (1995b) Pacific Region, 1995 management plan. Plankton – Euphausiids: Fisheries and Oceans, Canada.

Anon. (1996) Report of the Research Meeting on North Pacific Krill Resources 5, Tohoku National Fisheries Research Institute, Fisheries Agency (in Japanese).

Anon. (1998a) Report of the Research Meeting on North Pacific Krill Resources 6, Tohoku National Fisheries Research Institute, Fisheries Agency (in Japanese).

Anon. (1998b) Pacific Region, 1998 management plan. Plankton – Euphausiids: Fisheries and Oceans, Canada.

Baker, A. de C., Boden, B. P. and Brinton, E. (1990) *A Practical Guide to the Euphausiids of the World*. Natural History Museum Publications, London.

Butterworth, D.S. & Miller, D.G.M. (1987) A note on relating Antarctic krill catch-per-unit-effort measures to abundance trends. *S. Afr. J. Antarctic Res.* **17**, 112–16.

CCAMLR (1990a) *Statistical Bulletin*, Vol. 1 (1970–1979). CCAMLR, Hobart, Australia (CCAMLR-SB/90/1).

CCAMLR (1990b) *Statistical Bulletin*, Vol. 2 (1980–1989). CCAMLR, Hobart, Australia (CCAMLR-SB/90/2).

CCAMLR (1997) *Statistical Bulletin*, Vol. 9 (1987–1996). CCAMLR, Hobart, Australia (CCAMLR-SB/97/9).

Dolzhenkov, V.N., Lubimova, T.G., Makarov, R.R., Parfenovich, S.S. & Spiridonov V.A. (1988) Some peculiarities of the USSR krill fishery and possibilities to use fishery statistics in studies of krill biology and stocks. In *Selected Scientific Papers, 1988* (SC-CAMLR-SSP/5), Vol. Part I, pp. 237-252. CCAMLR, Hobart, Australia.

Ebisawa, Y. (1995) Characteristics of echo-traces of krill. In *Manual for species identification, measurements of body size and sampling of Euphausiid species off Sanriku and Joban areas*. (North Pacific Krill Resources Research Group, ed.). 32 pp. (in Japanese).

Eddie, G.C. (1977) The harvesting of krill. *Southern Ocean Fisheries Survey Programme*, FAO Rome. GLO/SO/77/2.

Endo, Y. & Ichii, T. (1989) CPUEs, body length and greenness of Antarctic krill during 1987/88 season in the fishing ground north of Livingston Island. In *Selected Scientific Papers, 1989* (SC-CAMLR-SSP/6), pp. 323–38, CCAMLR, Hobart, Australia.

Everson, I. (1978) Antarctic fisheries. *Polar Record* **19**, 233–51.

Everson, I. (1988) Can we satisfactorily estimate variation in krill abundance? In *Antarctic Ocean and Resources Variability* (D. Sahrhage, ed.), pp. 199–208. Springer-Verlag, Berlin.

Everson, I. & Goss, C. (1991) Krill fishing activity in the southwest Atlantic. *Antarctic Sci.* **3**, 351–58.

Fedulov P.P., Murphy, E. & Shulgovsky, K.E. (1996) Environment-krill relations in the South Georgia marine ecosystem. *CCAMLR Sci.* **3**, 13–30.

Fisher, L.R., Kon, S.K. & Thompson, S.Y. (1953) Vitamin A and carotenoids in some Mediterranean crustacea with a note on the swarming of *Meganyctiphanes*. *Bull. Inst. Oceanogr. Monaco* **1012**, 1–19.

Haig-Brown, A. (1994) Going for the krill. In *National Fishermen*, May 1994, pp. 18–19.

Ichii, T. (1987) Observations of fishing operations on a krill trawler and distributional behavior of krill off Wilkes Land during the 1985/86 season. In *Selected Scientific Papers, 1987* (SC-CAMLR-SSP/4), pp. 335–63. CCAMLR, Hobart, Australia.

Ichii, T. (1990) Distribution of Antarctic krill concentrations exploited by Japanese krill trawlers and minke whales. In *Proceedings of NIPR Symposium on* Polar Biology **3**, 36–56.

Ichii, T., Katayama, K., Obitsu, N., Ishii, H. & Naganobu, M. (1998) Occurrence of Antarctic krill (*Euphausia superba*) concentrations in the vicinity of the South Shetland Islands: relationship to environmental parameters. *Deep-Sea Res. I* **45**, 1235–62.

Ikeda, T. & Dixon, P. (1982) Body shrinkage: a possible over-wintering strategy of the Antarctic krill (*Euphausia superba*). *J. Exp. Mar. Biol. Ecol* **62**, 143–51.

Kawaguchi, S., Ichii, T. & Naganobu, M. (1996) CPUE and net towing depth and body length of krill during the winter operation of Japanese krill fishery around South Georgia. Working paper for CCAMLR; Working Group on *Ecosystem Monitoring and Management* (WG-EMM), 96/51. CCAMLR, Hobart, Australia.

Kawaguchi, S., Ichii, T. & Naganobu, M. (1997) Catch per unit effort and proportional recruitment indices from Japanese krill fishery data in Subarea 48.1. *CCAMLR Sci.* **4**, 47–63.

Kawaguchi, S., de la Mare, W.K., Ichii, T. & Naganobu, M. (1998) Do krill and salp compete ? Contrary evidence from the krill fisheries. *CCAMLR Sci.* **5**, 205–216.

Kock, K.-H. (1992) *Antarctic Fish and Fisheries*. Cambridge University Press, Cambridge.

Kock, K.-H. (1994) Fishing and conservation in southern waters. *Polar Rec.* **30**, 3–22.

Kodama, J. (1995a) Current status and problems of prediction of fishing condition for *Euphausia pacifica* in Miyagi Prefecture. *Bull. Japanese Soc. Fish. Oceanogr.* **59**, 145–47.

Kodama, J. (1995b) Commercial krill fishing net. In *Manual for species identification, measurements of body size and sampling of Euphausiid species off Sanriku and Joban areas.* (North Pacific Krill Resources Research Group, ed.), pp. 29–31 (in Japanese).

Kotani, Y. (1992) How to predict fishing conditions of *Euphausia pacifica*. Report of the Research Meeting on North Pacific Krill Resources **1**, 108–113 (in Japanese).

Kuroda, K. (1994) Euphausiid fishery in the Japanese waters and related research activities. *Gekkan Kaiyo* **26**, 203–209. (in Japanese).

Kuroda, K. & Kotani, Y. (1994) Euphausiid fishery and interannual variation of the fishing conditions in the Sanriku-Joban coastal areas. *Gekkan Kaiyo* **26**, 210–217 (in Japanese).

Laws, R.M. (1977) Seals and whales of the Southern Ocean. *Phil. Trans. Roy. Soc., London* **279**, 81–96.

Loeb, V., Siegel, V., Holm-Hansen, O., Hewitt, R., Fraser, W. & Trivelpiece, W. (1997) Effects of sea-ice extent and krill or salp dominance on the Antarctic food web. *Nature* **387**, 897–900.

Marin, V.H., Mujica, A. & Eberhard, P. (1991) Chilean krill fishery: analysis of the 1991 season. In *Selected Scientific Papers, 1991* (SC-CAMLR-SSP/8), pp. 273–87. CCAMLR, Hobart, Australia.

Marr, J.W.S. (1962) The natural history and geography of the Antarctic krill (*Euphausia superba*). *Discovery Rep.* **32**, 33–464.

Miyashita, K., Aoki, I., Seno, K., Taki, K. & Ogishima T. (1997) Acoustic identification of isada krill *Euphausia pacifica* Hansen, off the Sanriku coast, north-eastern Japan. *Fish. Oceanogr.* **6**, 266–71.

Murphy, E.J., Trathan, P.N., Everson, I., Parkes, G. & Daunt, F. (1997) Krill fishing in the Scotia Sea in relation to bathymetry, including the detailed distribution around South Georgia. *CCAMLR Sci.* **4**, 1–17.

Nakamura, T. (1992) Recent aspect of krill fishing grounds off Joban-Kashima area in relation to warming tendency. *Bull. Japanese Soc. Fish. Oceanogr.* **56**, 155–57.

Nicol, S. (1991) CCAMLR and its approaches to management of the krill fishery. *Polar Rec.* **27**, 229–36.

Nicol, S. & Endo, Y. (1997) Krill fisheries of the world. *FAO Fisheries Technical Paper* **367**, 1–100.

Odate, K. (1991) Fishery biology of the krill, *Euphausia pacifica*, in the northeastern coasts of Japan. *Fishery Science Library*, 40 (in Japanese).

Odate, K. (1992) History and present status of the krill (*Euphausia pacifica*) fishery. Report of the Research Meeting on North Pacific Krill Resources **1**, 72–82 (in Japanese).

Romaine, S.J., Mackas, D.L., Macaulay, M.C. & Saxby, D.J. (1995) Comparisons of repeat acoustic surveys in Jervis Inlet, British Columbia, 1994–1995. In *Harvesting krill: Ecological impact, assessment, products and markets*, Vol. **3**(3) (T.J. Pitcher & R. Chuenpagdee, eds). Vancouver: University of British Columbia Fisheries Centre Research Reports, pp. 48–50.

Sawadate, M. (1993) Current status and problem of the krill (*Euphausia pacifica*) fishery in the vicinity of Otsuchi. Report of the Research Meeting on North Pacific Krill Resources **2**, 87–88 (in Japanese).

SC-CAMLR (1992) *Report of the Eleventh Meeting of the Scientific Committee.* CCAMLR, Hobart, Australia (SC-CAMLR-XI).

SC-CAMLR (1997) *Report of the Sixteenth Meeting of the Scientific Committee.* CCAMLR, Hobart, Australia (SC-CAMLR-XVI).

Shimadzu, Y. (1984) A brief summary of Japanese fishing activity relating to Antarctic krill, 1972/73–1982/83. In *Scientific Papers 1982–1984*, pp. 439–71. CCAMLR, Hobart, Australia.

Siegel, V. (1987) Age and growth of Antarctic Euphausiacee (Crustacea) under natural conditions. *Marine Biology* **96**, 483–95.

Siegel, V. (1988) A concept of seasonal variation of krill (*Euphausia superba*) distribution and abundance west of the Antarctic Peninsula. In *Antarctic Ocean and Resources Variability* (D. Sahrhage, ed.), pp. 219–30. Springer, Berlin.

Siegel, V. & Loeb, V. (1995) Recruitment of Antarctic krill *Euphausia superba* and possible causes for its variability. *Mar. Ecol. Prog. Ser.* **124**, 45–56.

Siegel, V., de la Mare, W.K. & Loeb, V. (1997) Long-term monitoring of krill recruitment and abundance indices in the Elephant Island area (Antarctic Peninsula). *CCAMLR Sci.* **4**, 19–36.

Suzuki, M. (1986) On the distribution of *Euphausia pacifica* in the southern Joban and Kashima-nada areas. *Bull. Tohoku Branch Japanese Soc. Fish. Sci.* **37**, 30–31 (in Japanese).

Trathan, P.N., Everson, I., Murphy, E.J. & Parkes, G. (1998) Analysis of trawl data from the South Georgia krill fishery. *CCAMLR Sci.* **5**, 9–30.

Chapter 10
Products Derived from Krill

Stephen Nicol, Ian Forster and John Spence

10.1 Introduction

Small-scale krill fisheries, which have provided sources of fishing bait and feed for fish culture, have developed in a number of regions (Fisher *et al.*, 1953; Mauchline & Fisher, 1969). In some areas these fisheries have grown into larger-scale operations (Nicol & Endo, 1997, 1999). Lately, a fishery for Antarctic krill (*Euphausia superba*) has been carried out on a large scale (Miller, 1991). The Antarctic krill fishery developed primarily because of the large size of the krill population and its apparent ease of harvesting, but development has slowed recently because of the high cost of fishing, and the lack of a suitable product, or products, with a reliable and effective economic return (Bykowski, 1986). Because most of the northern hemisphere krill fisheries were developed to supply localised markets for bait and for aquaculture feed, there has been limited development of new products from northern fisheries, until recently. In contrast, there has been a considerable effort devoted to producing a range of products from the fishery for Antarctic krill, and in developing a market for its products. The products of the Antarctic krill fishery have been reviewed a number of times (Eddie, 1977; Everson, 1977; Grantham, 1977; Suzuki, 1981; Budzinski *et al.*, 1985; Suzuki & Shibata, 1990). We will concentrate primarily on recent developments in fisheries products and their uses.

10.2 Constraints to using krill

Introduction

Krill are generally smaller in size than most other organisms that have been commercially harvested. Antarctic krill, with a maximum wet weight of approximately 2 g, is the largest euphausiid that has been harvested, while the smallest harvested krill are of the order of 0.01–0.02 g (Nicol & Endo, 1997). Although other small species of crustaceans have been commercially harvested, and there are large fisheries for species such as sergestid shrimps (Parsons, 1972; Omori, 1978; Neal & Maris, 1985), the small size of krill, and some of their biological characteristics, have made their harvesting and processing particularly problematic.

Rapid spoiling

The digestive gland in the cephalothorax of krill contains powerful hydrolytic enzymes, including proteases, carbohydrases, nucleases and phospholipases, which begin to break down the body tissues immediately following death, particularly if crushing has occurred (Bykowski, 1986). Because these enzymes cohabit, they are mutually protected against their degrading effects, a property rare in nature (Anheller *et al.*, 1989). These enzymes are controlled by an inhibitor system which is disabled on death, facilitating the rapid process of autolysis (Sjodahl et al. 1998). The characterisation of these enzymes has been most complete for Antarctic krill (Sjodahl *et al.*, 1998), but there has been some research into the enzymes of *Meganyctiphanes norvegica* (Peters *et al.*, 1998) as well. Because of the different trophic niches of species of krill, it is likely that they will possess different suites of enzymes, but it appears that most species, certainly those that are currently harvested, all suffer from rapid spoiling as a result of autolytic processes (Bykowski, 1986). Although this property of krill has its drawbacks for commercialisation, it also has been utilised in the development of commercial products (see later sections).

The level of digestive enzymes in krill is highest in those individuals that have been actively feeding. These animals have a distinct greenness in the digestive gland, which is associated with a 'grassy' flavour in krill products manufactured for human consumption (Bykowski, 1986). Actively feeding krill also contain more acids and some ketones, alcohols and sulphur compounds than non-feeding krill, and the presence of volatile compounds affects the odour of krill (Gajowiecki, 1995). For these reasons, green krill are avoided by the Antarctic fishery and the catch is graded on its quality by reference to its greenness (Ichii, Chapter 9). Charts for this purpose have been produced (CCAMLR 1993; Plate 2, facing p. 182).

Krill proteins have a relatively high level of solubility, when compared to fish proteins, and this solubility increases with the degree of autoproteolysis (Kolokowski, 1989). This can present challenges in temporary storage prior to processing (Bykowski, 1986), when loss of soluble fractions can occur. The high solubility of krill proteins also has some advantages for producing certain types of end product, however (see hydrolysates in section on aquaculture feed additives).

The lipids of *E. superba* are subject to change during refrigerated storage (Kolakowska, 1988), with the critical factors being the time between capture and freezing and the temperature of freezing. Free fatty acids increase markedly following death, and rapid deep freezing is necessary to maintain product quality.

Bacteria

The bacterial flora of Antarctic krill have been described, and their activity in the gut may contribute significantly to spoilage (Donachie & Zdanowski, 1998), but their abundance and role as krill symbionts is uncertain (Fevolden, 1981; Virtue *et al.*, 1997). They do seem, however, to play a smaller role in the post-capture breakdown of krill than do auto digestive processes (Rakusa-Suszczewski & Zadanowski, 1980). Bacteria have also been shown to have a digestive function in *M. norvegica*, but the spoilage effect of bacteria on species of krill harvested at higher temperatures is unknown (Donachie *et al.*, 1995).

Parasites

Krill may be intermediate hosts of parasites, which can be passed on to other organisms that ingest them; this may be of particular importance when uncooked or unprocessed krill are used in aquaculture feeds. The issue of biosecurity has been viewed as the greatest threat to shrimp aquaculture, and there is particular sensitivity concerning the introduction of viruses in feed produced from other crustaceans (Lotz, 1997). Nothing is known about the viral infections of krill.

There is evidence that *M. norvegica* and *Thysanoessa raschii* are important intermediate hosts of the helminth *Anisakis simplex* (Hays *et al.*, 1998). North Pacific euphausiids have a low level of occurrence of larval digeneans, cestodes, nematodes and acanthocephalans (Kagei, 1985), and such parasites were looked for but not found in Antarctic krill (Kagei *et al.*, 1978). Other symbionts of krill with less harmful potential have also been reported (Nemoto, 1970; Kulka & Corey, 1984; Nicol, 1984; Rakusa-Suszczewski & Filcek, 1988), but reports on studies examining parasitism in krill are rare.

Fluoride

The high fluoride content of the exoskeleton of krill was first indicated by research into *E. superba* (Soevik & Breakkan, 1979), but all other species of krill so far examined have been found to have similarly high levels (Sands *et al.*, 1998). It seems likely that high exoskeleton fluoride concentration is a general feature of euphausiids and that fisheries on krill will have to take this feature into account when assessing potential products.

The fluoride in krill is localised in the exoskeleton, where it can reach concentrations of 3500 $\mu g\ F\ g^{-1}$ dry weight (Virtue *et al.*, 1995), but concentrations in the muscle and other internal tissues appear to be less than 100 $\mu g\ F\ g^{-1}$ dry weight (Sands *et al.*, 1998). Once krill die, however, there is very rapid leaching of the fluoride from the exoskeleton into the tissue (Adelung *et al.*, 1987), even if frozen to $-20°C$ (Christians & Leinemann, 1983). Freezing to temperatures lower than $-30°C$ is necessary to prevent migration of fluoride from the shell into the muscle tissue. Rapid peeling appears to be the most efficient way of separating the fluoride-rich shell from the flesh, although boiling also fixes fluoride in the shell (Bykowski, 1986). The chemical form of fluoride in the shell is unknown but there are techniques which can produce a low fluoride (5–21 $\mu g\ F\ g^{-1}$ dry weight) krill paste or krill protein concentrate by washing with organic acid or water (Tenuta-Filho, 1993).

Vertebrates that are fed on krill tend to accumulate fluoride to deleterious levels in their bones and tissues (Krasowska, 1989) and the levels of this element in krill meal are as much as four times the allowable levels in feed for the European Community (Bykowski, 1986). Those animals with a natural diet that contains krill, however, appear to be able to maintain low tissue fluoride levels and tolerate high bone fluoride levels (Schneppenheim, 1980; Oehlenschläger & Manthey, 1982). Many fish species naturally eat krill and do not appear to accumulate fluoride in their tissues or bones, consequently, whole krill can be used as an aquaculture feed for many species without long-term accumulation of tissue fluoride (Grave, 1981).

Despite many feeding trials on the use of krill meal for a variety of domestic animals, it seems that the high fluoride level of most krill meal, and its high cost, have prevented further developments in this area (Kotarbinska & Grosyzk, 1977; Oehlenschläger, 1979; Bykowski, 1986).

10.3 Current uses of krill

Introduction

There has been considerable effort expended in developing Antarctic krill products for human consumption, but, most of the krill catch has been used for domestic animal feed, and, particularly in recent years, for aquaculture feed. The Japanese Antarctic krill fishery, which takes most of the current catch, produces four types of product: fresh frozen (46% of the catch), boiled-frozen (10% of the catch), peeled krill meat (10% of the catch) and meal (34% of the catch). These products are used for aquaculture and aquarium feed (43% of the catch), for sport fishing bait (~ 45% of the catch) and for human consumption (~12% of the catch) (1999 figures, T. Ichii, Japan National Research Institute of Far Seas Fisheries, pers. comm.).

E. pacifica caught off Japan is used for sport fishing (~ 50% of the catch), feed for fish culture (particularly as a reddening agent), and a small amount is used for human consumption (Kuroda, 1994). Most of the *E. pacifica* from the Canadian fishery is frozen for export to the US, where it is used in the production of fish feed or pet food (Haig-Brown, 1994).

The proposed fisheries for *M. norvegica*, *T. inermis* and *T. raschii* off the East coast of Canada are aimed at producing frozen krill. In addition, it is intended to produce freeze-dried krill for ornamental fish and for public aquaria and freeze-dried krill as an ingredient in salmon feed and as a flavourant for food for human consumption (Nicol & Endo, 1997). *E. nana* caught off the Uwajima Bay, South East Japan, are used as feed for red sea bream (Y. Endo, pers. comm.).

Yields in the manufacture of products from Antarctic krill vary from nearly 100% for krill hydrolysates, to 80–90% for fresh-frozen and boiled-frozen (Plate 6, facing p. 182), to 8–17% for peeled krill and 10–15% for meal. There has been research into ways of improving the efficiency of recovery in the Antarctic krill fishery and reducing waste. Press waters and liquid by-products can contain significant amounts of protein and lipid, which can be removed by filtration, enzymatic or biotechnological methods (Dolganova, 1994).

Krill for human consumption

Individual krill products

Much of the available information on krill products for human consumption has been summarised in earlier reviews (Budzinski *et al.*, 1985; Suzuki & Shibata, 1990; Plate 7, facing p. 182). Currently, 43% of the Japanese Antarctic krill catch is processed for human consumption as boiled then frozen krill or peeled krill tail meat frozen in blocks on board. Canned tail meat is no longer produced from the Japanese catch. Information on products for human consumption from the Antarctic krill fisheries from nations other than Japan is not generally available. In the past, Antarctic krill has also been used for the production of fermented protein products, for spun protein products (i.e. surimi) (Suzuki & Shibata, 1990). A small amount of *E. pacifica* caught off Japan is also being used for human consumption (Kuroda, 1994).

Efforts have been made to produce low shell (hence fluoride) products. Krill paste produced by traditional methods (Budzinski *et al.*, 1985; Plate 8, facing p. 182) and alkaline and acid processed krill protein concentrates have been produced in a low fluoride form, by either organic acid washings or by simple water washings (Tenuta-Filho, 1993). Using either treatment, fluoride concentrations of less than 21 $\mu g\ g^{-1}$ (dry matter) were obtained, whereas untreated protein concentrates may have values of ~250 $\mu g\ g^{-1}$ (dry matter) (Oehlenschläger, 1981). These processes yield high protein recovery and a product with low enough fluoride concentrations for human consumption.

Considerable research has been carried out in Poland into producing krill precipitates using autoproteolysis, making use of krill's high level of proteolytic enzymes to produce a high yield (80% protein recovery) concentrate (Kolakowski & Gajowiecki, 1992). In this process, whole krill is mixed with water and heated. The hydrolysate is centrifuged to remove the shells and the precipitate is coagulated. The final product has low fluoride content (< 29 mg F kg^{-1}), a protein content of 18–22%, fat less than 7% and a high level of carotenoid pigments, giving the precipitate a pink-red colouration. This product is used mainly as a colorant and a flavourant additive to fish feeds and other products for human consumption.

Krill as a food additive

Freeze-dried krill concentrate prepared from peeled tail meat is currently being marketed as a food additive and as a health food supplement by a Spanish company.[1] This 'Antarctic Krill Concentrate' is advertised as having a number of useful properties such as: high n-3 fatty acid content, moderate caloric content, high

nutritional value and ease of digestion, and is advertised as having a major revitalising function on the body. Suggested uses of this product as a dietary supplement include during: pregnancy, lactation, pre- and post-menopausal stages, growth, post-operative procedures, cancer prevention, radiotherapy, chemotherapy, syndromes of immunodeficiency and treatment of various nutritional disorders. Antarctic Krill Concentrate is advertised by the manufacturer as containing important oligo-elements, including antioxidants and minerals required to prevent dental cavities and osteoporosis. The recommended dosage is approximately 5 g day^{-1}. Antarctic Krill Concentrate is promoted as being 100% natural and free of any side effects, even when taken at higher dosages. The n-3 fatty acids in dehydrated krill products are reported to remain unaltered even if stored for longer periods, retaining all their beneficial properties.

Antarctic Krill Concentrate is produced as flakes, or as a loose powder, with different degrees of granulation, and these have a light salmon-pink colour and an excellent shrimp-like taste. It is promoted as an excellent natural colouring and flavouring agent, which is effective even in small quantities when used in a variety of foods. It is claimed to be suitable for the production of special dietary meals and growth food products and requires no special storage conditions. Such specialty products are likely to be of high value, but will utilise only small volumes of krill.

Aquaculture feed

Products

The development of krill products for human consumption has been a focus of the Antarctic krill fishery in the past, but products for aquaculture are likely to dominate in the near future. Global aquaculture production more than doubled between 1986 and 1996 and currently accounts for over one-quarter of all fish consumed (Naylor *et al.*, 1998). Consequently, demand for quality aquaculture feed and feed ingredients is growing rapidly and supplies are uncertain (Rumsey, 1993).

The existing or proposed coastal krill fisheries in the northern hemisphere have been developed to provide local sources of feed for aquaculture and there have been similar proposals in other areas, e.g. *Nyctiphanes australis* in South Eastern Australian waters (Virtue *et al.*, 1995, 1996). It seems likely that stocks of other species of krill will also be investigated once krill has become a more established aquaculture feed. Currently, most of the krill caught in all the commercial fisheries is used for aquaculture feed. For Antarctic krill, 34% of the Japanese catch is fresh frozen and 20% of this is used for aquaculture and 32% is used to produce meal which is used in fish culture. Fifty per cent of the Japanese *E. pacifica* catch and much of the Canadian catch of this species is used as an ingredient in feed for fish culture.

One company, Specialty Marine Products, based in West Vancouver, Canada, is planning a major project to harvest Antarctic krill for the burgeoning aquaculture industry. The company currently markets limited quantities of krill hydrolysates (liquid and spray-dried) to the global livestock and aquaculture industries. The company expects to begin its sustained harvest programme during 2000. The krill will be processed on board the harvest vessels to ensure that the high standards for product quality demanded by the global aquafeed industry will be met.

The use of euphausiids in aquafeeds has steadily increased in recent years (Nicol & Endo, 1997; Storebakken, 1988). Aquafeed manufacturers have been including krill products in feeds primarily to enhance the palatability of the feed, but krill is also a good source of astaxanthin, a carotenoid pigment which is used to give a characteristic red colour to some species.

Krill is available commercially in a variety of forms for use in aquafeeds: as a meal, as frozen blocks of whole krill, and more recently as hydrolysates. Krill meal is produced by much the same methods used in the manufacture of fish-meal (Bykowski, 1986), whereas krill hydrolysates are prepared by the partial enzymatic digestion of whole krill under controlled conditions. Freeze-dried krill and krill hydrolysate have been used in fish feeds on an experimental basis, but their generally high costs have precluded large-scale use.

As with any ingredient, the freshness of the raw material and the type of processing influences the suitability of the final product for inclusion in feeds. This is particularly the case for krill, since many of the components that contribute to the flavour, as well as the pigments, are easily oxidised if exposed to excessive temperature during processing and drying. Typically, antioxidants are added prior to processing to reduce this loss and to preserve the lipid quality (D. Saxby, pers. comm.).

Krill species have a number of features that make them attractive ingredients for aquaculture feeds (Anderson *et al.*, 1997). These include: palatability enhancement of feeds; a source of carotenoid pigments; a source of essential fatty acids; a well-balanced amino acid profile; and an improvement of larval fish survival. To date, the suitability of adding krill to feeds has been tested with a wide variety of fish species, including: Atlantic and Pacific salmon (*Salmo salar* and *Onchorynchus* spp.); red sea bream (*Pagrus major*); largemouth bass (*Micropterus salmoides*), eels (*Anguilla* spp.); yellowtail (*Seriola quinqueradiata*); yellow perch (*Perca flavescens*); walleye (*Stizostedion vitreum*); whitefish (*Coregonus clupeaformis*); sea bass (*Dicentrarchus labrax*); sea bream (*Sparus aurata*) and Australian seabass (*Lates calcarifer*).

Adding krill hydrolysates to aquafeeds, even at levels of only a few per cent by weight, make them more palatable (Forster, 1998), increasing feed consumption, resulting in higher survival and weight gain. Using krill hydrolysate to improve palatability enables the production of feeds that include higher levels of inexpensive nutrient sources (e.g. plant and other protein sources) without affecting fish production (Oikawa & March, 1997).

There is considerable literature on the nutritional value of krill in cultured fish diets (Storebakken, 1988). All species of krill that have been examined are highly nutritious and can be used successfully as a source of protein, energy and flesh pigmenting carotenoids for aquaculture species. Krill also contain a good balance of amino acids and are effective feed stimulants as well. Krill-fed salmon were also found to have a superior taste, but did not significantly accumulate fluoride from the krill exoskeletons in their flesh. This paper (Storebakken, 1988), should be consulted for a review of the earlier literature.

Aquaculture feed additives

In recent years, several studies have been conducted to examine the value of krill products in aquafeeds. The primary focus of this research has been to demonstrate the effectiveness of krill as a feeding stimulant. Krill is known to have a positive effect on the feeding behaviour of some fish. Shimizu *et al.* (1990) showed that diets supplemented with krill meal stimulated feeding behaviour in sea bream (*Pagrus major*) and that this effect was probably due to the presence of proline, glycine and glucosamine. Not only does krill stimulate feeding, it promotes growth in some species of fish (Allahpichay & Shimizu, 1985). The growth promoting factors seem to be steroids located in the cephalothorax region, and thus are available in non-muscle meal.

The use of krill (*E. pacifica*) as a food source for hatchery-reared salmon smolts has contributed to increased disease resistance (Haig-Brown, 1994). This is attributed to the early development of the immune system when using krill as a food source. The nutritive value of *Nyctiphanes australis* has recently been assessed with regard to its possible use as an aquaculture feed by Virtue *et al.* (1995). These researchers found that *N. australis* contained, on average, 52% protein and up to 9.5% lipid on a dry weight basis. The lipid content of *N. australis* was marked by the presence of high quantities of unsaturated fatty acids with n-3 fatty acids accounting for 49% of the total fatty acid content. Carotenoids were present at levels of up to 320 $\mu g\ g^{-1}$ and were mainly (79.5%) in the form of astaxanthin. Fluoride levels were as high as those reported in other species of krill (up to 3507 $\mu g\ F\ g^{-1}$).

Krill hydrolysates are effective feeding stimulants in rainbow trout diets (Oikawa & March, 1997). A control diet containing a mixture of fish-meal (20% of diet), and plant protein sources (wheat (20%), soybean protein concentrate (10%), corn gluten meal (10%), and canola meal (22%)) was compared to the same mixture supplemented with 2% krill hydrolysate in replacement of fish-meal; and a third diet, which also contained 2% krill hydrolysate, but this was blended into the oil and coated on to the pellets. Krill hydrolysate significantly increased consumption of diets in this trial and reduced feed wastage. Fish fed the second diet, with 2% krill hydrolysate added, consumed 34% more feed and wasted less than half as much as those fed the control feed containing no krill hydrolysate, as a proportion of total feed input. This difference was even greater for fish fed the diet in which krill hydrolysate was coated on to the pellets. In this case, the fish consumed 69% more feed and again wasted less than half as much of the test feed, relative to those fed control feed. This study further indicated that, because fish fed diets containing the krill hydrolysates consumed the feed more rapidly, they required less time to reach satiation than those that were fed the control diet. The growth rates of the animals followed the pattern of feed intake.

Krill products have been shown to be very effective in stimulating feeding in fish that are in transition between life cycle stages. For example, weaning largemouth bass (*Micropterus salmoides*) from live feeds to commercial dry pellets is facilitated by feeding freeze-dried krill as a starter diet for four days (Kubitza & Lovshin, 1997). Following transfer to seawater, feed consumption by salmon smolts is often inhibited, and it can take several weeks for full feeding to resume. Atlantic salmon smolts fed a krill hydrolysate coated feed have higher feed intake, better feed conversion efficiency (0.74 vs 0.87), faster growth, and lower percentage of 'failed smolt' than those fed a regular smolt feed (Santosh Lall of the Department of Fisheries and Oceans Canada, pers. com.). The results of these trials were used in the development of commercial smolt transfer diets.

Protein and lipid content

The protein content of krill is generally 60–65%, on a dry weight basis. The amino acid profile of this protein is very well balanced with respect to fish-meal and the requirements of cultured fish and crustacea (Table 10.1).

The digestibility of the protein of krill products is variable. The controlled hydrolysis of many protein sources improves amino acid availability. A portion of the amino acids in krill hydrolysates is present either in free form, or as short-chain polypeptides, which tend to be more available than is intact protein. A comparison of *in vitro* protein digestibility of krill hydrolysate and two krill meals, one from a Japanese source and one from Russia, indicated considerable differences (I. Forster, unpublished results). The Modified Torry Pepsin Digestibility coefficients of the Japanese and Russian krill meals were 45 and 42%, respectively, while the digestibility coefficient of krill hydrolysate was 86%.

Lipid levels in krill products range from 10–20% on a dry weight basis. Krill oil is rich in highly unsaturated fatty acids, most notably of the n-3 fatty acids (Table 10.2). These fatty acids are especially prone to degradation by oxidation and care must be taken to ensure the freshness of the raw material and that appropriate conditions are maintained during processing, including the addition of suitable levels of antioxidants (Kolakowska, 1989, 1991b).

Pigments

Carotenoids levels in krill are around 30 $\mu g\ g^{-1}$ and these appear to deteriorate rapidly during storage if not refrigerated below 0°C (Czerpak *et al.*, 1980; Kolakowska, 1988). *E. pacifica* contains large amounts of carotenoid pigments, especially astaxanthin. For this reason, up to 50% of the Japanese *E. pacifica* catch is used as an ingredient in fish feeds to add a reddish colour to the skin and meat of fish species

Table 10.1 Recommended protein and amino acid contents for cultured fish and shrimp compared to the protein and amino acid levels of krill meal, krill hydrolysate and anchovy meal. Values are reported as percentage of protein.

	Channel catfish[1]	Rainbow trout[1]	Pacific salmon[1]	Common carp[1]	Tilapia[1]	Shrimp[2]	Krill meal[3]	Krill hydrolysate[4]	Anchovy meal[1]
Total protein (% dry wt)	32	38	38	35	32	40	55	60	65.4
Arginine	3.8	3.9	5.4	3.7	3.7	5.8	5.8	7.0	5.9
Histidine	1.3	1.8	1.6	1.8	1.5	2.1	1.9	2.8	2.5
Isoleucine	2.3	2.4	2.0	2.2	2.7	3.4	7.5	5.0	4.8
Leucine	3.1	3.7	3.5	2.9	3.0	5.4	4.9	8.3	7.7
Lysine	4.5	4.7	4.5	5.0	4.5	5.3	6.5	6.9	7.7
Methionine + cysteine	2.0	2.6	3.6	2.7	2.8	3.6	3.8	4.3	4.0
Phenylalanine + tyrosine	4.4	4.7	4.6	5.7	4.8	7.1	8.9	7.7	7.7
Threonine	1.8	2.1	2.0	3.4	3.3	3.6	4.0	5.0	4.3
Tryptophan	0.4	0.5	0.4	0.7	0.9	0.8	ND[5]	1.5	1.1
Valine	2.6	3.2	2.9	3.1	2.4	4.0	5.3	5.5	5.3

[1] NRC, 1993
[2] Guillaume, 1997
[3] Rehbein, 1981
[4] Biozyme Systems, Inc. internal data. Values are % of total amino acids of *E. superba*
[5] Not determined

Table 10.2 Fatty acid profile of krill hydrolysate (*E. superba*)[6].

Fatty acid	Liquid	Spray-dried
14:0	5.6	5.5
14:1n5	0.1	0.0
15:0	0.3	0.0
16:0	18.0	21.6
16:1n7	4.2	5.3
16:2	0.4	0.0
16:4n1	0.5	0.0
18:0	0.8	2.7
18:1n9	15.1	22.2
18:2n6	2.0	4.8
18:3n3	1.0	1.2
18:4n3	1.4	1.5
20:1n9	0.4	1.0
20:2n6	0.3	0.0
20:3n3	0.2	0.0
20:4n6	0.7	1.2
20:5n3	18.1	18.1
22:1n11	1.4	0.0
22:5n3	0.4	0.4
22:6n3	14.7	11.9

[6] Biozyme Systems, Inc. internal data

such as red sea bream, coho salmon, rainbow trout, yellowtail, and others (Kuroda, 1994). Extracts from Antarctic krill have also been successfully used as pigmenting agents for yellowtail and coho salmon (Fujita *et al.*, 1983a; Arai *et al.*, 1987). As Odate (1991) pointed out, red is traditionally viewed by the Japanese as an indication of good luck and red sea bream and lobsters are often used as offerings in celebrations. Moreover, a reddish colour in fish meat is thought to stimulate the appetite.

The principal pigment in krill is astaxanthin, although other carotenoid pigments are also found (Czerpak *et al.*, 1980; Czeczuga, 1981; Yamaguchi *et al.*, 1983; Kolakowska, 1988; Funk & Hobson, 1991). The level of astaxanthin can vary considerably among different krill products and species, but generally is between 150–200 ppm (dry basis) for *E. superba*, while *E. pacifica* can contain considerably more. Astaxanthin is present in free form, or esterified to either one or two fatty acids (mono- and di-ester, respectively). By contrast, synthetic astaxanthin, which is widely used in aquafeeds, is exclusively in the non-esterified form. The esterified forms of astaxanthin must be converted to the free form prior to being absorbed from the gut. Krill has been shown to be effective in pigmenting a variety of fish (Fujita *et al.*, 1983a, b; Arai *et al.*, 1987), although this has not always been found to be the case in salmonids (Foss *et al.*, 1987; Whyte *et al.*, 1998). Recent work with Atlantic salmon compared the bioavailability of astaxanthin from spray-dried krill hydrolysate, krill meal and synthetic astaxanthin at levels in the experimental diets up to 50 ppm. The astaxanthin from krill hydrolysate was 91% available as the synthetic source (I. Forster, unpublished data), whereas the bioavailability of krill meal was much less. Astaxanthin is very susceptible to degradation from high heat, such as occurs during the manufacture of aquafeeds. In general, under these conditions the destruction is expected to be about 10–15%, and is about the same for free and esterified forms.

Minerals

Krill products are a good source of minerals for aquatic animals (Table 10.3). Recent work has shown that rainbow trout fed feeds containing krill as the principal protein source had significantly less dorsal fin erosion than did those fed the fishmeal based control feed (Lellis & Barrows, 1997). Fin erosion is a common problem among cultured salmonids, and may negatively affect survival, disease resistance, desirability among sportfish anglers and the commercial value of the fish. This beneficial effect of krill is thought to be derived from the balance and availability of micro minerals (Rick Barrows, US Dept. of Fish and Wildlife, Bozeman, Montana, pers. comm.).

Table 10.3 The mineral content of krill hydrolysate (*E. superba*).

Mineral	Level[7]
Sodium	2.6%
Calcium	2.3%
Phosphorus	1.9%
Potassium	1.3%
Sulfur	1.3%
Magnesium	0.4%
Zinc	40 ppm
Iron	36 ppm
Copper	33 ppm

[7] Biozyme Systems, Inc. internal data. Values are on a dry weight basis.

Krill as shrimp feed

Little work has been done to date, to investigate the potential of krill products in feeds for shrimp. Krill products also provide many of the nutrients known to be important for shrimp; e.g. amino acid, fatty acid and minerals. In addition, they are a good source of astaxanthin, and it is believed that they possess excellent olfactory attractant characteristics, which is of particular relevance to companies involved in the development of penaeid shrimp feeds. Much work remains to be done in this area.

Uses of krill in aquafeeds

Krill products are added to feeds for aquatic animals in a variety of company and product specific ways. Some feed manufacturers add whole krill to the feed mixture prior to extrusion, in partial replacement of water. Other companies add dried krill by spraying a blend of the powder and fish oil on to the pelleted feed. Krill products (freeze-dried krill and krill hydrolysate) have been added directly to culture water to stimulate feeding of larval fish.

A new product which uses the enzymes of krill to hydrolyse and liquefy krill, before pelletising and drying is being developed in British Columbia for salmon farming and may be applicable to krill fisheries elsewhere (Haig-Brown, 1994). A Canadian firm, Biozyme Systems, Inc., is producing these high value krill hydrolysate products using a proprietary process it has developed. These products are offered in liquid, concentrated, dried or frozen forms. The markets for these products are in the animal and aquaculture feed industries (D. Saxby, pers. comm.[2]).

Sport fishing

A considerable percentage of the Japanese Antarctic krill catch is used for sport fishing bait. Fresh frozen krill comprises 34% of the total catch and of this 70% is sold whole as bait and 10% of this is used as chum for sport fishing. Kuroda (1994) suggests that there is little competition between *E. superba* and *E. pacifica* used for sport fishing, because smaller *E. pacifica* is used as a chum (about 50% of total catch), whereas the larger *E. superba* is mostly used as a bait.

Aquarium feed

A small quantity of Antarctic krill is freeze-dried for the home aquarium market. An estimated 50% of the catch of *E. pacifica* from the British Columbia fishery ends up in aquarium food (Dave Barrett, pers comm.[3]). Krill for ornamental fish feed are marketed on the basis of their high quality proteins and n-3 fatty acids[4]. Fish feeds using euphausiids are promoted as a protein source that is digestible, increases the colour, improves disease resistance, stimulates breeding and enhances growth of fish. Krill can be combined with other high biological-value nutrients (e.g. Spirulina algae), as a basis for top quality fish feeds.

Euphausiids for aquarium use are available in several forms: freeze-dried, frozen or as an ingredient in flake and pelleted feeds and are marketed as a feed to be used when the fish have not been eating, due to being moved or when recovering from a disease problem.

Chemicals and pharmaceuticals

Introduction

There has been a steady development of krill products for non-nutritional uses such as for pharmaceuticals and for industrial uses (Nicol & Endo, 1997). These include the production of chitin and chitosan from krill shells and the utilisation of krill enzymes for pharmaceutical (Melrose *et al.*, 1995; Mekkes *et al.*, 1997) and other purposes (Makes, 1992). These demands may not develop to the point where they become a major economic justification for krill fishing, but they may be instrumental in putting the fishery on a sound economic footing.

Chitin

Krill is a source of chitin (Anderson *et al.*, 1997) and considerable research has been carried out into the extraction of this chemical from Antarctic krill (Breski, 1989). The chitin content of krill and the annual production of chitin by krill has been reviewed by Nicol & Hosie (1993) and the chitin composition of whole Antarctic krill is between 2.4 and 2.7% of the dry weight.

Chitin and chitosan, which is derived from chitin, have a wide variety of current and potential uses, ranging from loudspeaker membranes to cholesterol lowering applications (Sandford, 1989; Nicol, 1991; Maezaki *et al.*, 1993; Peter, 1995), which might become lucrative by-products of the krill fishing industry. It is unlikely, however, that a krill fishery would be initiated solely to produce chitin.

Lipids

The lipid composition of Antarctic krill has been reviewed by Suzuki & Shibata (1990). More detailed analyses of lipids have been carried out recently. Antarctic krill caught in winter were analysed for their fatty acid composition by Kolakowska *et al.* (1994). Polyunsaturated n-3 fatty acids were found to comprise 19% of the total fatty acids and were stable during processing, making krill an attractive nutritional source of these fatty acids. Winter-caught krill had similar levels of fatty acids to their summer-caught counterparts but there was less variability, probably because of the lack of reproductive activity which can raise the lipid content of mature females to more than 8% of their wet weight (Kolakowska, 1991a). Winter season krill contained 3% wet weight lipids (Kolakowska *et al.*, 1994) and the highly unsaturated fatty acids, eicosapentaenoic acid (EPA) and dodecahexaenoic acid (DHA), accounted for about 19% of the total fatty acids. Kolakowska (1991b) concluded that *E. superba*, and even the waste products from the current Polish processing technology, is a valuable source of n-3 polyunsaturated fatty acids. The lipids are more stable, and contain much higher levels of carotenoid pigments than some fish-meals.

The chemical composition of *E. pacifica* has not been studied as intensively as *E. superba*. The fatty acid composition of *E. pacifica* is similar to that of *E. superba*, with 14:0 and 16:0 being the main saturated acids (10–20% of the total fatty acids), whereas 16:1 and 18:1 are the main monounsaturated fatty acid components (6% and 15%, respectively). *E. pacifica* contains higher amounts of EPA (*c.* 23%) and DHA (*c.* 14%) than does *E. superba* (Yamada, 1964).

The lipid composition of the three major north Atlantic species of krill, *M. norvegica*, *T. raschii* and *T. inermis*, have been examined in detail (Saether *et al.*, 1986). The fatty acid composition appears to be related to the diet, which varies seasonally, but these three species contain more depot lipids – triacylglycerides in *M. norvegica*, and wax esters and glycerophospholipids in the two *Thysanoessa* species – than does Antarctic krill.

N. australis from coastal Tasmania has lower lipid levels than do higher latitude species (8.5% dry weight vs 15–50%) and this level varies little with season (Virtue *et al.*, 1995). Such lower level of total lipid may have advantages when it comes to using this species as an aquaculture feed. In terms of lipid composition, *N. australis* contains very high levels of long chain n-3 polyunsaturated fatty acids, which are essential for healthy growth in salmonids, and has higher levels of the carotenoid pigment astaxanthin than does *E. superba*. Both these features favour its use as an aquafeed ingredient despite its lower protein value (52% dry weight) than traditional salmon feed meals (60–70%).

Enzymes

Krill contain very effective hydrolytic enzymes, including proteases, carbohydrases, nucleases and phospholipases. These enzymes appear to be concentrated in the digestive gland in the cephalothorax of the krill. These enzymes have found medical uses in debriding necrotic tissue (Anheller *et al.*, 1989) and as chemonucleolytic agents (Melrose *et al.*, 1995).

Three approaches have been taken to producing an enzymatic debriding agent for the treatment of necrotic wounds (Karlstam *et al.*, 1991). In the first process, whole krill are de-oiled and the proteolytic enzymes are isolated and purified by size-exclusion chromatography. The second process also de-oils the raw krill, followed by homogenisation of the residue, further de-oiling of the aqueous solution and precipitation using an organic solvent. This results in an enzyme powder with low specific activity. In the third process, the enzymes are purified from squeezed autolysed krill. The krill extract can be used directly, or after further purification of individual enzymes.

Krill digestive proteases have also been examined as potential chemonucleolytic agents (Melrose *et al.*, 1995) and proteases from Antarctic krill have shown considerable potential. Chemonucleolysis is a therapeutic procedure whereby a degradative agent is injected to reduce the height of vertebral discs and diminish disc pressure on inflamed nerve roots in cases of sciatica.

In another development of the medical application of enzymes from Antarctic krill, Phairson Medical has identified and purified a single enzyme from krill (PHM-101), and is moving through the lengthy steps of drug development with this compound. They have identified the molecular mechanism of action of this enzyme. It acts on a wide range of protein cell adhesion molecules and several other clinically relevant molecules, and it appears to readily cut CD4, CD8, ICAM 1 & 2, E & L Selectin and VCAM, all of which are important in inflammation. It also cuts an adhesion molecule on *Candida albicans*, as well as degrades IL-2, bradykinin, and staphylococcus enterotoxin B. This broad specificity may result from the intrinsic flexibility of the molecule. It is a serine protease, with properties common to chymotrypsin, trypsin, collagenase and elastase. It functions over a wide pH range and is relatively stable up to 55°C. Clinically, it has shown a response in the treatment of oral and vaginal candida infections, acne and wound care and may have even broader uses. (R. Franklin, Phairson Medical, pers. comm.[5]).

Krill enzymes have also found a use as agents for assisting the restoration of works of art (Makes 1992). Development of markets for high value products, such as krill enzymes, along with the production of food items or aquaculture feed ingredients, will stimulate the growth of krill fisheries.

Economics

Development of products from krill fisheries is highly cost-dependent, but the economics of krill fisheries has not been subject to close scrutiny recently. The market value of *E. superba* is difficult to ascertain accurately, but whole frozen krill, in 1996 prices, fetched approximately Aus.$0.32 kg^{-1}, whereas frozen tail meat was reported to fetch around Aus.$9.50 kg^{-1} (figures from industry sources). No published economic analysis on the Antarctic krill fishery has been attempted since the review in Budzinski *et al.* (1985). The low level of fishing for Antarctic krill in the last five years has been attributed to the high costs and low returns associated with standard catching and processing technology.

The value of euphausiids landed in British Columbia varied between Can$0.23 and $0.88 kg^{-1} between 1984 and 1994 and was Can$0.55 – $0.88 kg^{-1} in 1995. The total landed value of the euphausiid fishery in British Columbia is comparatively small, varying between Can$28 000 in 1985 and $415 000 in 1990. The 1995 catch was worth approximately Can$357 000.

The average price of *E. pacifica* on landing in Japan for 1989–1993 was ¥2.7 billion in total. The average unit price was ¥45.4 kg^{-1}, with a lower price of ¥21–36 kg^{-1} in the last 2 years, because of the very high catch in 1992 (Kuroda & Kotani, 1994). The availability of Antarctic krill does not seem to affect the price of *E. pacifica*, as long as the production level of the former species does not change drastically as shown by multiple regression analysis (Yoshida, 1995). Yoshida (1995) showed that the average unit price of *E. pacifica* (Y in ¥ kg^{-1}) can be predicted by the cumulative catch at the end of March (X_1 in kT) and the annual catch of the previous year (X_2 in kT) as follows:

$$Y = -1.15X_1 - 0.35X_2 + 99.41$$

with multiple correlation coefficient of 0.89. This means that the increase in X_1 by 10 kT causes prices to be reduced by 11.5 ¥ kg^{-1}, and the same increase in X_2 causes a price reduction of 3.5 ¥ kg^{-1}. There is a good agreement between the real average price and the calculated value during the years 1987–1994 (Nicol & Endo, 1997).

10.4 Trends and future developments

Considerable innovation has been displayed to develop products from krill that will provide a reasonable economic return. Harvesting and processing technology has improved markedly over the history of the fishery, but the costs of embarking on an Antarctic fishery are still largely prohibitive. The historic concentration of this fishery on producing products for human consumption may change as innovative aquafeed products, such as hydrolysates, gain acceptance and demand for aquafeeds grows. In 1993 12% of the world's fish-meal was used for aquaculture feeds and a projection to the year 2000 saw the requirement rise to 10–25% of fish-meal production (Rumsey, 1993). In 1997, about 1.8 million tonnes of wild fish were required to produce the fish-meal and oil used to produce 644 kT of Atlantic salmon (Naylor *et al.*, 1998). Other farmed species have similar conversion ratios. Between 1970 and 1990 world production of fish-meal had increased 27%, and was projected to decline by 5% between 1990 and 2000, despite an annual growth rate of aquaculture of about 12%. The development of aquafeeds using sources of protein other than fish-meal has received considerable attention in recent years. The growth rate of aquaculture, and the trend towards intensive rather than subsistence production methods, will put greater demands on existing fisheries for fish-meal – many of which are currently over-exploited – and may drive a move toward exploiting alternate sources of aquafeed ingredients, such as krill. All krill species have a number of features that make them attractive as source materials for the aquaculture feed industry. As these features become more well-known, the demand for krill will grow. The prices of krill will tend to decrease as increases in the size of krill fisheries improve the economies of scale. Although much of the development of new aquafeed products is occurring in the Northern Hemisphere (Haig-Brown, 1994), management caution in the development of so-called 'forage fisheries' is tending to prevent their development in coastal and shelf areas. Thus, if demand is generated for krill products for aquaculture, this will only be able to be met from the Antarctic region.

It is likely that krill products for human consumption will continue to be produced, but they will not be the driving force behind the fishery in the near future. In parallel with the development of krill products for aquaculture and human consumption, there will be a likely increase in demand for krill products for non-nutritional uses, such as for pharmaceuticals and for industrial uses. Some of these demands may develop to the point where they become a major economic justification for krill fishing, or at least will be instrumental in placing krill fisheries on a sound economic basis.

Notes

[1] Information provided by: C. Falkenberg, Monteclaro Asesores, S.L. Acacias 10, Monteclaro, 28223 Madrid, Spain.
[2] D. Saxby, Biozyme Systems, Inc., West Vancouver, BC V7V 1N6 Canada.
[3] D. Barratt, Murex Aqua Foods, Inc.
[4] Aquatrol inc. http://www.aquatrol.com/aboutus.html
[5] Phairson Medical Ltd., 602 The Chambers, Chelsea Harbour, London, SW10 0XF, UK.

References

Adelung, D., Buchholz, F., Culik, B. & Keck, A. (1987) Fluoride in tissues of krill *Euphausia superba* Dana and *Meganyctiphanes norvegica* M. Sars in relation to the moult cycle. *Pol. Biol.* **7**, 43–50.

Allahpichay, I. & Shimizu, C. (1985) Separation of growth promoting factors from non-muscle krill meal of *Euphausia superba. Bull. Japanese Soc. Sci. Fish.* **51**, 945–51.

Anderson, S., Richardson, N.L., Higgs, D.A. & Dosanjh, B.S. (1997) The evaluation of air-dried whole krill meal as a dietary protein supplement for juvenile Chinook salmon (*Oncorhynchus Tshawytscha*). Canadian Technical Report of Fisheries and Aquatic Sciences 2148.

Anheller, J.-E., Hellgren, L., Karlstam, B. & Vincent, J. (1989) Biochemical and biological profile of a new enzyme preparation from Antarctic krill *Euphausia superba* Dana suitable for debridement of ulcerative lesions. *Arch. Dermatol. Res.* **281**, 105–110.

Arai, S., Mori, T., Miki, W., Yamaguchi, K., Konsu, S., Satake, M. & Fujita, T. (1987) Pigmentation of Juvenile Coho salmon with carotenoid oil extracted from Antarctic krill. *Aquaculture* **66**, 255–64.

Breski, M.M. (1989) Production and application of chitin and chitosan in Poland. In *Chitin and Chitosan. Sources, chemistry, biochemistry, physical properties and applications* (G. Skjak-Braek, T. Anthonsen & P. Sandford, eds). Elsevier Science, New York, USA.

Budzinski, E., Bykowski, P. & Dutkiewicz, D. (1985) Possibilities of processing and marketing of products made from Antarctic krill. *FAO Fisheries Technical Paper* 268. pp. 1–46.

Bykowski, P.J. (1986) Possibilities of utilising Antarctic krill. Infofish. *Marketing Digest* 6/86, 11–13.

CCAMLR (1993) *Scientific Observer's Manual (Pilot Edition).* Commission for the Conservation of Marine Living Resources, Hobart, Australia, 68 pp.

Christians, O. & Leinemann, M. (1983) Investigations on the migration fluoride from the shell into the muscle flesh of Antarctic krill (*Euphausia superba*) in dependence of storage temperature and time. *Archiv fur Fischereiwisschaffen* **34**, 87–95.

Czeczuga, B. (1981) The presence of carotenoids in some invertebrates of the Antarctic coast. *Comp. Biochem. Physiol.* **69B**, 611–15.

Czerpak, R., Jackowska, H. & Mical, A. (1980) Qualitative analysis of carotenoids in particular parts of body of males and females of *Euphausia superba* Dana (Crustacea). *Polish Polar Res.* **1**, 139–45.

Dolganova, N. (1994) Comprehensive processing of krill and fish press waters and other liquid by-products. *Acta Ichthyologica et Piscatoria* **XXIV**, 171–77.

Donachie, S.P. & Zdanowski, M.K. (1998) Potential digestive function of bacteria in krill *Euphausia superba* stomach. *Aquat. Micro. Ecol.* **14**, 129–36.

Donachie, S.P., Sabrowski, R., Peters, G. & Buchholz, F. (1995) Bacterial enzyme activity in the stomach and hepatopancreas of *Meganyctiphanes norvegica* (M. Sars, 1857). *J. Exp. Biol. Ecol.* **188**, 151–65.

Eddie, G.C. (1977) *The harvesting of krill. Southern Ocean Fisheries Survey Programme.* FAO Rome. GLO/SO/77/2, 1–76.

Everson, I. (1977) *The living resources of the Southern Ocean. Southern Ocean Fisheries Survey Programme.* FAO Rome. GLO/SO/77/1, 1–156.

Fevolden, S.E. (1981) Bacteriological characteristics of Antarctic krill (crustacea, euphausiacea). *Sarsia* **66**, 77–82.

Fisher, L.R., Kon, S.K. & Thompson, S.Y. (1953) Vitamin A and carotenoids in some Mediterranean crustacea with a note on the swarming of *Meganyctiphanes. Bull. Inst. Oceanogr. Monaco* **1012**, 1–19.

Forster, I. (1998) Krill hydrolysates. *Int. Aquafeeds* **4**, 21–24.

Foss, P., Storebakken, T., Austreng, E. & Liaaen-Jensen, S. (1987) Carotenoids in diets for salmonids. V. Pigmentation of rainbow trout and sea trout with astaxanthin and astaxanthin dipalmitate in comparison with canthaxanthin. *Aquaculture* **65**, 293–305.

Fujita, T., Satake, M., Hikichi, S. *et al.* (1983a) Pigmentation of cultured yellowtail with krill oil. *Bull. Japanese Soc. Sci. Fish.* **49**, 1595–1600.

Fujita, T., Satake, M., Watanabe, T. *et al.* (1983b) Pigmentation of cultured red sea bream with astaxanthin diester purified with krill oil. *Bull. Japan. Soc. Sci. Fish.* **49**, 1855–61.

Funk, V.A. & Hobson, L.A. (1991) Temporal variations in the carotenoid composition and content of *Euphausia pacifica* Hansen in Saanich Inlet, British Columbia. *J. Exp. Mar. Biol. Ecol.* **148**, 93–104.

Gajowiecki, L. (1995) Effects of feeding intensity of Antarctic krill (*Euphausia superba* Dana) on qualitative and quantitative composition of its volatile compounds. *Adv. Agr. Sci.* **4**, 75–82.

Grantham, G.J. (1977) *The Southern Ocean: The Utilization of Krill. Southern Ocean Fisheries Survey Programme.* FAO Rome. GLO/SO/77/3, 1–61.

Grave, H. (1981) Fluoride content of salmonids fed on Antarctic krill. *Aquaculture* **24**, 191–96.

Haig-Brown, A. (1994) Going for the krill. *National Fisherman*, May 1994, 18–19.

Hays, R., Measures, L.N. & Huot, J. (1998) Euphausiids as intermediate hosts of *Aniaskis simplex* in the St. Lawrence estuary. *Can. J. Zool.* **76**, 1226–35.

Kagei, N. (1985) Krills and their parasites (abstract only). *Bull. Mar. Sci.* **37**, 768.

Kagei, N., Asano, K. & Kihata, M. (1978) On the examination against the parasites of Antarctic krill *Euphausia superba. Sci. Repts. Whale Res. Inst. Tokyo* **30**, 311–13.

Karlstam, B., Vincent, J. Johansson, B. & Bryno, C. (1991) A simple purification method of squeezed krill for obtaining high levels of hydrolytic enzymes. *Prep. Biochem.* **21**, 237–56.

Kolakowska, A. (1988) Changes in lipids during the storage of krill (*Euphausia superba* Dana) at 3°C. *Z. Lebensm. Unters. Forsch.* **186**, 519–23.

Kolakowska, A. (1989) Krill lipids after frozen storage of about 1 year in relation to storage time before freezing. *Die Nahrung* **33**, 241–44.

Kolakowska, A. (1991a) The influence of sex and maturity stage of krill (*Euphausia superba* Dana) upon the content and composition of its lipids. *Polish Pol. Res.* **12**, 73–78.

Kolakowska, A. (1991b) The oxidability of krill lipids. *Polish Pol. Res.* **12**, 89–104.

Kolakowska, A., Kolakowski, E. & Szczygielski, M. (1994) Winter season krill (*Euphausia superba* Dana) as a source of n-3 polyunsaturated fatty acids. *Die Nahrung* **38**, 128–34.

Kolakowski, E. (1989) Comparison of krill and Antarctic fish with regard to protein solubility. *Z. Lebensm. Unters. Forsch.* **188**, 419–25.

Kolakowski, E. & Gajowiecki, L. (1992) Optimization of autoproteolysis to obtain an edible product 'precipitate' from Antarctic krill (*Euphausia superba* Dana). In *Seafood Science and Technology* (E.G. Bligh, ed.). Fishing News Books, Oxford.

Kotarbinska, M. & Grosyzk, K. (1977) Evaluation of krill meal as a feed for growing pigs. *Nowe rolnice* **26**, 27–29. (In Polish.)

Krasowska, A. (1989) Influence of low-chitin krill meal on reproduction of *Clethrionomys glareolus* (Schreber, 1780). *Comp. Biochem. Physiol.* 94C(1), 313–20.

Kubitza, F. & Lovshin, L.L. (1997) Effects of initial weight and genetic strain on feed training largemouth bass *Micropterus salmoides* using ground fish flesh and freeze dried krill as starter diets. *Aquaculture* **148**, 179–90.

Kulka, D.W. & Corey, S. (1984) Incidence of parasitism and irregular development of gonads in *Thysanoessa inermis* (Kroyer) in the Bay of Fundy (Euphausiacea). *Crustaceana* **46**, 87–94.

Kuroda, K. (1994) Euphausiid fishery in the Japanese waters and related research activities. *Gekkan Kaiyo* **26**, 210–217.

Kuroda, K. & Kotani, Y. (1994) Euphausiid fishery and interannual variation of the fishing conditions in the Sanriku-Joban coastal areas. *Gekkan Kaiyo* **26**, 210–217.

Lellis, W.A. & Barrows, F.T. (1997) The effect of diet on dorsal fin erosion in steelhead trout (*Oncorhynchus mykiss*). *Aquaculture* **156**, 229–40.

Lotz, J.M. (1997) Viruses, biosecurity and specific pathogen-free stocks in shrimp. *World J. Microbiol. Biotech.* **13**, 405–413.

Maezaki, Y., Tsuji, K., Nakagawa, Y. *et al.* (1993) Hypocholesterolemic effect of chitosan in adult males. *Biosci. Biotech. Biochem.* **57**, 1439–44.

Makes, F. (1992) Enzymatic examination of the authenticity of a painting attributed to Rembrant. Krill enzymes as diagnostic tool for identification of 'The Repentant Magdlene', Institute of Conservation of the University of Gothenburg, Gothenburg.

Mauchline, J. & Fisher, L.R. (1969) The biology of euphausiids. *Adv. Mar. Biol.* **7**, 1–454.

Mekkes, J.R., Le Poole, I.C., Das, P.K., Kammeyer, A. & Westerhof, W. (1997) *In vitro* tissue-digesting properties of krill enzymes compared to fibrinolysin/DNAs, papaine and placebo. *Int. J. Biochem. Cell Biol.* **29**, 703–706.

Melrose, J., Hall, A., Macpherson, C., Bellenger, C.R. & Ghosh, P. (1995) Evaluation of digestive proteinases from the Antarctic krill *Euphausia superba* as potential chemonucleolytic agents. *In vitro* and *in vivo* studies. *Arch. Orthop. Trauma Surg.* **114**, 145–52.

Miller, D.G.M. (1991) Exploitation of Antarctic marine living resources: A brief history and a possible approach to managing the krill fishery. *S. Afr. J. Mar. Sci.* **10**, 321–39.

Naylor, R.L., Goldburg, R.J., Mooney, H.M., Beveridge, M., Clay, J., Folke, C., Kautsky, N.,

Lubchenko, J., Primavera, J. & Williams, M. (1998) Nature's subsidies to shrimp and salmon farming. *Science* **282**, 883–84.

Neal, R.A. & Maris, R.C. (1985) In *The biology of crustacea. Fisheries biology of shrimps and shrimplike organisms* (A.J. Provenzano, ed.). Academic Press, London.

Nemoto, T. (1970) On the parasitic organisms in a krill, *Euphausia similis*, from Suruga Bay. *J. Oceanogr. Soc. Japan* **26**, 283–95.

Nicol, S. (1984) *Ephelota* sp. a Suctorian found on the euphausiid *Meganyciphanes norvegica. Can. J. Zool.* **62**, 744–46.

Nicol, S. (1991) Life after death for empty shells. *New Scientist* **1755**, 36–38.

Nicol, S. & Endo, Y. (1997) *Krill Fisheries of the World.* FAO, Rome, 110 pp.

Nicol, S. & Endo, Y. (1999) Krill fisheries: development, management and ecosystem implications. *Aquat. Liv. Resour.* **12**, 1–17.

Nicol, S. & Hosie, G.W. (1993) Chitin production by krill. *Biochem. System. Ecol.* **21**, 181–84.

NRC (1993) *National Research Council. Nutrient Requirements of Fish.* National Academy Press, Washington, D.C., 114 pp.

Odate, K. (1991) Fishery biology of the krill, *Euphausia pacifica*, in the Northeastern coasts of Japan. *Suisan Kenkyu Sosho* **40**, 1–100.

Oehlenschläger, J. (1979) First attempt to feed krill to mink. *Informationen fur die Fischwirtschaft* **26**, 48. (In German.)

Oehlenschläger, J. & Manthey, M. (1982) Fluoride content of Antarctic marine animals caught off Elephant Island. *Polar Biol.* **1**, 125–27.

Oehlenschlager, J.a.S. (1981) A functional protein concentrate (KPC) from Antarctic krill (*Euphausia superba*, Dana 1850). Preparation, chemical composition and functional properties. *Z. Lebensm. Unters. Forsch.* **172**, 393–98.

Oikawa, C.K. & March, B.E. (1997) A method for assessment of the efficacy of feed attractants for fish. *Prog. Fish Culturist* **59**, 213–17.

Omori, M. (1978) Zooplankton fisheries of the world: A review. *Mar. Biol.* **48**, 199–205.

Parsons, T.R. (1972) Plankton as a food source. *Underwater J.* **4**, 30–37.

Peter, M. (1995) Applications and environmental aspects of chitin and chitosan. J.M.S. – *Pure Appl. Chem.* **A32**, 629–40.

Peters, G., Saborowski, R., Mentlein, R. & Buchholz, F. (1998) Isoforms of an n-acetyl-beta-D-glucosaminidase from the Antarctic krill, *Euphausia superba*: purification and antibody production. *Comp. Biochem. Physiol. B-Biochem. Mol. Biol.* **120**, 743–51.

Rakusa-Suszczewski, S. & Filcek, K. (1988) Protozoa on the body of *Euphausia superba* Dana from Admiralty Bay (the South Shetland Islands). *Acta Protozoologica* **27**, 21–30.

Rakusa-Suszczewski, S. & Zadonowski, M. (1980) Decomposition of *Euphausia superba* Dana. *Pol. Arch. hydrobiol.* **27**, 305–311.

Rehbein, H. (1981) Amino acid composition and pepsin digestibility of krill meal. *J. Agric. Food Chem.* **29**, 682–84.

Rumsey, G.L. (1993) Fish meal and alternate sources of protein in fish feeds. Update 1993. *Fisheries* **18**, 14–19.

Saether, O., Ellingsen, T.E. & Mohr, V. (1986) Lipids of North Atlantic krill. *J. Lipid Res.* **27**, 274–85.

Sandford, P. (1989) Chitosan: commercial uses and potential applications. In *Chitin and Chitosan. Sources, chemistry, biochemistry, physical properties and applications* (G. Skjak-Braek, T. Anthonsen & P. Sandford, eds). Elsevier Science, New York.

Sands, M., Nicol, S. & McMinn, A. (1998) Fluoride in Antarctic marine crustaceans. *Mar. Biol.* **132**, 591–98.

Schneppenheim, R. (1980) Concentration of fluoride in Antarctic animals. *Meersforchung. Reports on Marine Research* **28**, 179–82.

Shimizu, C., Ibrahim, A., Toroko, T. & Shirakawa, Y. (1990) Feeding stimulation in sea bream, *Pagrus major*, fed diets supplemented with Antarctic krill meals. *Aquaculture* **89**, 43–53.

Sjodahl, J., Emmer, Å., Karlstam, B., Vincent, J. & Roeraade, J. (1998) Separation of proteolytic enzymes originating from Antarctic krill (*Euphausia superba*) by capillary electrophoresis. *J. Chromatography B* **705**, 231–41.

Soevik, T. & Breakkan, O.R. (1979) Fluoride in Antarctic krill (*Euphausia superba*) and Atlantic krill (*Meganyctiphanes norvegica*). *J. Fish. Res. Bd. Can.* **36**, 1414–16.

Storebakken, T. (1988) Krill as a potential feed source for salmonids. *Aquaculture* **70**, 193–205.

Suzuki, T. (1981) *Fish and Krill Protein: Processing Technology*. Allied Science Publishers, pp. 193–251.

Suzuki, T. & Shibata, N. (1990) The utilization of Antarctic krill for human food. *Food Reviews International* **6**, 119–47.

Tenuta-Filho, A. (1993) Fluorine removal during production of krill paste and krill protein concentrates. *Acta Alimaentaria* **22**, 269–81.

Virtue, P., Johannes, R.E., Nichols, P.D. & Young, J.W. (1995) Biochemical composition of *Nyctiphanes australis* and its possible use as an aquaculture feed source: lipids, pigments and fluoride content. *Mar. Biol.* **122**, 121–28.

Virtue, P., Johannes, R.E., Nichols, P.D. & Young, J.W. (1996) The biochemical composition of *Nyctiphanes australis* and its possible use as an aquaculture feed source: lipids, pigments and fluoride content pp. 1–4. CSIRO Divisions of Oceanography and Fisheries, Marine Laboratories, Hobart, Tasmania, Australia.

Virtue, P., Nichols, P.D. & Nicol, S. (1997) Dietary related mechanisms of survival in *Euphausia superba*: changes during long term starvation and bacteria as a possible source of nutrition. In *Antarctic Communities – species, structure and survival* (B. Battaglia, J. Valencia & D.W.H. Walton, eds). Cambridge University Press, Cambridge.

Whyte, J.N.C., Travers, D. & Sherry, K.L. (1998) Deposition of astaxanthin isomers in chinook salmon (*Oncorhynchus tshawytscha*) fed different sources of pigment. Fisheries and Oceans, Canada, Nanaimo, B.C.

Yamada, M. (1964) On the lipids of plankton. *Bull. Jap. Soc. Sci. Fish.* **30**, 673–81.

Yamaguchi, K., Miki, W., Toriu, N., Kondon, Y., Murakami, M., Konosu, S., Satake, M. & Fujita, T. (1983) The composition of carotenoid pigments in the Antarctic krill *Euphausia superba. Bull. Japan. Soc. Sci. Fish.* **49**, 1411–15.

Yoshida, H. (1995) A study on the price formation mechanism of *Euphausia pacifica. Bull. Jap. Soc. Fish. Oceanogr.* **59**, 36–38.

Chapter 11
Management of Krill Fisheries in Japanese Waters

Yoshi Endo

Three euphausiid species are commercially exploited in the Japanese waters: *Euphausia pacifica* Hansen, *E. nana* Brinton and *Thysanoessa inermis* (Kroyer) (Fig. 11.1). The most important is *E. pacifica* or, in the Japanese vernacular, 'tsunonashi-okiami' (krill without projection on the head), with an average annual catch of 62 192 t having been taken for the past 10 years. The total landed value of

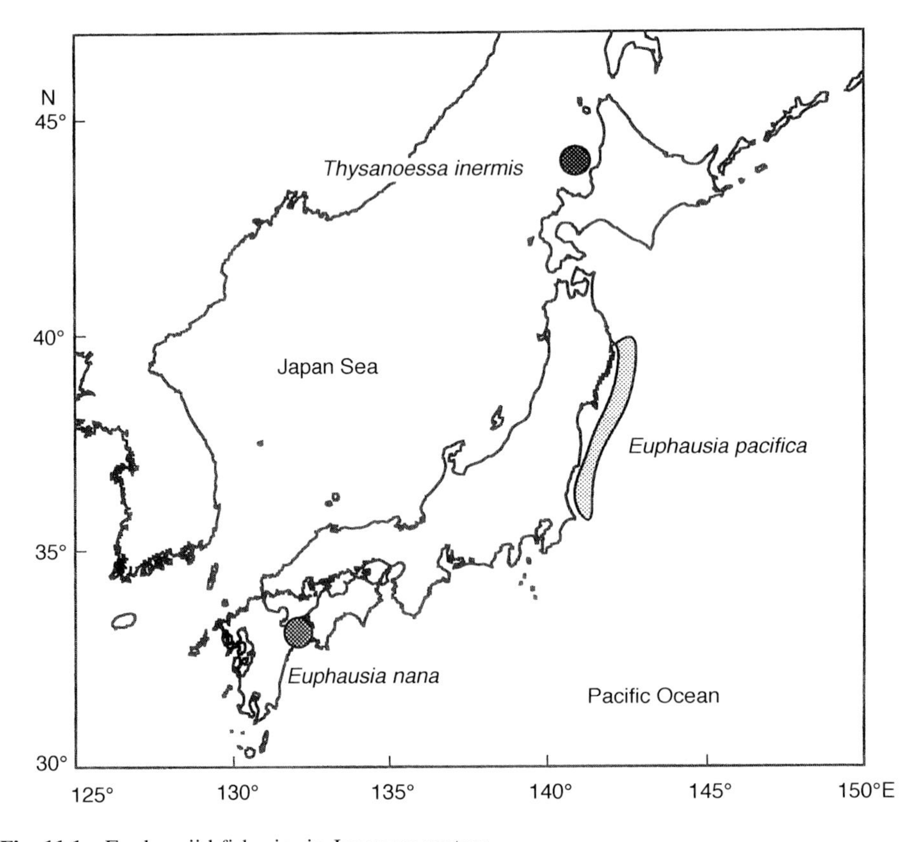

Fig. 11.1 Euphausiid fisheries in Japanese waters.

E. pacifica is 1.5–4.4 billion yen per year and the fishery is now very important in the Japanese coastal fisheries. The *E. pacifica* fishery is the second largest krill fishery in the world following *E. superba* fishery in the Antarctic Ocean (Nicol & Endo, 1997).

In Iwate Prefecture, annual catch of *E. pacifica* ranked second of all the sea fisheries following salmonid fish in 1994, 1995 and 1997, and ranked third following salmonid fish and squids in 1996; the landed values of *E. pacifica* ranked in the top seven following salmonid fish, tunas, squids, abalones and so on (Iwate Prefecture, 1999). In Miyagi Prefecture, annual catch of *E. pacifica* ranked 4th–6th following tunas plus marlins, squids, bonito and so on during the period 1993–1996; the landed value ranked in the top nine following tunas plus marlins, squids, bonito, gadoid fish, flatfish, salmonid fish and so on (Miyagi Prefecture, 1998). In the southern prefectures, only small amounts of *E. pacifica* have been landed during these years because of warmer oceanographic conditions, as will be discussed in Section 11.3.

11.1 History of the *Euphausia pacifica* fishery

The *Euphausia pacifica* fishery developed during the Meiji Era, approximately the last 100 years, from a fishery for sand lance (*Ammodytes personatus*) using a bow-mounted trawl in Sanriku coastal waters (Odate, 1991). The same method was used to harvest *E. pacifica* in the mid-1940s by the fishermen of the Oshika Peninsula in Miyagi Prefecture. Fishery statistics for *E. pacifica* have been available since 1953 from the Onagawa Fish Market.

The fishing season runs from January to June for sand lance, and from March to April for *E. pacifica*. When fishing conditions for sand lance were poor the fishermen changed the net and targeted *E. pacifica*. An increased requirement for sea bream culture and bait for sport fishing in the late-1960s caused an expansion in the fishery to the north as well as along the southern coasts of Miyagi Prefecture. In the Ibaraki Prefecture, fishing for *E. pacifica* began in 1972. A very large catch was made in 1974 when, for the first time in the history of the fishery in the prefecture, the first branch of the Oyashio Current intruded as far south as Inubo-zaki. A small amount of *E. pacifica* had been landed in the southern part of Ibaraki Prefecture previously, but large amounts were not landed until 1974. Shortly after that Fukushima Prefecture began an *E. pacifica* fishery, to be followed in 1975 by Iwate Prefecture, both becoming fishing prefectures.

Thus, the *E. pacifica* fishery which began in the Miyagi Prefecture developed into an important fishery in the Sanriku (Aomori, Iwate and Miyagi Prefectures from north to south) and the Joban (Fukushima and Ibaraki Prefectures) coastal waters (see Fig. 3.1.4).

A small-scale fishery for *E. pacifica* had also been conducted along the coast of the Japan Sea over the last 20 years in waters off the Akita, Yamagata, Shimane and Yamaguchi Prefectures and off the eastern coast of Tsushima Islands in the Nagasaki Prefecture (Kuroda, 1994). Catches from all of these areas have been small.

11.2 Present status of the *Euphausia pacifica* fishery

The *Euphausia pacifica* fishery is categorised as a licensed fishery, licenses being issued by the prefectural governor. Small boats of less than 20 t are engaged in the fishery. One- or two-boat seines are used in all the prefectures. In the Miyagi Prefecture, the original bow-mounted trawls are still used, but since 1991 both one-boat seine and bow-mounted trawls have been used (Kuroda & Kotani, 1994). Photographs of these fishing boats can be seen in Nicol & Endo (1997).

A bow-mounted trawl can only catch surface swarms within 8 m of the surface. On the other hand, one- and two-boat seines catch subsurface swarms, detected by echo-sounder, as deep as 150 m. Over the period 1991 to 1997 in the Miyagi Prefecture, boat seines had a higher CPUE (4.7–7.8 t/boat/day) than the CPUE achieved using bow-mounted trawls (0.3–4.0 t/boat/day). These differences being 1.5 to 15.7 times over the same fishing periods.

The fishing season lasts generally from February to July, but varies from area to area and from year to year (Fig. 11.2). The main fishing season in Sanriku waters is from March to April. That of Joban waters was similar until 1986, when it shifted to later in the year, from May to June (Odate, 1991; Kuroda & Kotani, 1994), a change which seems to reflect the warmer oceanographic condition prevailing since 1986 (Fig. 11.3). In warmer years, annual temperature minimum tends to occur later in the year. The fishing grounds lie over the continental shelf (< 200 m) within 10–20 nautical miles of the shore. The fishing depth is usually less than 50 m in Sanriku waters, but deeper (0–150 m) in the Joban coastal area (Kuroda & Kotani, 1994).

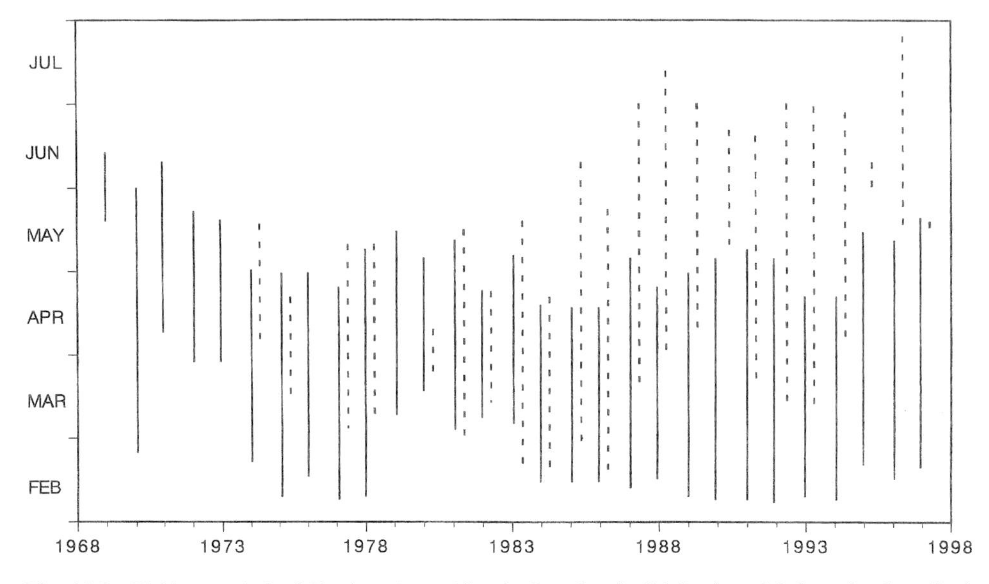

Fig. 11.2 Fishing period of *Euphausia pacifica* in Sanriku (solid line) and Joban (broken line) waters from 1969 to 1998 (redrawn from Odate (1991) and Tohoku Natl. Fish. Res. Inst. (1998)).

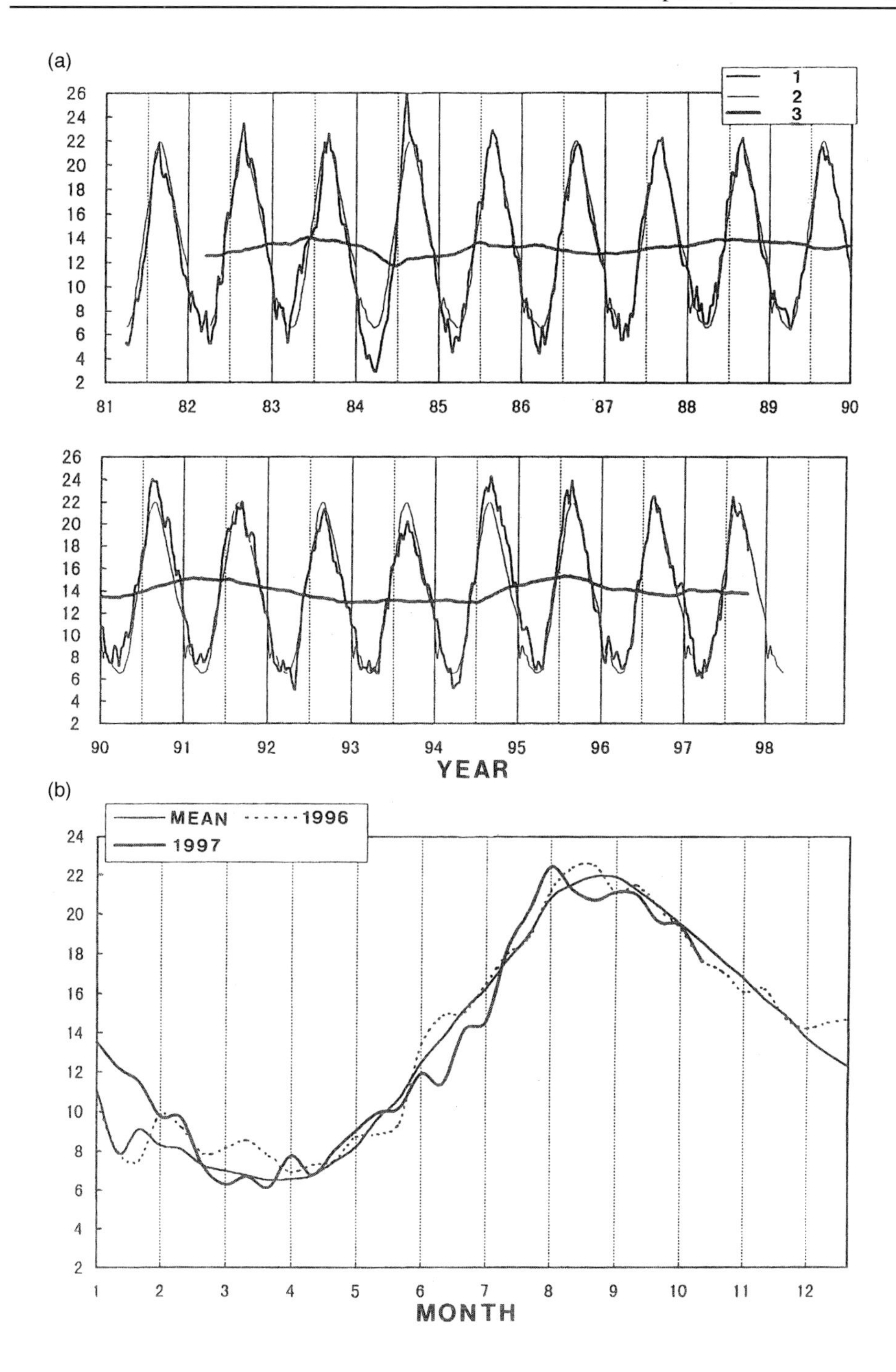

Fig. 11.3 Sea surface temperature (°C) in Sanriku waters (Honda, 1998). (a): interannual variation from 1981 to 1997. 1: raw data; 2: mean values in the last 17 years; 3: 12-month running mean. (b) close-up of 1996 and 1997 with mean values in the last 17 years.

The total annual catch of *E. pacifica* has increased steadily over the last 20 years (Fig. 11.4) exceeding 40 000 t in 1978, 80 000 t in 1989 and 100 000 t in 1992. This increase was aided in about 1975 by the introduction of plastic containers to reduce the rate of deterioration; each of these containers holds about 30 kg of *E. pacifica*. Previously, krill were stored in the fish hold in bulk. A further innovation was the introduction of fish pumps in the 1980s (Kuroda & Kotani, 1994). In 1993, the total catch decreased to 60 881 t, following the introduction of catch regulations which were imposed in Miyagi and Iwate Prefectures in order to control the market and obviate a decline in value of the catch (Kuroda & Kotani, 1994).

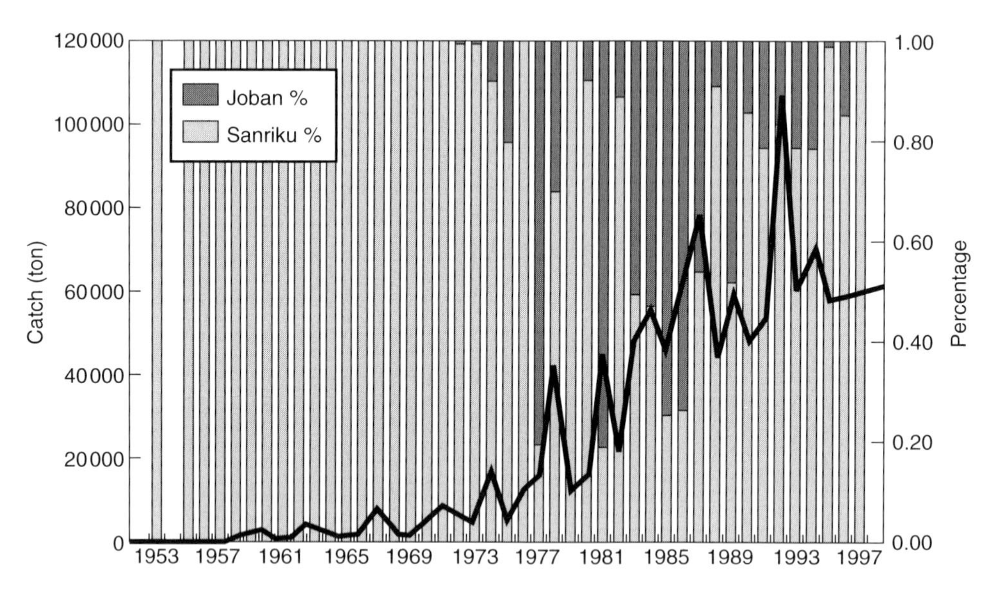

Fig. 11.4 Annual variations in the catch of *Euphausia pacifica* (solid line) and the percentage composition (bars) from Sanriku and Joban waters (redrawn from Tohoku Natl. Fish. Res. Inst. (1998)).

11.3 Variation of annual catch of *Euphausia pacifica* and its causative factors

Most of the catch in the 1970s was landed in Miyagi Prefecture. In the early to mid-1980s, following the commencement of the fishery in Ibaraki Prefecture in 1972 and in Iwate Prefecture in 1975, Ibaraki Prefecture expanded its catch taking more than 60% of the total catch in 1986. After 1988, Iwate and Miyagi Prefectures have been the main fishing prefectures with similar catch levels (Odate, 1991; Kuroda & Kotani, 1994); this again reflects a warmer oceanographic condition after 1988.

The fishing grounds are close to the front between the first branch of the Oyashio Current and the coastal waters with optimal surface water temperatures of 7–9°C (Odate, 1991). Conditions suitable for fishing are therefore highly dependent on the strength of the Oyashio Current (for general current system in Japanese waters see Endo, Chapter 3). Odate (1991) found an inverse relationship between the southernmost latitude of the 5°C sea-surface temperature isotherm and the annual *E. pacifica* catch. As a more reliable indicator of the fishing condition, she calculated the area of the water mass with a temperature < 5°C at 100 m depth between 35° and 42°N and west of 145°E over the period 1970–1978. A very high correlation coefficient, 0.93, was obtained between area and annual catch.

Odate (1991) classified oceanographic patterns into three types based on the *E. pacifica* fishing conditions (Fig. 11.5).

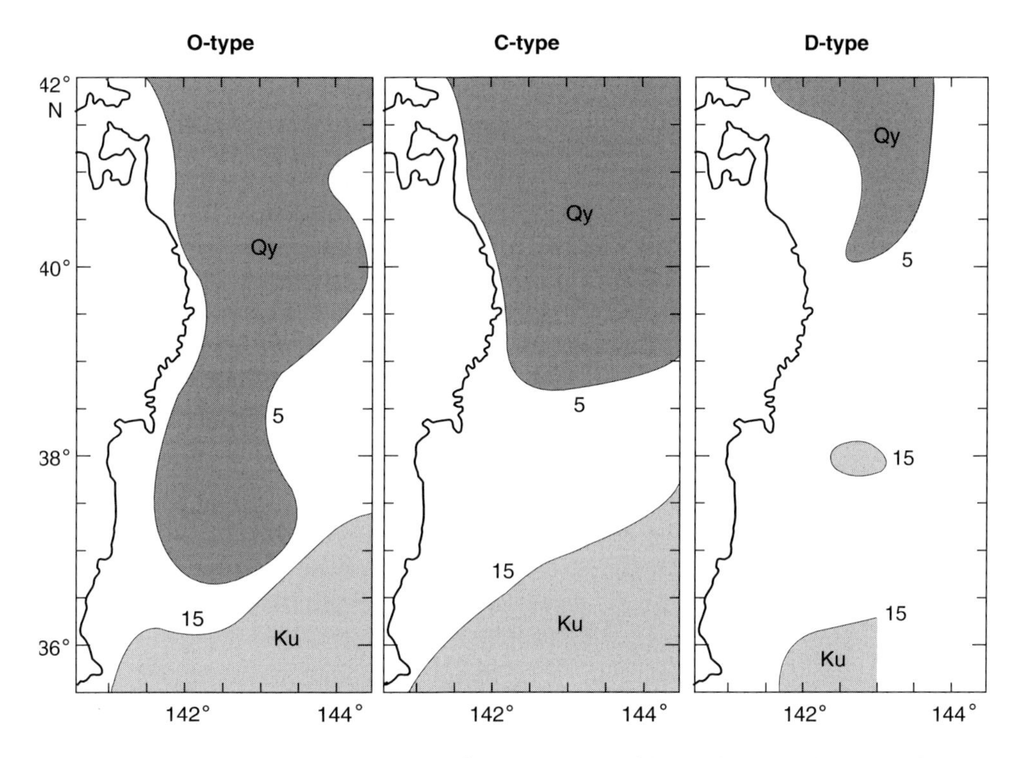

Fig. 11.5 Three oceanographic conditions (O, C and D types) in Sanriku and Joban coastal waters which relate to the formation of fishing ground for *Euphausia pacifica*. Figures indicate the surface water temperature (°C) (redrawn from Odate (1991)).

(1) O-type (good catch years, 1977, 1978, 1981, 1983–1987)
The first branch of the Oyashio Current intrudes along the Sanriku and Joban coastal areas as far south as Kashima-nada (about 36°30′N). This forms a marked front against the warm water associated with the Kuroshio Current. However, in extremely cold years, a water mass much colder than 5°C prevails in Sanriku coastal area, and a larger catch is landed in the Joban area compared to the Sanriku area.

(2) C-type (intermediate catch years, 1980, 1982, 1988, 1989)
The Oyashio Current prevails in Sanriku waters, and warm water related to the Kuroshio Current is present in the Joban area, causing a relatively narrow transition zone between the two currents. In this case, good catches are restricted to Sanriku waters.

(3) D-type (poor catch years, 1976, 1979)
The Oyashio Current is weak and only present in the northern part of Sanriku waters. Coincidentally the Kuroshio Current is also weak leaving a very wide transition zone between these two currents.

Kodama & Izumi (1994) undertook a similar analysis with a classification based largely on the strength of the Oyashio Current. Their A-type is similar to the D-type of Odate (1991). They subdivided Odate's O-type into two categories (B and C) designated by their distance from the shore to the first branch of the Oyashio Current. When the first Oyashio branch is strong but farther offshore, classified as B-type, suitable water temperatures $>5°C$ for fishing ground formation still exist in the southern Sanriku and Joban coast, arising from which good catches are expected in the coastal areas of Miyagi, Fukushima and Ibaraki Prefectures. When the Oyashio first branch is very strong and comes close to the shore, classified as C-type, suitable water temperatures $>5°C$ are restricted to the Joban–Kashima-nada areas, and good catches are expected only in those areas.

11.4 Prediction of fishing conditions of *Euphausia pacifica*

Of the various factors related to the *E. pacifica* fishery, the most important for the fishermen to be able to predict are the onset and length of the fishing season, and the area of occurrence of the fishery. At present the prediction is not carried out by public organisations, but information on the oceanographic and fishing conditions are provided by prefectural fisheries experimental stations in Miyagi, Fukushima and Ibaraki (Kuroda, pers. comm.).

Kotani (1992) undertook a correlation analysis between krill catches in various sea areas of the Pacific coast and the southernmost position of the first branch of the Oyashio Current. A significant correlation ($p < 0.01$) was found between the annual catch in the whole area, Ibaraki and Iwate coasts, and the southernmost extent of the first branch of the Oyashio Current. In other words, the further south the Oyashio Current intrudes, the more catch is taken from these areas. Kotani also found a good correlation ($p < 0.01$) between the southernmost position of the first Oyashio branch in January and the southernmost position in that year. This indicates that the krill catch can be predicted by the southernmost latitude of the first branch of the Oyashio Current in January, about two months prior to the beginning of the fishing season.

Kotani (1992) assumed the optimum temperature for *E. pacifica* swarming to be

5–10°C and tried to predict the first fishing day of the year by correlation analyses. He measured the distance from the fishing grounds of Onagawa and Otsu to the water mass with a temperature of 10°C at 100 m depth for each 10-day period from December to January and examined the relationship between this distance and the days from the mid-day of each period to the first fishing day of the year. He found a significant correlation ($p < 0.05$) between the distance from the Onagawa fishing grounds to the water mass and the days from mid-December and January to the first fishing day within the Miyagi Prefecture in the year. A highly significant correlation ($p < 0.01$) was also found between the distance from the Otsu fishing grounds to the water mass and the days from late January to the first fishing day of Ibaraki Prefecture in the year.

Southward shifts of the Oyashio Current seem to depend on the wind field over the North Pacific Ocean. Sekine (1988a, b) examined the relationship between the North Pacific winds and the extent of southward intrusion of the Oyashio Current and concluded that the Aleutian low pressure system was more strongly developed and shifted southward during years of abnormal southward intrusion of the Oyashio. In relation to the El Niño, Sekine & Suzuki (1991) pointed out that the anomalous southward intrusion of the Oyashio has a tendency to occur in one year after or one year before the occurrence of the El Niño, when the sea-surface temperature appears warm in the west to central region of the equatorial Pacific.

Tomosada (1994) examined the relationship between atmospheric circulation indices and water temperature at several depths in the transition zone (see Section 11.3) off northeastern Japan. He found that there is a negative correlation between the 4th-year preceding Far East polar vortex index and the water temperature at 200 m depth in the transition zone, and a positive correlation between the 4th-month preceding Far East zonal index and water temperature at 50 m depth in the transition zone. When the Far East polar vortex index is large, cold air flows down to high latitudes and seawater is cooled by the cold air. When Far East zonal index is small, the westerly wind meanders and stimulates southward flow of the cooled water, resulting in a southward shift of the Oyashio Current. He showed the possibility of predicting long-term trends in the fishing condition of *E. pacifica* through latitudinal shifts of the Oyashio Current.

Warm-core rings cut off from the Kuroshio Extension sometimes greatly affect the fishing conditions for *E. pacifica* by preventing the first branch of the Oyashio Current from intruding southward (see Endo, Chapter 3). The fishing conditions in 1990 were a good example, where a warm-core ring existed off Kinkasan Island from January to March (Fig. 11.6). The catch of *E. pacifica* was 31 000 t in Iwate Prefecture but only 10 000 t in Miyagi Prefecture, the lowest reported catch in the last 12 years. Another example was in 1996 when a long-lived warm-core ring, 93A, existed in the northern part of Sanriku waters and blocked the southward intrusion of the Oyashio Current (Fig. 11.6). As a result, substantial catches were restricted to the coastal waters of Iwate and Miyagi Prefectures, namely, the northern part of Sanriku waters. Sekine (1988a) reported that the Tsugaru Warm Current water and

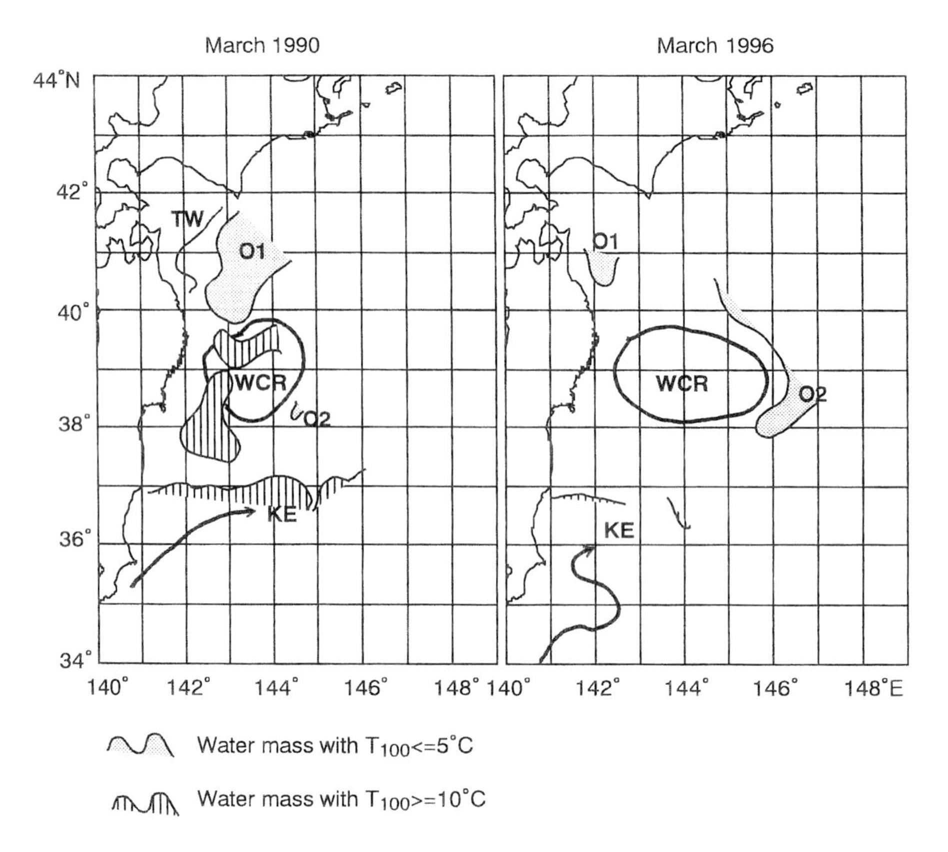

Fig. 11.6 Two examples of the oceanographic condition in which the warm-core ring prevented the first Oyashio branch from intruding southward and the fishing grounds of *Euphausia pacifica* were restricted to the northern Sanriku waters (redrawn from the map of Tohoku Natl. Fish. Res. Inst.).

the Kuroshio warm-core rings may block the southward intrusion of the Oyashio Current, as apparently occurred in the years when the Aleutian low pressure system was strongest. So not only the first branch of the Oyashio Current, but also the presence or absence of warm-core ring(s) and the strength of the Tsugaru Warm Current affect the fishing conditions for *E. pacifica*. These oceanographic conditions must therefore be taken into account in order to predict more effectively the fishing conditions for *E. pacifica*.

11.5 Management

The *Euphausia pacifica* fishery is designated as a licensed fishery with licenses issued by a prefectural governor. Small boats less than 20 t are engaged in the fishery (Table 11.1). The composition of the fleets has already been described in Section

Table 11.1 Contents of licence and regulation on *Euphausia pacifica* fishery in 1997 (Anon., 1998). A full container weighs 30 kg in Iwate and Miyagi Prefectures and 28 kg in Fukushima and Ibaraki Prefectures.

Prefecture		Contents of Licence	Regulation
Iwate	No. boat seine	298	Total catch 28 500 tonnes
	Fishing period	all year round (until catch limit is attained)	Max. number of containers per day
	Landing time	until 15:30	400 for boat > 9 tonnes
	Fishing ground	sea area off Iwate Prefecture	300 for boat < 9 tonnes
	Boat size	< 20 tonnes	
Miyagi	No. boat seine	145	Total catch 32 500 tonnes
	No. bow mounted trawl	208	28 500 tonnes for boat seine
	Fishing period	15 Feb–31 May for boat seine*	4000 tonnes for bow mounted trawl
		1 Jan–31 Jul for bow mounted trawl**	Max. number of containers per day
	Landing time	until 15:00	350 for boat > 10 tonnes
	Fishing ground	sea area off Miyagi Prefecture	330 for boat < 10 tonnes
	Boat size	< 20 tonnes	
Fukushima	No. boat seine	283	Total catch 10 500 tonnes
	Fishing period	1 Feb–31 Jul	Max. catch 6 tonnes per boat per day
	Operation time	dawn–dusk	
	Fishing ground	sea area off Fukushima Prefecture	
	Boat size	< 7 tonnes	
Ibaraki	No. boat seine	498	Total catch 10 500 tonnes
	Fishing period	11 Feb–31 Jul	Max. catch 6 tonnes per boat per day
	Operation time	dawn–dusk	
	Fishing ground	sea area off Ibaraki Prefecture	
	Boat size	< 5 tonnes	

* The period shortened to 75 days from the first fishing day by self-regulation
** The period shortened to 1 Feb–30 Jun by self-regulation

11.2. Several hundred fishing boats are permitted to catch *E. pacifica* from each prefecture with the overall number of krill fishing vessels totalling around 1500 in the years 1994–1997. Over this period the total number of landings by all boats in Miyagi Prefecture was 4500–6000 each season. In 1996 and 1997, there were less than 10 landings by vessels using bow-mounted trawls, a reflection of the fact that surface swarms did not appear in those years. In 1996, the surface-water temperature in Sanriku waters was 0.6 to 2.0°C higher than the 17-year average from late January to late March and in 1997, it was 1.5 to 4.6°C higher from early January to mid February (Honda, 1998). This suggests that the higher surface-water temperature may have prevented surface aggregations from forming.

Fishery regulations are set separately for each prefecture. The licence of the prefectural governor defines the fishing period, the time allowed at sea before the vessel must return to port (also referred to as the operational time), fishing area, boat size and other factors (Table 11.1). Other regulations include total catch limit for the season and a maximum number of plastic containers permitted for each boat per day. Such regulations are not based on scientific rationale. Feasibility studies have been initiated to estimate the standing stock of *E. pacifica* employing two different methods. These are direct assessment using quantitative acoustics and indirectly from egg counts in plankton surveys (see Endo, Chapter 3). There is a clear need to develop a fisheries management regime based on a scientific assessment of the krill populations.

11.6 Economics

The price of *E. pacifica* on landing in 1989–1993 averaged 2.7 billion yen for the entire area. The mean unit price per kilogram ranged from 29.6 yen in 1993 to 79.8 yen in 1996 in Miyagi Prefecture. The average unit price seems to depend on the cumulative catch at the end of March in any year and the annual catch of the previous year.

Yoshida (1995) showed that the average unit price of *E. pacifica* (Y in yen per kg) can be predicted by the cumulative catch at the end of March (X_1 in 1000 t) and the annual catch of the previous year (X_2 in 1000 t) as follows:

$$Y = -1.1\,5X_1 - 0.35X_2 + 99.41$$

with a multiple correlation coefficient of 0.89. This means that the increase in X_1 by 10 000 t reduces the price by 11.5 yen, and the same increase in X_2 reduces price by 3.5 yen. There is a reasonably good agreement between the real average price and the calculated value in 1987–1994.

Availability of Antarctic krill does not seem to affect the price of *E. pacifica*, as long as the harvested level of the former species does not change drastically (Yoshida, 1995).

11.7 Usage of landed *Euphausia pacifica*

The main commercial uses of *E. pacifica* are for sport fishing, feed for fish culture, and human consumption, with the last item being by far the smallest (Kuroda & Kotani, 1994). Kuroda and Kotani suggest that there is little competition between Antarctic krill, *E. superba* and *E. pacifica* used for sport fishing. Whereas *E. pacifica*, being smaller, is used as chum bait and accounts for about 50% of the total catch, *E. superba* being somewhat larger are used as a bait. The remaining 50% of *E. pacifica* catch is used as an ingredient in aquaculture feeds to add reddish colour to the skin and meat of fish such as red sea bream (*Pagrus major*), coho salmon (*Oncorhynchus kisutch*), rainbow trout (*Salmo gairdnerii*), yellowtail (*Seriola quinqueradiata*) and others (Kuroda, 1994), since *E. pacifica* contains a high level of carotenoid pigments (5.4 mg/100 g, Sakurai & Omori 1985). As Odate (1991) has pointed out, Japanese people favour red as a colour since it is thought to bring good luck, consequently red sea bream and lobsters are used as offerings in celebrations.

E. pacifica is rich in essential amino acids (Kayama *et al.*, 1976) and in n-3 series unsaturated fatty acids such as EPA (eicosa pentaenoic acid) and DHA (docosahexaenoic acid) (Yamada, 1964), fatty acids which are known to be effective against circulatory diseases such as arterial sclerosis (Saito, 1994). There are higher concentrations of EPA and DHA than are found in *E. superba* and since they constitute more than 30% of the total fatty acids, this is likely to ensure that *E. pacifica* continues to be an important food source for humans.

11.8 Other Euphausiid species commercially exploited in the Japanese waters

Thysanoessa inermis has been commercially exploited since the early 1970s in the inshore waters of Shakotan Peninsula and Yagishiri Island, off the western coast of Hokkaido (Fig. 11.7, Hanamura *et al.*, 1989; Kotori, 1994). Surface aggregations of this species are fished in the daytime usually from early March to early April. A spoon net, with a diameter of 2 m and a 3–4 m handle, is used to target the aggregations. The yearly catch varies from very little up to 200 t. The price has varied from 75 to more than 3000 yen per kg and was reasonably stable at around 100 yen per kg, in 1994 and 1995. There was no catch in 1996 (Kotori, pers. comm.).

The *T. inermis* fishery is categorised as a recognised fishery by the Fisheries Regulation Committee of Ishikari-Shiribeshi waters. There are no regulations affecting the catch, although the fishing period is limited to February–June. Biological characteristics of the aggregating individuals have been investigated by Hanamura *et al.* (1989) and the aggregations have been shown to comprise fully mature individuals with males possessing spermatophores and females having spermatophores attached to the thelycum. Hanamura *et al.* concluded that *T. inermis* were engaged in reproductive activities during the swarming season.

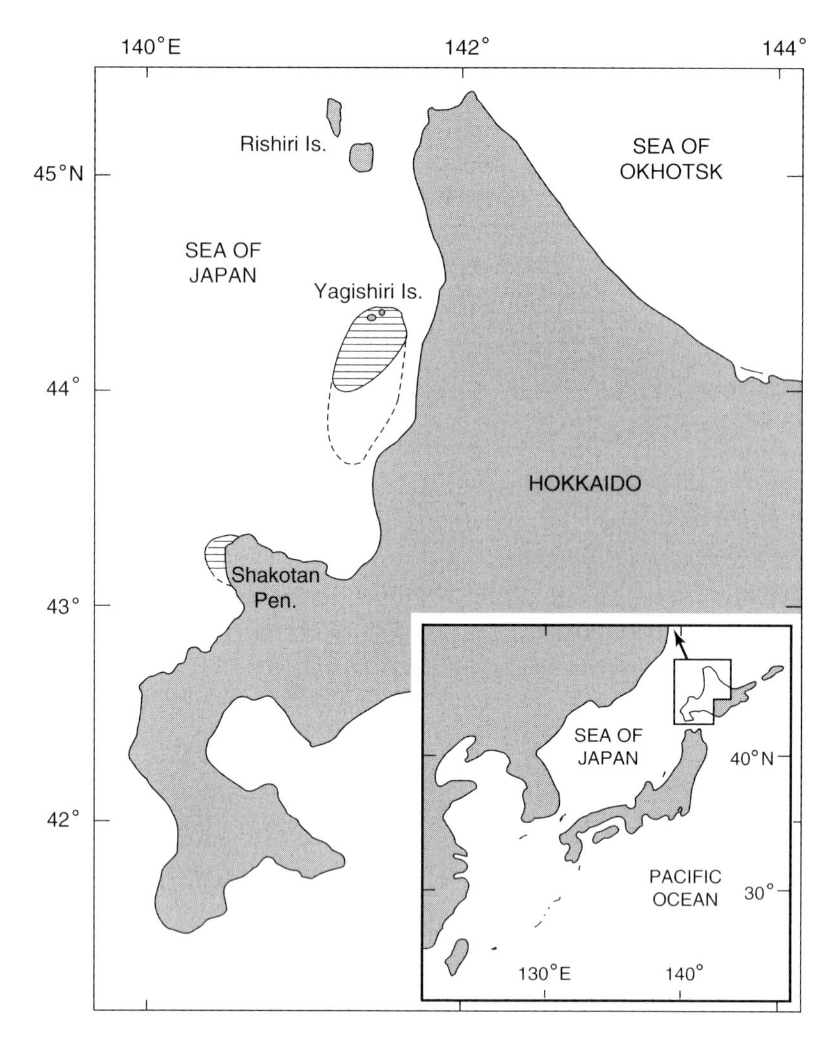

Fig. 11.7 Location of fishing grounds (hatched area) for *Thysanoessa inermis* off the western coast of Hokkaido (Hanamura *et al.*, 1989).

'Ami-ebi' or *E. nana*, an allied species of *E. pacifica*, has been commercially exploited for 26 years in Uwajima Bay, Ehime Prefecture, Shikoku (Fig. 11.8, Hirota & Kohno, 1992). Fishery statistics have been available since 1976. Over the period 1980 to 1991, the yearly catch varied from 2000 to 5000 t. Since this period, there have been no substantial catches reported (Hirota, pers. comm.). Two fishing methods are employed to catch *E. nana* aggregations. One is purse seine operated at night during the period March to July by a group of vessels comprising a netting boat, a transport-boat, and three or less light-boats using fish gathering lamps. These seines usually target sardines and mackerel, but catch *E. nana* when the conditions for catching these species are poor. The other fishing method uses a boat seine of

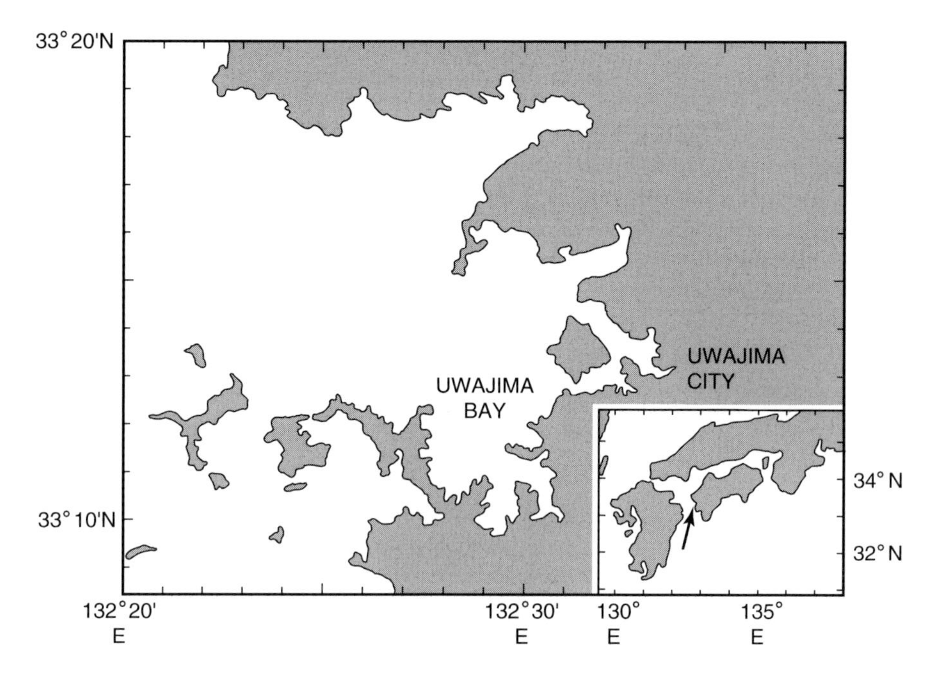

Fig. 11.8 Map of the Uwajima Bay area, Ehime Prefecture where *Euphausia nana* is commercially exploited (adapted from Hirota & Kohno (1992)).

drawers type operated in the daytime by a troop which consists of two netting boats, a boat with fish sounder and a transport-boat, from spring to early summer.

The catch of *E. nana* is used mainly as a feed in the aquaculture industry for the production of red sea bream. The price is about 50 yen per kg. When supply exceeds demand the price falls, but, to protect the fishing industry the catch is then controlled taking into account the *E. pacifica* catch in Sanriku and Joban waters.

The distribution of *E. nana* is neritic with a range extending from the southern Japanese coasts to the East China Sea and Taiwanese waters (Hirota & Kohno, 1992). The species does not seem to be endemic to Uwajima Bay because the fishing season is restricted and adults are reported to perform diel vertical migration with daytime depth of 300–400 m in Sagami Bay (Hirota *et al.*, 1983).

References

Anon. (1998) Fishing condition of *Euphausia pacifica* in Sanriku and Joban coastal waters in 1997. Rep. Res. Meeting on Krill Resources and Oceanogr. Conditions in the Northwest. Pacific 57–62 (in Japanese).

Hanamura, Y., Kotori, M. & Hamaoka, S. (1989) Daytime surface swarms of the euphausiid *Thysanoessa inermis* off the west coast of Hokkaido, northern Japan. *Mar. Biol.* **102**, 369–76.

Hirota, Y. & Kohno, Y. (1992) A euphausiid fishery in Uwajima Bay, Ehime Prefecture.

Fisheries Biology and Oceanography in the South-western Waters of Japan **8**, 89–95 (in Japanese).

Hirota, Y., Nemoto, T. & Marumo, R. (1983) Vertical distribution of euphausiids in Sagami Bay, central Japan. *La mer* **21**, 151–63.

Honda, O. (1998) Characteristic features in the surface and mid water temperatures in Tohoku waters. Rep. Res. Meeting on Krill Resources and Oceanogr. *Conditions in the Northwest Pacific* 32–41(in Japanese).

Iwate Prefecture (1999) *Fisheries of Iwate Prefecture*. 31 pp. (in Japanese).

Kayama, M., Ikeda, Y. & Komaki, Y. (1976) Studies on the lipids of marine zooplankton, with special reference to wax ester distribution in crustaceans and its *in vivo* formation. *Yukagaku (Lipid Chemistry)* **25**, 329–34 (in Japanese with English abstract).

Kodama, J. & Izumi, Y. (1994) Factors relevant to the fishing ground formation of *Euphausia pacifica* and the relation to the demersal fish resources. *Gekkan Kaiyo (Kaiyo Monthly)* **26**, 228–35 (in Japanese).

Kotani, Y. (1992) What aspects and how to predict fishing conditions of *Euphausia pacifica*? Rep. Res. Meeting on North Pacific Krill Resources No. 1, 108–113 (in Japanese).

Kotori, M. (1994) A euphausiid fishery in the west coast of Hokkaido. *Gekkan Kaiyo (Kaiyo Monthly)* **26**, 248–50 (in Japanese).

Kuroda, K. (1994) Euphausiid fishery in the Japanese waters and related research activities. *Gekkan Kaiyo (Kaiyo Monthly)* **26**, 203–209 (in Japanese).

Kuroda, K. & Kotani, Y. (1994) Euphausiid fishery and interannual variation of the fishing conditions in the Sanriku-Joban coastal areas. *Gekkan Kaiyo (Kaiyo Monthly)* **26**, 210–217 (in Japanese).

Miyagi Prefecture (1998) *White book on fisheries in Miyagi Prefecure*. 74 pp. (in Japanese).

Nicol, S. & Endo, Y. (1997) Krill fisheries of the world. *FAO Fish. Tech. Paper* **367**, 1–100.

Odate, K. (1991) Fishery biology of the krill, *Euphausia pacifica*, in the Northeastern Coasts of Japan. *Suisan Kenkyu Sosho (Library of Fisheries Study)* **40**, 1–100 (in Japanese with English abstract).

Saito, H. (1994) Element composition of euphausiids, with special reference to lipids. *Gekkan Kaiyo (Kaiyo Monthly)* **26**, 224–28 (in Japanese).

Sakurai, S. & Omori, A. (1985) On the body color of *Euphausia pacifica*. Business Rep. of the 1985 fiscal year, *Ibaraki Pref. Fish. Exp. St.*, 169 (in Japanese).

Sekine, Y. (1988a) Anomalous southward intrusion of the Oyashio east of Japan, 1. Influence of the seasonal and interannual variation in the wind stress over the North Pacific. *J. Geophys. Res.* **93**, 2247–55.

Sekine, Y. (1988b) A numerical experiment on the anomalous southward intrusion of the Oyashio east of Japan, Part I. Barotropic model. *J. Oceanogr. Soc. Japan* **44**, 60–67.

Sekine, Y. & Suzuki, Y. (1991) Occurrence of anomalous southward intrusion of the Oyashio and the variation in global atmospheric circulation. *Umi to Sora (Ocean and Sky)* **67**, 11–23 (in Japanese with English abstract).

Tohoku National Fisheries Research Institute (1998) Interannual variation in the catch and fishing period of *Euphausia pacifica*. Rep. Res. Meeting on *Krill Resources and Oceanogr. Conditions in the Northwest. Pacific* 51–56 (in Japanese).

Tomosada, A. (1994) A feasibility study of predicting water temperature variation in the Transition area off northeastern Japan. Rep. Res. Meeting on *North Pacific Krill Resources* No. 3, 100–101 (in Japanese).

Yamada, M. (1964) On the lipids of plankton. *Bull. Jap. Soc. Scient. Fish.* **30**, 673–81 (in Japanese).

Yoshida, H. (1995) A study on the price formation mechanism of *Euphausia pacifica*. *Bull. Jap. Soc. Fish. Oceanogr.* **59**, 36–38 (in Japanese).

Chapter 12
Management of Krill Fisheries in the Southern Ocean

Denzil Miller and David Agnew

12.1 Introduction

Over the past thirty years, the growing demand for protein from the sea along with increasingly restricted access to historical fishing grounds has resulted in the development of many 'unconventional' fisheries (Robinson, 1982; Budzinski *et al.*, 1985; Miller, 1990). The Antarctic krill *(Euphausia superba* Dana, hereinafter referred to as 'krill' in this chapter) fishery is a clear example of such development.

Exploitation of Antarctic marine living resources has been characterised by intense and sporadic cycles (Gulland, 1977, 1983) often resulting in severe depletion of harvested stocks (especially seals, whales and finfish) (Laws, 1977, 1985, 1989; Kock, 1991, 1992). This has raised serious concerns about the future effective management and sustainable utilisation of resources (Mitchell & Sandbrook, 1980) with krill, as a key species, assuming prominence (Nicol, 1989; Miller, 1991).

Krill has long been recognised as an integral component of many Antarctic marine food webs (e.g. Marr, 1962; Everson, 1977, 1984; Knox, 1984; Miller & Hampton, 1989 among others). Several important attributes enhance its potential as an exploitable resource (Everson, 1977, 1978; El-Sayed, 1988; Nicol, 1989) as well as its 'availability' to many predators. These are, *inter alia*, high global abundance (Everson, 1977; Gulland, 1983), high nutritional value (Grantham, 1977; Budzinski *et al.*, 1985), and a tendency to aggregate which increases the ease of location and capture (Eddie, 1977; El-Sayed & McWhinnie, 1979).

Exploratory fishing for krill commenced in the early 1960s with catch levels being initially low (Table 12.1). The build-up of catches was slow and it was not until the 1973–1974 season that the fishery assumed commercial significance. In this review, catch statistics are used to indicate areas where the fishery has been most active and from where the largest catches have been taken. Trends in catches are analysed to provide insights into possible relationships between the fishery, its operational constraints and krill distribution/abundance (i.e. availability). Earlier reviews by Miller (1990) and Agnew & Nicol (1997) are updated to include recent efforts by the 23-nation Commission for the Conservation of Antarctic Marine Living Resources (CCAMLR) and its Scientific Committee (SC-CAMLR) to develop a scientifically

Table 12.1 Annual krill catches prior to 1973/74 (from Everson, 1978; Bengtson, 1984; Miller, 1990).

Season	Catch (t)	Comments	Reference
1961/62	4	Krill	Burukovskiy & Yaragov (1967)
1963/64	70	Krill	Stasenko (1967)
1964/65	306	Krill	Nemoto & Nasu (1975)
1966/67	?	Krill	Nemoto & Nasu (1975)
1968/69	>140	Krill	Ivanov (1970)
1969/70	100	UMC*	FAO (1976)
1970/71	1300	UMC*	FAO (1976)
1971/72	2100	UMC*	FAO (1976)
1972/73	7459	Krill	FAO (1976)
			Nemoto & Nasu (1975)

* Catches of 'Unspecified Marine Crustacea' for areas closely adjacent to the Antarctic area assumed by Everson (1978) to be krill.

based management regime for the krill fishery. Advice on the management of Antarctic marine living resource exploitation has been developed by specialist working groups of SC-CAMLR. For krill these comprised the Working Group on Krill (WG-Krill) (1987–1994) and the Working Group for the CCAMLR Ecosystem Monitoring Programme (WG-CEMP) (1986–1994). In 1994 the two Groups were combined into the Working Group for Ecosystem Monitoring and Management (WG-EMM).

12.2 Trends in the fishery

Available data

Krill catches are reported from three major statistical areas in the Antarctic, which are each divided into a number of subareas or divisions (Fig. 12.1). Summary catch and effort data are submitted annually to CCAMLR, with catches being attributed to split-years – a split-year being the twelve-month period from 1 July in one year to 30 June in the next. Since 1990, CCAMLR fisheries statistics have been published annually in a *Statistical Bulletin* (Statistical Bulletins 1990–1998). The former Soviet Union was responsible for the bulk of catches taken during the period under review. Soviet catch data in all areas were only separated by month from 1983 and effort data (e.g. hours fished) are only available from that year. From 1991–1992, krill catches and the reporting thereof by the former Soviet Union were split between the Russian Federation (a CCAMLR Member) and Ukraine (non-CCAMLR Member prior to 1995). For the bulk of the analyses post-1992, we have combined the catches of these two nations, considering them synonymous with those of the Soviet Union.

The data used to investigate relationships between sea-ice and the krill fishery were derived from ice concentration data held in the database of the National Snow

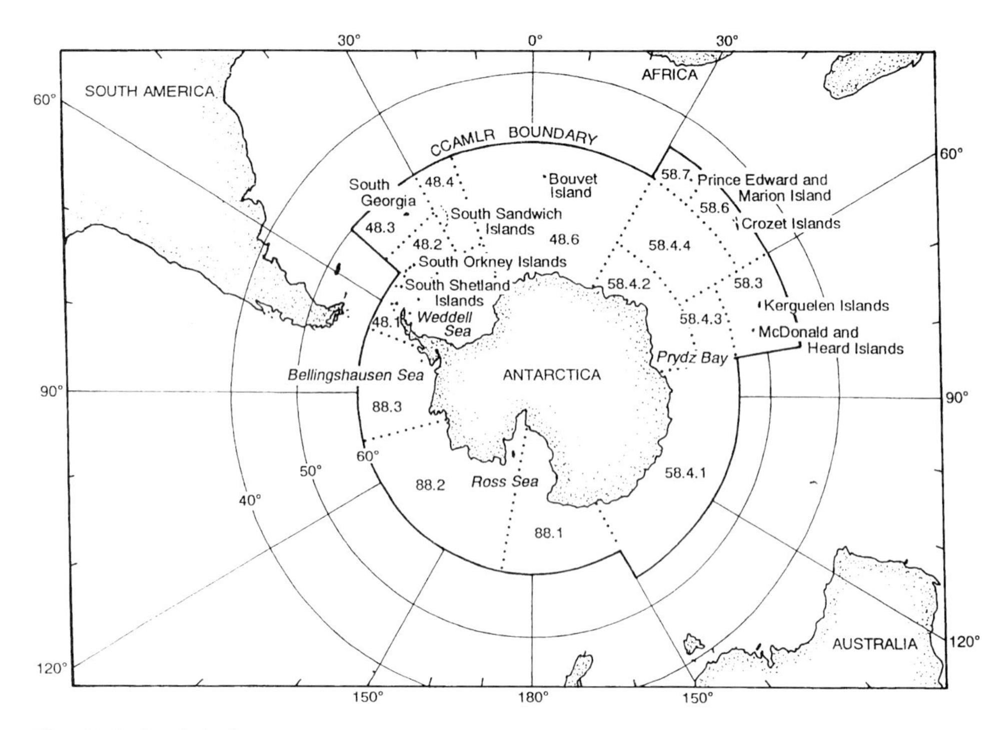

Fig. 12.1 Statistical Areas, Subareas and Divisions of the CAMLR Convention Area (from Anon., 1995). With permission from the CCAMLR.

and Ice Centre, Colorado, USA and from the CCAMLR Statistical Bulletin (Statistical Bulletins 1990–1998). The possible association between sea-ice and krill catch-per-unit effort (CPUE) was investigated by comparing ice cover in October of one split-year with CPUE in that split-year. For example, ice cover in October 1988 (i.e. from the 1988–1989 split-year) was compared with krill CPUE in 1988–1989. In Subareas 48.1 and 48.2, most fishing takes place in summer and autumn (see Section on Seasonal distribution (i.e. 3–6 months after the October calculation date)). In Subarea 48.3, fishing takes place in the winter (June–September) and so the correlation was performed between ice in October and krill CPUE of the following split-year (e.g. ice in October 1988 from the 1988–1989 split-year was compared with CPUE from the 1989–1990 split-year).

Catch trends

Annual catches

From the first commercial takes in the early 1970s, krill catches rose steadily from 19 785 t in 1973–1974 to a peak of 528 201 t in 1981–1982 (Fig. 12.2). Catches then declined sharply until 1983–1984 due to marketing and processing problems possibly associated with the discovery of high levels of fluoride in the exoskeleton of krill

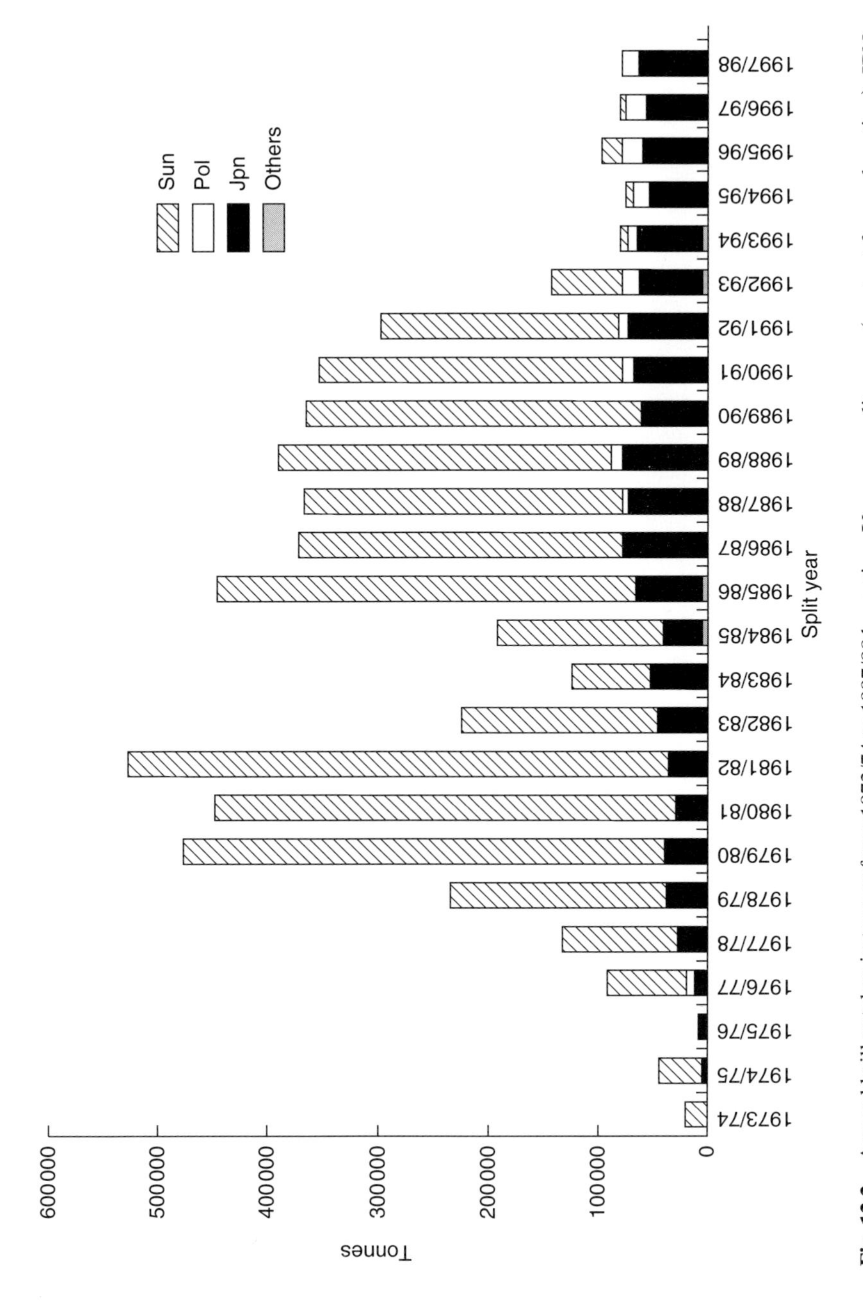

Fig. 12.2 Annual krill catches in tonnes from 1973/74 to 1997/98 by nation. Years are split years (see text for explanation). JPN = Japan, POL = Poland, SUN = Soviet Union (including Russia and Ukraine post-1990/91) and OTHERS = Other krill fishing nations (Bulgaria, Chile, Korea and United Kingdom).

(Nicol & Endo, 1997; Ichii, Chapter 9). Following the solution of these technical problems by rapid processing and the development of improved peeling machines, catches increased again, levelling off at around 300 to 400 000 t. The next major impact on the krill fishery was triggered by the breakup of the Soviet Union in 1991 with the subsequent removal of fuel subsidies for Russian and Ukrainian vessels (Nicol & Endo, 1997). The subsequent drop in catches is graphically illustrated in Fig. 12.2.

Other key nations (mainly Poland, Chile and Korea) have accounted for about 4% of the total catch. Korea entered the fishery in 1978–1979, fishing sporadically in Subarea 58.4 from 1978–1979 to 1983–1984 and took about 9000 t during that time. It re-entered the fishery in 1986–1987, moving its operations to Subareas 48.1 and 48.2, and taking about 1500 t annually until 1991–1992. Both Chile and Poland undertook exploratory fishing for krill between 1976 and 1977, but did not enter the fishery in earnest until 1982–1983 and 1985–1986 respectively. Chile built up to annual catches of about 5000 t throughout the 1980s, leaving the fishery after the 1993–1994 season. Poland has taken variable catches between 2000 and 20 000 t to date. The United Kingdom (UK) has been a recent entry into the fishery taking 308 and 634 t in 1996–1997 and 1997–1998 respectively. Very recently there has been a resurgence of interest in fishing for krill, with the entry of some new players such as the UK, Argentina and USA (CCAMLR, 1998; SC-CAMLR, 1998, 1999), with some interest also being shown by Canada, a Convention Acceding State but not a Member of CCAMLR.

Initially, the krill fishery was distributed in all three CCAMLR Statistical Areas (48, 58 and 88; Fig. 12.1). Japan started fishing in Division 58.4.2 which was logistically convenient. However, following comprehensive exploration of most areas around the Antarctic continent, the fishery became concentrated in the Atlantic sector (Area 48) where the most predictable concentrations of krill are to be found (Fig. 12.3). Catches in Areas 58 and 88 were located along the continental shelf break, relatively close to the Continent (Ichii, 1988, 1990). Catches in Area 48 have also been concentrated on shelf breaks, to the north and west of the South Shetland Islands, in a rather more diffuse area north of the South Orkney Islands and to the north of South Georgia (Agnew 1993; Ichii, Chapter 9).

There has been one reported catch of krill outside the CCAMLR Convention Area in 1992–1993 by Poland immediately north of Subarea 48.1 (FAO Statistical Division 41.3.2) just to the south of the Antarctic Polar Front (Fig. 12.1). Occasionally patches of krill have been reported to drift in a north-easterly direction from Elephant Island to north of 60°S (i.e. out of the CCAMLR Area) with catches ultimately being made in the vicinity of 57–60°S, 50–60°W (Sushin & Myskov, 1993). Drifter buoys released close to the South Shetland Islands have also been observed to transit Division 41.3.2 *en route* to South Georgia (Ichii *et al.*, 1993), while the First International BIOMASS (Biological Investigations of Marine Antarctic Systems and Stocks) Experiment (FIBEX)(Anon., 1986) found significant quantities of krill in the same region (Trathan *et al.*, 1998). Division 41.3.2 thus appears to be bio-

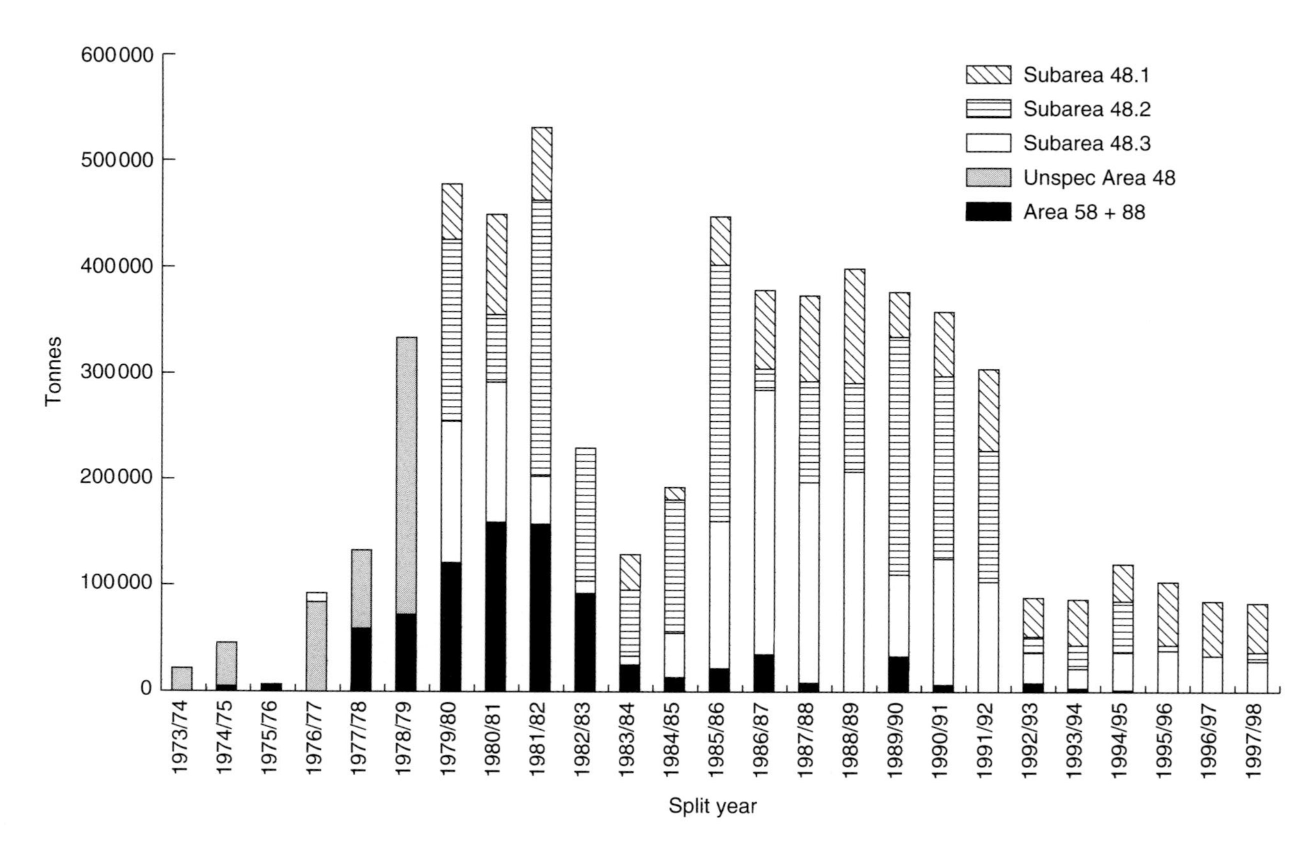

Fig. 12.3 Annual krill catches in tonnes by CCAMLR Statistical Area or Subarea. Years are split years (see text for explanation), A58&88 = Areas 58 and 88 combined, Unspec area 48 = Unspecified locations within Area 48, S48.1 = Subarea 48.1, S48.2 = Subarea 48.2 and S48.3 = Subarea 48.3.

geographically contiguous with Subarea 48.1 and could be considered as part of the Scotia Sea system, along with Subareas 48.1, 48.2 and 48.3.

Seasonal distribution

The seasonal pattern of krill catches around the Antarctic is, as one would expect, critically linked to seasonal sea-ice distributions. In areas adjacent to the Antarctic Continent fishing generally commences in the early austral spring as the ice edge moves southwards exposing the krill-rich waters of the shelf break, and ceases in about May with krill appearing to remain over the shelf break under the ice during the winter (Ichii, 1990). In Subarea 48.2, which is slightly more oceanic than other close-continent areas, fishing commences at a similar time but catches are taken over a longer period (January to May) and usually peak slightly later than in Subarea 48.1 (March–April). By contrast, the waters around South Georgia (Subarea 48.3) remain ice-free in winter, and so constitute the only waters Antarctic-wide which contain fishable krill concentrations at this time of year. The positions of krill catches in the Atlantic sector have therefore been clearly linked to the progressive encroachment of ice northwards during the autumn and winter so forcing the fishing fleet also to move north (Dolzhenkov *et al.*, 1989; Miller, 1990; Everson & Goss, 1991; Agnew, 1993).

There has been a general trend over the last decade for fishing in Subareas 48.1 and 48.2 to show a shift in median catch date to later in the season (Fig. 12.4). In 48.1, the fishing period extended from about 4 months (December–March) in the early 1980s to 8 months (November–June) now. However, in 48.2, although it expanded from 6 months (December to May) in the early 1980s to 9 months in the early 1990s, the fishing season has since contracted to about 4 months (March–June). In 48.3, by contrast, the median catch date (of the winter fishery) has remained relatively unchanged, and as in Subarea 48.2 the duration of the fishery has contracted from 10 (January–October) to 5 months (May–September).

There are a number of explanations for these observed changes in fishing pattern, both environmental and operational. In Subarea 48.2, the trend in median catch date may have been a response to increasing availability of ice-free areas late in the season, allowing fishing to continue to June in many years. The number of ice-free days around Signy Island (Subarea 48.2), calculated by SC-CAMLR (1996) as the duration from ice retreating in the spring to ice returning in the autumn, displays a well-documented cyclic pattern (Murphy *et al.*, 1995), but also shows a general trend of increasing duration in the ice-free period over the last decade (Fig. 12.5). In contrast, oceanographic conditions around the Antarctic Peninsula mean that although the cyclic phenomenon is still apparent in Subarea 48.1 there is little long-term trend. With the exception of the period 1981–1982 to 1984–1985, no clear linking of the later median catch date to the availability of ice-free areas is possible (Figs 12.4 and 12.5). Furthermore, Agnew (1993) has noted the close dependence of the distribution of catches in Subarea 48.2 on variable sea-ice conditions, in contrast

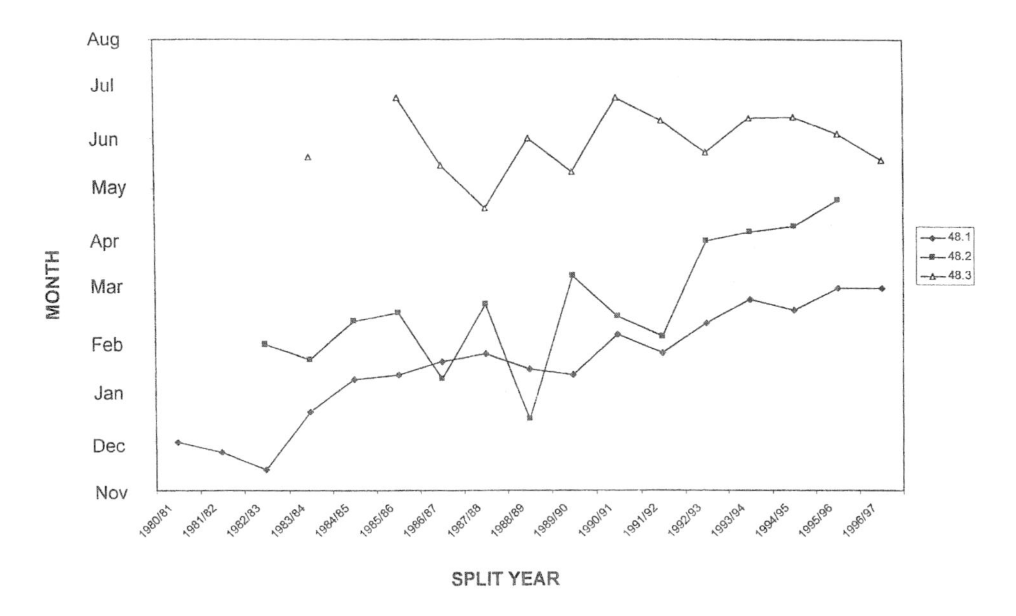

Fig. 12.4 Median catch date by month. For krill fisheries in CCAMLR Subareas 48.1 and 48.2 the median date refers to the summer fishery of the split year. For Subarea 48.3, the median date refers to the winter fishery which commenced on 1 January of a split year. For instance, for 1986–87 this is the winter fishery of the year 1987, taking place between 1 January 1987 and 31 December 1987. 48.1 = Subarea 48.1, 48.2 = Subarea 48.2 and 48.3 = Subarea 48.3.

to Subarea 48.1 where a remarkably consistent pattern of catches from year to year was found. The trend to later autumn fishing in Subareas 48.1 and 48.2 may therefore only partially be driven by decreasing ice coverage (see Section 12.5).

12.3 Developing a management regime

Background

The CAMLR Convention was negotiated between 1978 and 1980, entering into force in 1982. The substantial krill fishery throughout this time, coupled with growing interest from countries such as Poland, Korea and Chile, imparted an urgency to CCAMLR's establishment. As already indicated, this was particularly attributable to krill's perceived pivotal role as a key ecosystem component and concern that it not be over-exploited (Edwards & Heap, 1981). Therefore, on its entry into force CCAMLR was faced with the daunting task of developing an effective strategy to meet the Convention's primary objectives and to manage an anticipated expansion in krill exploitation levels (Everson, 1988a; Miller, 1991).

The fundamental management objectives of CCAMLR are set out in Article II of the Convention (Table 12.2). From the outset the ecological importance of krill as a key species in the Antarctic marine ecosystem, and the consequent need to manage

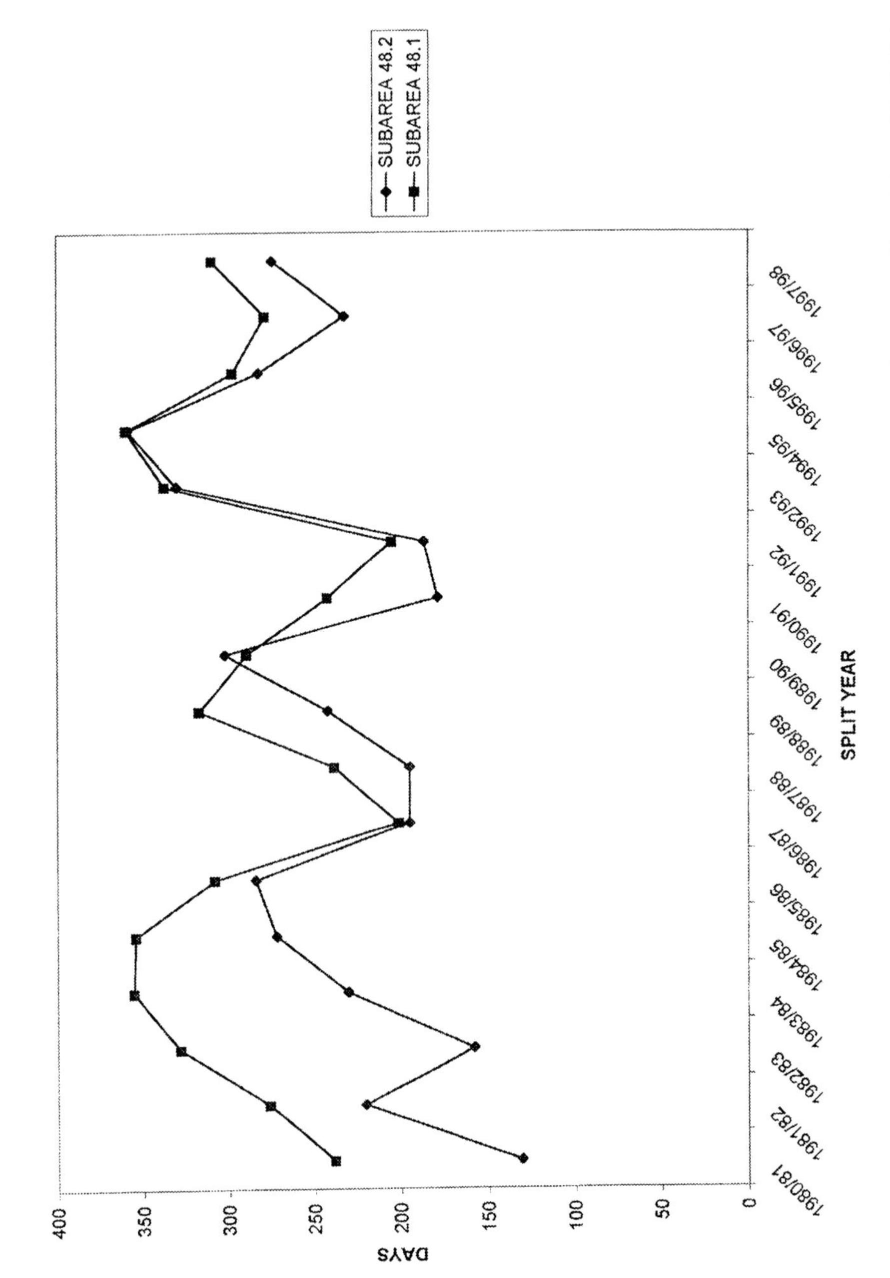

Fig. 12.5 Number of ice-free days (duration of the period between ice retreat and ice return) calculated by SC-CAMLR (1997) for sites in Subareas 48.1 and 48.2.

Table 12.2 Article II of the CAMLR Convention (from Anon., 1995).

(1) The objective of this Convention is the conservation of Antarctic marine living resources

(2) For the purposes of this Convention, the term 'conservation' includes rational use

(3) Any harvesting and associated activities in the area to which this convention applies shall be conducted in accordance with the provisions of this Convention and with the following principles of conservation:

(a) prevention of decrease in the size of any harvested population to levels below those which ensure its stable recruitment. For this purpose its size shall not be allowed to fall below a level close to that which ensures the greatest net annual increment;
(b) maintenance of the ecological relationships between harvested, dependent and related populations of Antarctic marine resources and the restoration of depleted populations to the levels defined in sub-paragraph (a) above, and
(c) prevention of changes or minimization of the risk of change in the marine ecosystem which are not potentially reversible over two or three decades, taking into account the state of available knowledge of the direct and indirect impact of harvesting, the effect of the introduction of alien species, the effects of associated activities on the marine ecosystem and of the effects of environmental changes, with the aim of making possible the sustained conservation of Antarctic marine living resources.

stocks in such a way as to minimise potential ecological risks both to krill and to krill predators, was recognised (Everson, 1977; Knox, 1984; Butterworth, 1986; Miller, 1991; Nicol & de la Mare, 1993). CCAMLR was the first international fisheries arrangement to explicitly assume both precautionary and ecosystem approaches as basic management principles. As a result, CCAMLR has addressed management of the krill fishery from two perspectives: a bottom-up, 'direct', approach (i.e. examining krill as exploitable populations with predictable dynamics), and a top-down, 'indirect', approach (i.e. considering the potential effects of krill exploitation on the ecosystem as a whole).

Formulating operational management objectives

Butterworth (1986) was among the first to recognise that more traditional approaches to fisheries management, such as $F_{0.1}$ are unlikely to be applicable to krill. This is attributable to such factors as potential density-dependent fishing mortality arising from the species' aggregating behaviour, problems associated with accurately ageing krill and the general inapplicability of $F_{0.1}$ in multi-species applications. $\mathbf{F}_{0.1}$ is often used as a reference point in fisheries management. If unexploited fisheries are subject to gradual increases in fishing mortality (**F**) the yield (catch) per individual recruit rises to either a maximum peak or asymptote $\mathbf{F}_{0.1}$ which is that fishing mortality at which the increase in yield-per-recruit is 10% of the increase in yield-per-recruit at the origin. As a management reference point it is at a lower **F**, which is more conservative than the often quoted $\mathbf{F}_{max}$. Subsequently the need for 'operational management principles' (after Butterworth, 1990) to manage krill stocks in specified 'management' areas was recognised by CCAMLR. Four inter-related elements were identified as important (SC-CAMLR 1990):

(1) Requiring a basis for assessing the status of krill stocks (an estimator).
(2) Developing an algorithm to specify appropriate regulatory procedures [i.e. a 'Catch control law' such as a Total Allowable Catch (TAC)].
(3) Simulating and testing the performance of management procedures [i.e. components of both (1) and (2)] and
(4) Developing operational definitions of Article II (i.e. quantities measurable from field observations) to provide criteria against which management performance procedures may be assessed.

The operational management principles accepted by CCAMLR in the development of krill management goals (SC-CAMLR, 1990; CCAMLR, 1990) were:

(1) To aim to keep krill biomass at a level higher than would be the case for single-species harvesting considerations.
(2) Given that krill dynamics have a stochastic component, to focus on the lowest biomass that might occur over a future period, rather than on the average biomass at the end of that period, as might be the case in a single-species context.
(3) To ensure that any reduction of food to predators which may arise out of krill harvesting is not such that land-breeding predators with restricted foraging ranges are disproportionately affected compared to predators in pelagic habitats and
(4) To examine what levels of krill escapement are sufficient to meet the reasonable requirements of predators.

These principles are applied so as to account for sustained consistency in catch levels with time while taking into account uncertainties (i.e. krill demand of predators) so that, on available information, at least, the possibility of violating the objectives of Article II is reduced (Butterworth *et al.*, 1994).

In a practical sense, CCAMLR's first task was to identify management areas in which to apply the above principles. However, the delineation of krill stock boundaries has proved extremely difficult since the extent to which krill are resident in various areas, or move between them as passive particles in the eastward flowing Antarctic Circumpolar Current, is largely unknown and likely to be quite variable (Miller & Hampton, 1989; see Section 12.5, Krill flux). For practical purposes, therefore, CCAMLR has focused its efforts on areas where the fishery is primarily located (i.e. Subareas 48.1, 48.2, 48.3, and to a lesser degree 58.4). These areas, particularly 48.1 and 48.3, contain sites where large colonies of land-based, krill predators breed or are located (Bengtson, 1984; Croxall *et al.*, 1988; Croxall, 1990).

Consequently, all elements of the management paradigm (fishery, krill and land-based predators) overlap in space and/or time to varying degrees.

Second, methods for estimating appropriate levels of harvesting (i.e. yield) were needed. A number of different approaches have been developed for this purpose – both direct methods of estimating yield, based on surveys of stock biomass, and indirect methods of estimating trends in biomass. Finally, the effect of krill harvesting on dependent predators has also been incorporated at various levels, with explicit account being taken of such effects in the yield models in association with subsequent monitoring of potential changes likely to be associated with krill harvesting. The following sections describe these various approaches.

12.4 Estimating stock size and yield

Direct approaches

Estimation of krill yield

At the outset, CCAMLR recognised that a substantial amount of data would be required before traditional feedback assessments might be used for krill. Given the size of most management areas for the species and difficulties inherent to determining age (Siegel & Nicol, Chapter 5), it is doubtful whether traditional annual assessments of krill will ever be achievable (Butterworth, 1986; Everson, Chapter 3). CCAMLR therefore concentrated on developing calculations of long-term yield based on the approach of Beddington & Cooke (1983), as initially modified by Butterworth *et al.* (1992, 1994). In this approach, estimates of krill recruitment variability, growth and natural mortality are used in a stochastic simulation to determine the effects of various levels of harvesting on the population. For each level of harvesting, the parameter γ in the equation $Y = \gamma B_o$ is calculated and the stock is tracked over a 20-year period (where Y = yield and B_o = unexploited biomass). Because this stochastic model is run a large number of times, the distribution of various management quantities can be determined and the probability of various outcomes of both management and fisheries-associated actions can be calculated. To select an appropriate level of yield, a three part decision rule, described in Table 12.3, was developed to select an appropriate value for γ.

This approach requires an estimate of pre-exploitation biomass (B_o) from acoustic survey information. In its first application (SC-CAMLR, 1991), acoustic estimates of krill biomass from the 1981 FIBEX survey (Anon., 1986) were used to estimate B_o in Statistical Area 48. The FIBEX estimate was favoured since it covered Area 48 as a whole and there was little necessity for adjustment to account for krill flux (i.e. movement). A year after initiating the above, CCAMLR adopted its first krill Conservation Measure (CM)(CM 32/X)(Table 12.4) in 1991 using results from the yield model (γ was estimated as 0.1 and the FIBEX B_o estimate for Area 48 was 15 million t).

Table 12.3 Three-part decision rule to select value of γ (proportionality coefficient) used by CCAMLR in setting precautionary catch limits for the Antarctic krill fishery (SC-CAMLR, 1994). See text for full details of management procedures and their development.

(i)	γ (γ_1) is chosen so that probability of spawning biomass dropping below 20% of pre-exploitation median level over a 20-year harvesting period is 10%;
(ii)	γ (γ_2) is chosen so that median krill escapement over a 20-year period is 75%, and
(iii)	The lower of γ_1 and γ_2 is selected as the level of γ for the calculation of krill yield.

CM 32/X incorporated two 'limitations' on the uncontrolled expansion of krill fishing. Firstly, the precautionary krill catch limit of 1.5 million t for Area 48 was seen as being sufficiently above prevailing historic catch levels to allow for reasonable growth of the fishery, but low enough to minimise the possibility of a detrimental impact on krill stocks (i.e. some allowance was made for uncertainty in parameter estimates used in the yield model) (SC-CAMLR, 1991). Secondly, the 'catch trigger' (620 000 t) for Subareas 48.1, 48.2 and 48.3 was a little higher than the largest annual krill catch to date. This was perceived to be the level at which rapid expansion of catch is most likely so necessitating suitable subdivision of catch by Subarea (i.e. paragraph 3 of CM 32/X) to avoid possible unacceptable concentration of catch within the foraging areas of vulnerable predators (see below and Section 12.6, Overlap of krill fishery and predators)(SC-CAMLR, 1991).

Initially the krill yield model used a fixed natural mortality, M. However, there is considerable uncertainty concerning a suitable level for this parameter, so in later models it was ascribed to a uniform distribution between 0.6 and 1.0. Other parameters, such as variable age-at-recruitment, seasonal growth and catch history, as well as age-at-first-capture were also added to the model. Prior distributions for a number of parameters were refined, particularly recruitment variability and the relationship between this and likely variability in M (de la Mare 1994a, b). Recruitment variability is now estimated from survey data as variation in the ratio of 1- and 2-year-old krill in the older population. There is a considerable time series of surveys from which this parameter can be estimated, especially around the Antarctic Peninsula region (Siegel *et al.*, 1998).

Recently an age-based krill population model has been constructed for the Antarctic Peninsula. This confirms that the recruitment, growth and natural mortality estimates used in the general yield model are internally consistent and are capable of reproducing population trends similar to those observed in surveys (Murphy *et al.*, 1999). The effects of serial correlation in krill recruitment have also been investigated, especially following observations of linkages between recruitment and sea-ice cover (Loeb *et al.*, 1997), and cyclicity in sea-ice dynamics (Murphy *et al.*, 1995). Finally, B_o values were recalculated for various Statistical Subareas (Trathan *et al.*, 1993, 1995) using a revised acoustic target strength for krill (after Everson *et al.*, 1990; Foote *et al.*, 1990; Greene *et al.*, 1990). Subsequently, CCAMLR

Table 12.4 CCAMLR Conservation Measures for krill.

CONSERVATION MEASURE 32/X

The total catch of *Euphausia superba* in Statistical Area 48 shall be limited to 1.5 million tonnes in any fishing season. A fishing season begins on 1 July and finishes on 30 June of the following year.

This limit shall be kept under review by the Commission, taking into account the advice of the Scientific Committee.

Precautionary limits to be agreed by the Commission on the basis of the advice of the Scientific Committee shall be applied to subareas, or on such other basis as the Scientific Committee may advise, if the total catch in Subareas 48.1, 48.2, and 48.3 in any fishing season exceeds 620 000 tonnes.

For the purposes of implementing this Conservation Measure, the catches shall be reported to the Commission on a monthly basis.

**CONSERVATION MEASURE 45/XI*

The total catch of *Euphausia superba* in Statistical Division 58.4.2 shall be limited to 390 000 tonnes in any fishing season. A fishing season begins on 1 July and finishes on 30 June of the following year.

This limit shall be kept under review by the Commission, taking into account the advice of the Scientific Committee.

For the purposes of implementing this Conservation Measure, the catches shall be reported to the Commission on a monthly basis.

CONSERVATION MEASURE 45/XIV

The total catch of *Euphausia superba* in Statistical Division 58.4.2 shall be limited to 450 000 tonnes in any fishing season. A fishing season begins on 1 July and finishes on 30 June of the following year.

This limit shall be kept under review by the Commission, taking into account the advice of the Scientific Committee.

For the purposes of implementing this Conservation Measure, the catches shall be reported to the Commission on a monthly basis.

***CONSERVATION MEASURE 46/XI*

If the total catch of *Euphausia superba* in Statistical Subareas 48.1, 48.2 and 48.3 in any fishing season exceeds 620 000 tonnes, then catches in the following Statistical Subareas shall not exceed the precautionary catch limit prescribed below:

Antarctic Peninsula	Subarea 48.1	420 000 tonnes
South Orkney Islands	Subarea 48.2	735 000 tonnes
South Georgia	Subarea 48.3	360 000 tonnes
South Sandwich Islands	Subarea 48.4	75 000 tonnes
Weddell Sea	Subarea 48.5	75 000 tonnes
Bouvet Island region	Subarea 48.6	300 000 tonnes

Notwithstanding these subareal limits, the total sum of catches in any fishing season in all Subareas shall not exceed the precautionary catch limit of 1.5 million tonnes for the whole of Statistical Area 48 prescribed by Conservation Measure 32/X. A fishing season begins on 1 July and finishes on 30 June of the forthcoming year.

The above precautionary catch limits shall apply to the fishing seasons 1992/93 and 1993/94 after which time they will be reviewed by the Commission, taking into account the advice of the Scientific Committee.

* revised in 1995

** lapsed in 1994

has agreed to re-survey krill B_o in Area 48 during the austral summer of 1999–2000 (CCAMLR, 1998).

Alongside development of the above model, two important additional management principles were elaborated. Firstly, CCAMLR continually requested consideration of the possible allocation of the area precautionary catch limit of 1.5 million t to subareas within Area 48. A variety of methods were identified (e.g. based on historic catch levels, predator needs and various proportionate combinations of results from the yield model). It was acknowledged that there is some need to define krill 'management regions' in place of more *pro rata* designated Statistical Areas/Subareas (CCAMLR, 1992). The demarcation of such regions was recognised as needing to account for equivocal evidence on the potential effects of localised krill fishing on land-based predators (Agnew, 1993; Sushin & Myskov, 1993; Kerry *et al.*, 1993)(Section 12.6, Overlap of krill fishery and predators). Following work in 1992, CM 46/XI (Table 12.4) was adopted as an interim allocation of catches by Statistical Subarea within Area 48 for the 1992–1993 and 1993–1994 seasons (CCAMLR, 1992). However, this measure subsequently lapsed and its replacement awaits the results of attempts to refine alternative subarea allocations based on krill, fishery, predator and environmental considerations (e.g. along the lines outlined by Watters & Hewitt, 1993 and SC-CAMLR, 1999).

The second, far-reaching development was the elaboration of a decision rule to address the requirements of predators (CCAMLR, 1994). In early application of the krill yield model, allowance for predators had been discussed and a 'discount factor' to proportionately reduce calculated yield was proposed. The difficulty with this suggestion, of course, was that there was no way of determining an appropriate level for the discount factor, and in any case it was argued that krill-predator needs were taken into account implicitly through estimation of krill M. As an alternative, the population projections of the model were used to define a second decision rule to complement the rule safeguarding against critical reductions in stock biomass (Table 12.3). Together, both decision rules act in a conservative manner to ensure that krill spawning stock as well as predator (ecosystem) needs are safeguarded. By adopting such rules, CCAMLR finally achieved an explicit formulation of the operational criteria (Section 12.3, Formulating operational management objectives) it had originally identified as crucial for management of the krill fishery. Comparable rules have since been applied to species other than krill (e.g. Patagonian Toothfish, *Dissostichus* spp.) (SC-CAMLR, 1998).

In 1994, CCAMLR accepted the revised yield model, final FIBEX B_o estimates, and three-part decision rule, thereby completing the most important phase in the development of a krill management procedure within four years (1991–1994). The value of γ now accepted for krill, following the three-part decision rule, is 0.116. Catch limits for Divisions 58.4.2 and 58.4.1 have been set using this γ and estimates of B_o from Australian acoustic surveys (Pauly *et al.*, 1996; see also Everson, Chapter 3).

However, although a revision of the FIBEX results in 1994 using new target strengths for krill suggested that the precautionary yield in Area 48 should be 4.1

million tonnes, a new Conservation Measure has not yet been adopted. This is attributable to:

(1) a current lack of guidelines on which to base an allocation of catch limits to Subareas within Area 48, and
(2) a growing concern that the estimate of B_o from the 1981 FIBEX survey was made in a year of high krill biomass that is no longer characteristic of the South Atlantic.

Siegel *et al.* (1998) have demonstrated that, at least in the Antarctic Peninsula area, krill biomass was an order of magnitude lower for most of the period 1985–1995 than in 1981–1982. Although the biomass may have increased again in 1996 this is thought to be temporary (SC-CAMLR, 1998 – Annex 4, Appendix D, Fig. 12). Therefore, as already mentioned, a synoptic survey of Area 48 was planned for early 2000 to establish a new B_o for krill in the Area. Adoption of any replacement for CM 32/X also awaits the outcome of deliberations aimed at developing allocation procedures for catch limits to subareas (SC-CAMLR, 1998). As the current catch is perceived to be low (about 100 000 t and therefore less than 10% of the precautionary yield), there appears to be no immediate urgency to revise CM 32/X for Area 48 (Table 12.4) until improved estimates of krill precautionary catch limits are available based on, for example, the synoptic survey alluded to above.

Taking account of functional interaction(s) between krill and predators

Although the decision rules (Table 12.3) take explicit account of the needs of predators, arbitrary levels of krill escapement have been set to meet such needs commensurate with the rules. However, CCAMLR has realised that existing data on predator performance and krill variability might facilitate explicit modelling of the functional relationship between predator and krill populations. Consequently, this may facilitate more objective definition of the levels of escapement necessary to meet predator needs (Butterworth & Thomson, 1994, 1995).

The above approach initially utilised preliminary information on the population dynamics of Adélie penguins (*Pygoscelis adelie*), Antarctic fur seals (*Arctocephalus gazella*), crabeater seals (*Lobodon carcinophagus*) and blackbrowed albatross (*Diomedea melanophrys*) to identify functional relationships between juvenile and adult predator survival rates with krill abundance (Butterworth & Thomson, 1995). From the 'one-way' interaction model developed (i.e. where fluctuations in krill abundance impact on predator populations, but not vice versa), variability in annual krill recruitment was shown to render predator populations less resilient to krill harvesting than deterministic evaluations would suggest (Butterworth & Thomson, 1995). These initial results focused subsequent discussion on interpreting preliminary adult survival rate estimates for some of the predator populations used in the model (SC-CAMLR, 1993).

Further model developments (Butterworth & Thomson, 1994), introduced the concept of krill 'availability' to provide a random component in the relationship between biomass and availability (i.e. with biomass as a function of spatial as well as temporal variability). For the blackbrowed albatross and Antarctic fur seal in particular, an almost unrealistically poor resilience was observed for these two species in relation to krill fishing, and by implication to changes in krill availability (even without variability in krill recruitment being taken into account) (Butterworth & Thomson, 1994). Currently, CCAMLR is now striving to ascertain whether the modelling technique is inappropriate or whether there are inherent negative biases in available estimates of predator survival rates from field data (SC-CAMLR, 1995, 1997).

In the absence of more quantitative assessments of predator responses to changes in krill availability in general, and to different levels of krill escapement in particular, SC-CAMLR has accepted a target level of escapement of 0.75 as an initial value on which to base management recommendations (SC-CAMLR, 1994). The current krill escapement value of 0.75 lies between the customary 0.5 level applicable in the single species context and the 1.0 level commensurate with no fishing. CCAMLR has acknowledged that this value may be revised as the modelling initiatives described here are further refined.

Indirect approaches

Tracking indices in krill biomass

In its early years, and following realisation that age-based assessments of krill were unlikely to be possible, CCAMLR considered using CPUE indices to monitor trends in krill population biomass and in biomass dynamic models. In 1986, simulation studies were commissioned of the Soviet and Japanese krill fisheries respectively to investigate the possibility of using CPUE as an index of krill biomass (SC-CAMLR, 1986, 1987). These studies (Butterworth, 1989a, b; Mangel, 1989; SC-CAMLR, 1988) concluded that certain catch-dependent indices, particularly those containing some element of fishery search-time, could be used to assess krill abundance. The CPUE simulation studies supported Shimadzu's (1985) and Everson's (1988a) earlier conclusions that catch-per-fishing time (i.e. CPH) provides the most useful index of local krill abundance. They also indicated that various catch and effort data might be utilised to derive a *Composite Index of Krill Abundance* (SC-CAMLR, 1988).

A recent experiment to investigate the properties of a *Composite Index* found that the frequency distributions of commercial catch-per-fishing time and krill density from acoustic surveys showed similar forms, although non-random movement by a fishing vessel may obscure such comparisons (SC-CAMLR, 1994). A further difficulty in using search-time is that fishing operations are limited by catch processing efficiency rather than krill availability. Unfortunately, the conclusion of this work is that CPUE is not a very good estimator of biomass and to date there have been no further attempts to utilise it. Nevertheless, CCAMLR has continued to encourage development of approaches to use CPUE data to monitor fishing activities in relation to krill biomass. More specifically, it has endorsed the need to collect information on fishing vessel activities at random times in an effort to quantify search-time (SC-CAMLR, 1993).

Estimating krill potential yield from predator consumption

As early as 1990, CCAMLR discussed the use of predator consumption rates to bound estimates of B_0 and, consequently, krill potential yield (SC-CAMLR, 1990; Agnew, 1992). Everson & de la Mare (1996) have recently refined these ideas, to provide a method for estimating B_0 from predator consumption rates around South Georgia with:

$$B_o = \frac{PT(M^2 + V[M])}{M^3} \tag{12.1}$$

where: P = annual krill consumption by land-based predators; M = annual krill mortality rate; T = krill retention time within predator foraging area(s); $V[M]$ = variance of M estimate.

For example, Everson & de la Mare estimate that, at South Georgia, 9.76 million t of krill are taken by predators annually, mostly within 100 km of the Island. With a turnover rate of once a year, and estimates of M and $V[M]$ of 0.6 and 0.1 respectively, the calculated instantaneous standing stock in the area of 9.5 million t is considerably greater than the revised FIBEX estimate of 4.1 million t for Area 48 as a whole (SC-CAMLR, 1994). This figure, when used with the estimate of γ from the potential yield model (0.116) infers an annual krill yield of 1.1 million t in the vicinity of South Georgia.

12.5 Implications on krill management of links between krill and environmental parameters

Krill flux

CCAMLR has recognised that application of both direct and indirect approaches is crucially dependent on movement (flux) of krill between various localities. Subsequently, developments in 1993 focused on the effect of krill flux (either passively as a result of water movement or actively through migration) in estimating yield, primarily as a consequence of potential impacts on the value of B_0 (SC-CAMLR, 1993).

A CCAMLR Workshop on Krill Flux in 1994 used data from satellite-tracked drifters to investigate passive transport of krill by prevailing easterly currents through the Scotia Arc (see Everson, Chapter 3). Results indicated water transport times between the Antarctic Peninsula and South Georgia of about six months (Ichii & Naganobu, 1996) and, as originally postulated by Marr (1962), the effects of this circulatory system may link krill populations around the South Shetlands, South Orkneys and South Georgia. Despite this work and general support for its conclusions from the results of the *CCAMLR Workshop on Area 48* (SC-CAMLR, 1998), the extent to which krill are exchanged between areas (through passive transport around the Antarctic Continent) or remain resident in highly productive areas (i.e. the South Shetland Islands) remains unknown.

The consequences of uncertainty about flux hold significance for management, especially in the setting of precautionary limits for subareas within Area 48 – effectively the 'management areas' used by CCAMLR (see Section 12.3). Should flux be important, the available krill stock in any one area will be greater than that estimated by a 'snapshot' survey. If krill are essentially resident in one area, then stock size will be equivalent to that seen in a survey; assuming that there is no flux results in a conservative yield estimate (SC-CAMLR, 1994), and that has been the approach adopted by CCAMLR to date. However, this approach does assume that, should flux be important, there is no fishing in upstream areas. Obviously, heavy fishing in upstream areas could remove all krill from downstream areas if flux was important, but would have no effect if the contrary was true.

In contiguous areas, such as Subareas 48.1, 48.2 and 48.3, flux-modified catch limits might necessitate further division of catch between subsidiary areas within an overall areal catch in order to take into account possible fluxes (i.e. movement) of krill biomass into such areas from a single upstream source (i.e. Subarea 48.1). Thus, the total yield estimate for Area 48 would remain the same, but the catch limits for individual subareas would change. Balanced against the current approach that not taking flux into account remains conservative, CCAMLR is increasingly aware that in the case where krill flux rates are not constant between areas, a simple *pro rata* allocation of some total precautionary catch for Area 48 to Subareas could lead to inappropriately high local catches in some smaller areas (SC-CAMLR, 1992 onwards).

Sea-ice

Apart from water circulation, the environmental factor most likely to influence krill distribution and abundance is sea-ice (Mackintosh, 1972, 1973). A recent examination of long-term data on ice concentration has established that although total Antarctic sea-ice cover may remain fairly constant from year to year, anomalies in the northern ice extent progress clockwise around the Continent with a periodicity of approximately 7–9 years (Murphy *et al.*, 1995; Stammerjohn & Smith, 1997); a situation reflected further in local sea-ice distribution.

Variability in krill recruitment is thought to be linked to such cycles in the Antarctic Peninsula region. Kawaguchi & Satake (1994), Siegel & Loeb (1995) and

Loeb *et al.* (1997) have reported that winters with a substantial extent of sea-ice are followed by summers of high krill recruitment, as shown by high proportions of juvenile krill (i.e. animals spawned the previous summer) taken in both commercial and research net hauls. Such observations may be attributable to winter sea-ice offering an important feeding/nursery ground and refugium for both adult and larval/juvenile krill (following suggestions by Hamner *et al.*, 1989; Daly, 1990). These results also intimate that krill recruitment is largely independent of spawning stock size in the region. However, Siegel & Loeb (1995) suggest that prolonged winter sea-ice cover leads to early krill maturation and spawning with subsequent enhanced recruitment to the next season's ice. Consecutive years of heavy sea-ice cover therefore act in concert to increase krill recruitment (Loeb *et al.*, 1997) while the absence of sea-ice may influence the dominance of either krill or salps at different times (Loeb *et al.*, 1997).

Possible associations between sea-ice and krill CPUE are investigated by comparing ice cover in October of one split-year with catch-per-hour (CPH) in that split-year (see Section 12.2 for description of analysis). In Subarea 48.1, a strong positive correlation is observed between CPH during the period January to March, when fishing usually takes place (see Section 12.2), and sea-ice conditions the previous winter (Table 12.5). This may be due to such effects as noted in the previous paragraph, but it is not possible to confirm whether the observed increase in CPH in Subarea 48.1 is due to increased density of adult krill or the occurrence of large numbers of recruiting krill. Nevertheless, and irrespective of how it is mediated, any correlation between krill CPH and ice may have operational consequences for the fishery.

By contrast, negative or weak associations are evident between CPH and sea-ice in Subareas 48.2 and 48.3 with Fedulov *et al.* (1996) finding a correlation between

Table 12.5 Correlation coefficients comparing CPH (catch/hours fished) and sea-ice cover in October. For description of analysis refer to Section 12.2, Available data, of text. CHL = Chile, JPN = Japan, POL = Poland and SUN = Soviet Union (* – Significant at 0.05 level and ** – significant at 0.01 level).

Subarea	Nation	Ice concentration			n
		Subarea			
		88.3.1#	48.1	48.2 & 48.3	
48.1	JPN	0.36	0.75*	0.87**	14
48.1	CHL	0.10	0.58	0.78**	10
48.2	JPN@	−0.66*	−0.54	−0.22	12
48.2	SUN+	−0.21	−0.51	−0.06	9
48.3	SUN	−0.52	0.24	0.44	9
48.3	POL	−0.10	0.30	0.00	9

@ Excluding 1990 when only 1 tonne of krill caught
# Bounded by 75°S, 70°W; 60°S, 90°W
+ Combined Russian & Ukrainian Fisheries Data from 1991/92

krill catch-per-day fished and sea-ice for the Russian fishery around South Georgia during the period June–August. One implication of this finding is that sea-ice conditions may influence CPUE throughout Area 48. Although the data available in the CCAMLR database do not allow a detailed breakdown of CPH for the Russian fishery, some positive correlation is evident between CPH in the Soviet fishery and ice concentration the previous October (Table 12.5).

The cause of negative correlations between sea-ice and krill CPH in Subarea 48.2 shown in Table 12.5 remains unclear. Agnew (1993) has previously reported that the fishery in Subarea 48.2 is erratic, perhaps as a result of a variable current structure around the South Orkneys. It may be that years of heavy sea-ice concentration are associated with highly disrupted current flow fields around the Islands thereby resulting in different krill retention times in the region and hence variable catch rates.

From these results, linking sea-ice and the performance of the krill fishery in the forthcoming year offers an intriguing possibility for predicting catch rates based on a relatively simple environmental index (i.e. sea-ice cover). The matter is obviously one that warrants serious and further consideration.

12.6 Monitoring the effects of krill harvesting

Predator indices

As long ago as 1981, Everson (1981) indicated the likely effects on krill-dependent predators of intensive krill fishing in restricted area(s) (Table 12.6); a concern specifically addressed by CCAMLR Article II, Sub-paragraph (3).(b) (Anon., 1995). Coupled with possible effects of fishing on krill availability to predators, it is assumed that it will be possible to detect changes in krill availability through some index of predator performance (Croxall, 1990). Predator data would therefore need to be incorporated into the management paradigm (see also Section 12.3). Apart from the points already discussed in Section 12.4, CCAMLR has chosen to address this problem in various, but not mutually exclusive, ways. CCAMLR's first step was to develop an Ecosystem Monitoring Programme (CEMP)(SC-CAMLR, 1984, 1985) aiming:

> 'To detect and record significant changes in critical components of the ecosystem to serve as the basis for the conservation of Antarctic marine living resources. The monitoring system should be designed to distinguish between changes due to the harvesting of commercial species and changes due to environmental variability, both physical and biological.'

A system to regularly record selected life-history parameters of key seabird and seal populations has been in place since 1986 (SC-CAMLR, 1986). The programme involves using standard procedures to monitor a number of parameters of selected

Table 12.6 Postulated effects of intensive krill fishing on predators in restricted area(s) (after Everson, 1981 and Miller, 1991).

Impacted stocks	Single krill stock	Several areal krill stocks
Species close to/in fishing area (e.g. breeding seabirds)		
(i) Heavy fishing prior to critical predation period each season	Significant in fishing area, elsewhere minimal	Significant: Increasing with time to level-off eventually
(ii) Intensive fishing during and after critical predation period	Small	Significant in subsequent years
Species not tied to fishing area (e.g. whales)		
(i) Intensive fishing before predator normally present in fishing area	Small overall	Reduced density (i.e. feeding elsewhere)
(ii) Intensive fishing during and after predator present in fishing area	None or slight reduction in density	Slight short-term reduction Major long-term reduction

predator species (generally relating to breeding and foraging success), parameters estimating krill abundance and environmental parameters such as sea-ice extent (Anon., 1997, see also Everson, Chapter 7). Since 1992 SC-CAMLR has undertaken annual 'ecosystem assessments' using the data collected by CEMP. Agnew & Nicol (1997) have provided a comprehensive review of the Programme and research into multivariate analyses of CEMP parameters, especially, in detecting 'unusual' years (of krill availability or otherwise), continues to be a high CCAMLR priority (SC-CAMLR, 1997, 1998).

Overlap of krill fishery and predators

Despite calculations of krill yield that now take into account krill and predator requirements, CCAMLR has been acutely aware of the potential for local competition between predators and the krill fishery. From a global perspective (e.g. within a Subarea or Area), fishing mortality (F) might remain within the limits set by the yield model and so provide sufficient escapement for predator needs. However, at a more localised scale, F might be much greater and krill escapement therefore too low to sustain predators with restricted foraging ranges. This concern is exacerbated by the knowledge that the foraging of many bird and seal predators is restricted during breeding to close to land for about four months in the austral summer (Croxall, 1990; Croxall *et al.*, 1984, 1992).

CCAMLR's debate on possible competition between the krill fishery and krill predators has been largely influenced by an improvement in the availability of fine-scale, fisheries information for Subarea 48.1. From Fig. 12.3, it is apparent that although only about 16% of the total krill catch has come from Subarea 48.1, the annual catch has been relatively consistent at about 40 000 to 80 000 t. Based on an earlier description of haul-by-haul data from the Chilean fishery (Marin *et al.*, 1992), Agnew (1993) has shown that this catch is highly concentrated on the shelf areas of the South Shetland Islands, mostly between December and March, with patterns of high catch being largely predictable from year to year. The areas concerned support some of the largest concentrations of seabirds, including penguins, in the CCAMLR Area as well as substantial seal and whale populations (Everson, 1977; Laws, 1989).

Agnew & Nicol (1997) have emphasised that krill's important ecological status and contagious distribution result in a variety of possible interactions with other biota as well as with the environment over a wide range of spatial/temporal scales (see also Murphy *et al.*, 1988; Murphy, 1995). It is the functional relationship between predators, prey and the fishery at each scale which is crucial to determining whether potential overlap between fishing and predator foraging is important. Progress has been made in modelling functional relationships between krill and predators (Mangel & Switzer, 1998) following earlier work by Mangel (1994).

At the local scale (< tens of kilometres), Ichii *et al.* (1994a) have indicated that the fishery usually targets aggregation centres while penguins and the fishery may exploit the same krill aggregations around the South Shetland Islands. On the other hand, Bengtson *et al.* (1993) have shown that Chinstrap penguins *(Pygoscelis antarctica)* feed in the upper levels of such aggregations. Ichii *et al.* go on to suggest that the fishery and penguins exploit different krill size classes.

At a slightly larger scale (tens to fifties of kilometres), Agnew (1993) found patterns of fishing from 1988 to 1992 within Subarea 48.1 to be consistent between years. Seventy to 90% of the total catch for the Subarea was taken between January and March, within 100 km of shore-based penguin breeding colonies each year. The period December to March is the peak breeding time, when penguins are restricted to foraging distances within 100 km of their breeding colonies (Croxall *et al.*, 1985). Agnew's (1993) results prompted Ichii *et al.* (1994b) to consider the spatial distribution of penguin colonies and krill catches in more detail. They showed that even though a high percentage of the krill catch in Subarea 48.1 may be taken within the foraging ranges of land-based penguin colonies, the largest of these colonies are some distance from the fishing grounds. More specifically, the Japanese krill fishery is generally concentrated close to Livingston and Elephant Islands, while the largest penguin colonies are located on Low, Nelson and King George Islands (Wöehler, 1993).

The conclusions of Ichii *et al.* (1994b) are largely supported by studies showing that the foraging ranges of Chinstrap and Adélie penguins are also generally confined to within 100 km of their breeding colonies (Lishman, 1985; Everson, 1981, 1988b; Trivelpiece *et al.*, 1987). However, Kerry *et al.* (1993) have shown that Adélie penguins on Bechervaise Island (within Subarea 58.4) forage several hundreds of kilometres from land to feed on krill aggregations over the continental shelf break (i.e. where krill fishing took place in the early 1980s). If Adélie penguins are also able to exhibit similar foraging ranges in the Antarctic Peninsula (Subarea 48.1) area then some overlap with the fishery is still possible.

A further difficulty in interpreting the effects of potential overlaps between krill fisheries and predator foraging areas may be related to the krill flux issue. If important, this could lead to significant variability in the availability of krill to predators. For instance, based on modelling studies of krill availability, predator demands and water flow around South Georgia, Murphy (1995) has shown the importance of the magnitude and timing of horizontal fluxes in determining krill abundance available to predators. The situation at the Antarctic Peninsula is not so clear. While there may be longshore movements, and a krill flux in and out of various locations, there is also now considerable evidence to suggest that some krill stocks may remain resident in specific areas and that adult animals migrate on to the shelf to spawn (Ichii & Naganobu, 1996). The likely impact of overlap between fishing and predators as a function of different levels of flux and residency has yet to be adequately investigated. However, Mangel & Switzer (1998) have examined such a possibility in a model they have recently developed.

An area which remains unexplored pertains to the extent to which overall predation on krill can be attributed to land-based predators. There are a number of marine predators (fish, squid and whales) which have not yet been taken into account in fishery, krill and krill predator overlap studies. At least in the past, this predation has been substantial (Everson, 1977; Miller & Hampton, 1989).

Irrespective of the ultimate interaction between predators and fishery, CCAMLR has developed a suite of calculations to track the extent of the potential overlap between the two. Agnew & Phegan (1995) developed a model of land-based predator foraging demands at a small spatial scale. This model was used to develop an overlap index which expressed krill catch relative to predator foraging demands. The results (Fig. 12.6) show a progressive decrease in the index in Subarea 48.1 during the critical period when predators are tied to land-based breeding sites. The decrease is most apparent from 1991 onwards, the period during which CCAMLR has focused its attention increasingly on the fishery-landbased predator overlap problem at critical times for the latter (SC-CAMLR, 1991–1994). As shown in Section 12.2, the above can only partially be explained by increasing the availability of ice-free areas later in the season (Fig 12.5). Recently, CCAMLR has developed a variety of other overlap indices for predators and fishery, including the Schröeder index which reflects relative degrees of overlap (SC-CAMLR, 1998, 1999; Everson, Chapter 8).

At current levels of fishing, in Subarea 48.1 at least, krill escapement from the fishery is probably adequate to meet predator demands. Chinstrap penguins constitute by far the most numerous krill predator in the Subarea, and the most significant in terms of krill consumption in most of Area 48 (Croxall et *al.*, 1984, 1985). One and a half million Chinstrap pairs populate the South Shetland Islands, compared with 60 000 Adélie and 34 000 Gentoo penguins (*Pygoscelis papua*); over two million pairs of all three species are located in Subarea 48.1 (Wöehler, 1993). Total

Fig. 12.6 Krill catch in tonnes by split year taken within the critical distances of various land-based predator colonies within Subarea 48.1 during the breeding period (CPD) compared with the total catch in the Subarea (TCS).

krill consumed by these species has been calculated as 370 000 t for the South Shetland populations (Agnew, 1993) during the period (December to March) when foraging is restricted close to breeding sites. With the current krill catch being of the order of 40 000 to 80 000 t annually, and the krill biomass in Subarea 48.1 over the austral summer probably being greater than two million tonnes (Siegel, 1992; Trathan *et al.*, 1993, 1995), there is little suggestion that the latter is unable to support both penguins and other krill predators such as fish. However, should the catch in Subarea 48.1 increase substantially, the issue of overlap of predators and the fishery will once again assume prominence.

Effects of management measures on fishing

Concomitant with developing precautionary catch limits for krill, CCAMLR has identified a number of other approaches (e.g. reactive management, predictive management, open/closed areas, indirect methods, pulse fishing and feedback management) which may serve to manage krill fishing (SC-CAMLR, 1991–1993). The scope of these approaches and the resulting dialogue with fisherman have served to emphasise the need to evaluate objectively the effects of any management measures which might be introduced. CCAMLR has therefore accepted (SC-CAMLR, 1993) that while adequate protection should be afforded to krill-dependent predators at critical times and in specific areas, such protection should not exert unnecessary or unreasonable restrictions on the fishery. Various models have therefore been developed to assess the potential effects of selected management measures on the fishery.

Agnew & Marin (1994) constructed a spatial model based on the Chilean krill fishery to simulate the behaviour of the fishery in Subarea 48.1. The model takes into account fishing duration, fishing vessel performance, CPH and prevailing management measures in force. Agnew (1994) subsequently extended the model to include Japanese data, an estimate of encounter probability with fishable aggregations and to account for variability in CPUE as well as other parameters. In addition, the model was applied to fleets rather than single vessels and the performance of management measures was assessed by effects on:

(1) the total catch of the fleet over the whole fishing season, and
(2) potential interaction(s) between catches and foraging predators at specific times (i.e. during critical periods when predators are breeding).

Out of a range of management measures, Agnew (1994) found that a two-month closure (January to February) offered the most successful strategy for both the Chilean and Japanese fleets within 75 km of selected predator areas. This impacted least on the fishery, although some small overlap between predators and fishery remained.

The continued development of management models (SC-CAMLR, 1997) has exciting strategic implications since it provides a means to evaluate the potential effects of specific management action on both predators and fishery. Such models also serve to identify key parameters which may be used to test management procedures and to advance consideration of how specific indices of predator status, or performance, may be taken into account in management measure formulation (de la Mare, 1986).

12.7 Discussion

The emergence of many Antarctic fisheries in the early 1970s was undoubtedly predicated by developments at the Third United Nations Conference on the Law of The Sea (UNCLOS) (Boczek, 1983). As a consequence, control of most of the world's fisheries was gradually transferred from international arrangements to Coastal State jurisdiction within established national, 200-mile Exclusive Economic Zones (EEZs) (May *et al.*, 1979). However, unresolved sovereignty issues in the Antarctic have generally precluded the establishment of universally accepted EEZs (Freestone, 1995).

Antarctic fisheries have therefore been generally 'open to access' and have had much in common with the 'boom and bust' scenarios demonstrated by Antarctic whaling. Under such circumstances, and in the absence of effective international control, Antarctic fisheries have tended to move from one depleted species to the next (Kock, 1992). This pattern is clearly illustrated by trends in finfish catches following severe depletion of Antarctic cod *(Notothenia rossii)* stocks post-1972 (Fig. 12.7). With an overall steep decline in catches of finfish, a shift of focus to serious development of the krill fishery thus became inevitable from about 1978 onwards.

However, it should be noted that the mackerel icefish (*Champsocephalus gunnari*) fishery was a notable exception from the general boom or bust scenario associated with Antarctic fisheries. From 1978, catches of this species demonstrated a clear five-yearly cycle (Fig. 12.7). The first two of these (1978 and 1983) were followed by significant increases in krill catches within one or two years. In the latter case, the increase in icefish catches between 1982 and 1984 was associated with a marked reduction in krill catches. Independently, there was also a marked decline in krill catches just prior to, and immediately following, the break up of the Soviet Union in 1991. Together, these examples clearly suggest, like the finfish fishery, operational considerations (i.e. selective access to resources) have exerted a profound influence on krill fishery development.

As emphasised, considerable concern came to be expressed about possible over-exploitation of krill almost as soon as the fishery began to flourish. Proponents of moderation warned that given krill's perceived high abundance, the potential for rapid and uncontrolled growth of the fishery was high following precedent set by the explosive, and essentially unsustainable, expansion in the fisheries for a number of finfish (Fig. 12.7) in combination with the demise of the Peruvian anchovy harvest between 1958 and 1970 (Reid, 1988).

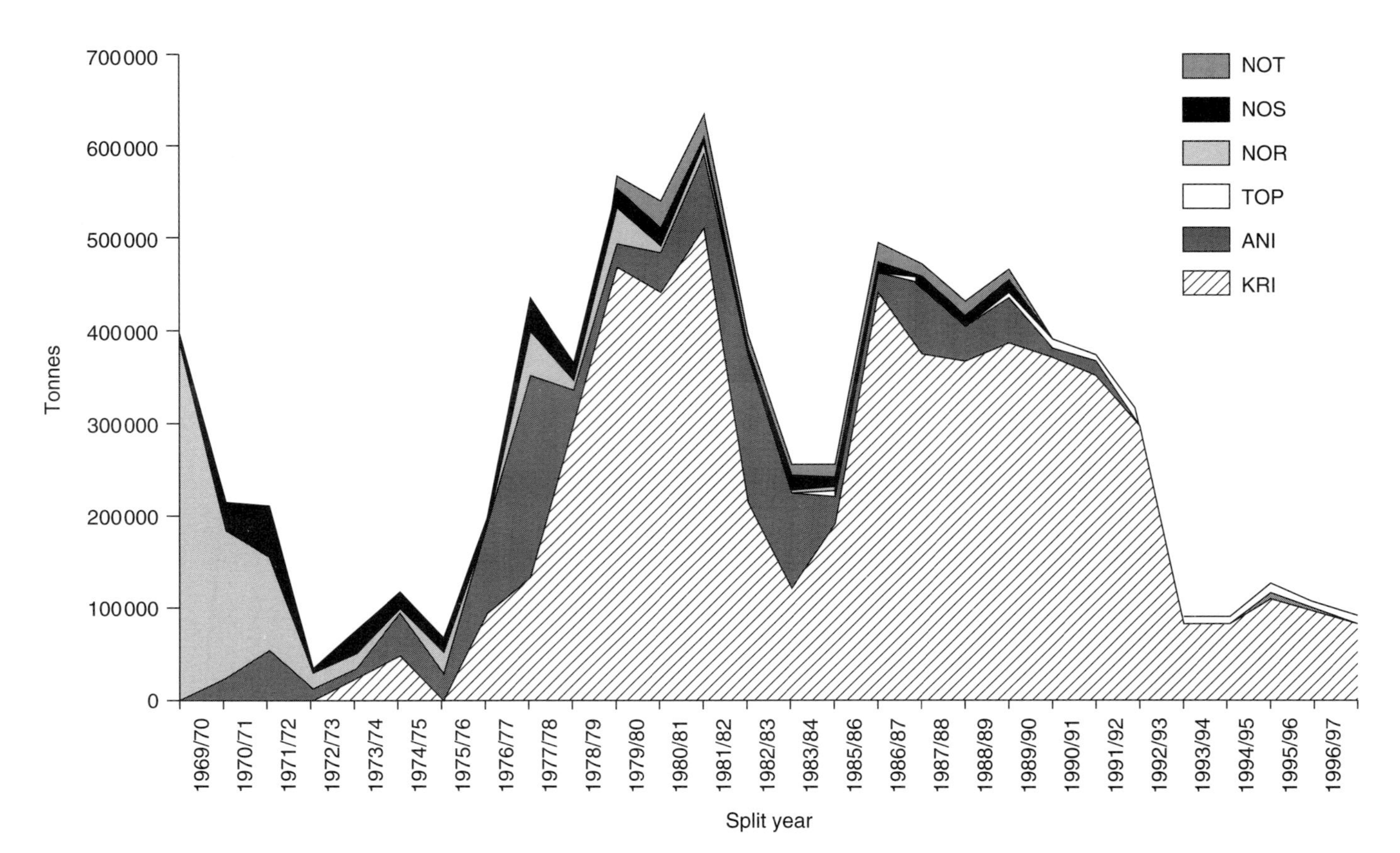

Fig. 12.7 Annual catches of major species in the CAMLR Convention Area from 1969/70 to 1996/97. ANI = *Champsocephalus gunnari* (Mackerel icefish); KRI = *Euphausia superba* (Krill); NOR = *Notothenia rossii* (Marbled rockcod); NOS = *Notothenia squamifrons* (Grey rockcod); NOT = *Patagonotothen guntheri* (Patagonian rockcod) and TOP = *Dissostichus eleginoides* (Patagonian toothfish).

The emergence of CCAMLR was the response to these concerns. Despite its notable successes in formulating a scientifically defensible, management policy for the krill fishery (Section 12.3), various factors have retarded the CCAMLR's efforts to address issues associated with krill management (Nicol, 1991). Croxall *et al.* (1992) have emphasised that CCAMLR's relatively slow progress was due to conceptual as much as practical difficulties. These necessitated a new approach to managing the fishery and were compounded by krill's contagious distribution, the fishery's potentially large geographical range and the assertions of a number of krill-fishing nations that:

- krill catches are small in relation to the available stock;
- there is no intention to increase krill catches dramatically in the near future;
- krill management must be based on the 'best scientific information' available and hence on scientifically formulated assessments. Such assessments are currently limited as data are inadequate, particularly on:
  - (a) krill abundance, distribution and flux;
  - (b) functional relationships between krill and predators, and
- historical catch levels do not offer a scientific basis for managing the krill fishery.

With an improved flow of data from the fishery, these restrictions have been largely circumvented by CCAMLR's acceptance of guiding principles, recognising SC-CAMLR's scientific advice as being the 'best available', adopting a precautionary approach to management in the face of uncertainty and exercising feedback rather than reactive management control (Section 12.3).

A second major factor which has undoubtedly influenced progress on management issues has been CCAMLR's requirement that decisions on matters of substance (i.e. conservation measures) be taken by consensus [Convention Article XII.(1)] (Croxall *et al.*, 1992). While this may have slowed down the making of decisions, it has ensured that all parties to the Convention are included in the decision-making process and that, in the absence of objections, decisions enter into force within a period of six months. Since CCAMLR, like so many international organisations, has no legal personality of its own, members are morally and strongly obligated to enforce management measures, especially when these have been agreed through a consensus approach.

Despite the above limitations, CCAMLR's progress in establishing formal conservation measures (including precautionary catch limits for krill) has compared favourably with other fisheries commissions and conventions elsewhere in the world (Miller, 1991) (Table 12.7). Indeed, CCAMLR has adopted various landmark measures (e.g. first measure, size limit, catch limit, closed areas etc.) more rapidly than most other international fisheries arrangements. Furthermore, CCAMLR remains the only international fisheries commission to date to have formally adopted any precautionary limit on catch in the interests of minimising the risk of over-exploitation and to reduce possible ecosystem effects.

Table 12.7 Development and enforcement of Conservation Measures by CCAMLR compared with the International Commission for the Southeast Atlantic Fisheries (ICSEAF, 1984) and the International Commission for the Conservation of Atlantic Tunas (ICCAT, 1985). Values in parentheses represent the time elapsed since the specific convention came into force while those in bold pertain directly to the krill fishery.

Attribute/Measure	CCAMLR	ICSEAF	ICCAT
Number of members	22	17	22
Date of entry into force	1982	1971	1969
First management measure	1984 (2) (**10**)	1973 (2)	1973 (3)
Fish size limit	1994 (13)	–	1974 (5)
Target fishing mortality	1987 (5)	?	1975 (6)
Total allowable catch (quota)	1987 (5)	1976 (5)	1982 (13)
Mesh size regulation	1984 (2)	1973 (2)	–
Close season(s)	1987 (5)	1977 (6)	1994 (25)
Closed area(s)	1984 (2)	1979 (8)	1982 (13)
Catch reporting system	1987 (6) (**8**)	1982 (11)	?
Inspection scheme	1987 (6)	1976 (5)	1974 (6)
Observation scheme	1992 (**11**)	?	?
Precautionary catch measure	1991 (**10**)	–	–
Catch trigger/subareal catch allocation	1991/92 (**10/11**)	–	–
Notification of new fishery	1991 (10)	–	–
Ecosystem Protection	1991 (10)	–	–
Prohibition of Fishing Without a Licence	1998 (16)	–	–
Satellite Vessel Monitoring	1998/2000 (16)	–	–
Trade Documentation	1999 (17)	–	?

Notwithstanding CCAMLR's obvious successes, uncertainties still surround the future of the Antarctic krill fishery. The first concerns the fishery's possible course of development. Obviously, the initial anticipated rapid expansion of the fishery has not occurred and it seems that with the break up of the heavily subsidised Soviet fishery the potential for such expansion has probably diminished. Such uncertainty is further compounded by a lack of complete economic analyses of the fishery to date.

Although economic analyses (e.g. Grantham, 1977; McElroy, 1980; Budzinski *et al.*, 1985) have focused on technological problems associated with krill processing and product quality, thorough analyses of krill's potential market demand remain limited. Both Anon. (1978) and Budzinski *et al.* (1985) have emphasised that determining the prices of, and demand for, krill are the most persistent problems in forecasting the fishery's future development. In this context, CCAMLR's on-going call for industrial information from krill fishing nations (e.g. SC-CAMLR, 1991), along with the precautionary catch limits in place, may be considered as a tactic to improve knowledge of the fishery's economic driving forces while ensuring that the risk of uncontrolled expansion is minimised. Even so, there is little information to explain the economic motives for the observed increase in krill catches by Ukraine during 1994–1995 compared with the 1993–1994 season, or for the stated intention to cease krill fishing by Russia and Chile from 1995 onwards (SC-CAMLR, 1995).

These examples serve to stress the urgency for economic studies of the fishery's potential in furthering the development of effective management measures.

The second, and in terms of the convention's objectives probably more relevant, uncertainty is to what extent, as well as how, ecosystem (i.e. multi-species) considerations may be more formally subsumed into management measures. While some areas where progress has been made (Section 12.3) have been highlighted, the topic is one which obviously warrants priority attention and development in the future.

Finally, to address these various uncertainties still requires more information (e.g. on search-time by fishing vessels) than is currently forthcoming from the krill fishery. Only with such information will it be possible to assess objectively the potential utility of regularly monitoring selected indices or fishery's factors (e.g. length frequency composition of catches) crucial to developing realistic management procedures. The placement of suitably qualified observers aboard krill fishing vessels would go a long way to improving the flow of data necessary to effectively monitor the fishery (e.g. SC-CAMLR, 1990; CCAMLR, 1991) and to assess its impact on available stocks as well as to evaluating future management action. CCAMLR initiatives in this regard (e.g. Ichii *et al.* 1993; 1994a, b; Mujica *et al.*, 1993; Vagin *et al.*, 1993) are therefore to be encouraged and their results are awaited with interest.

Acknowledgements

We thank our many CCAMLR colleagues who have given their time, wisdom and data to developing a sensible krill management regime. In particular, we greatly appreciate useful comments from, and discussions with, Professors D.S. Butterworth and J. Croxall, Drs I. Everson, T. Ichii, K-H. Kock, V. Marin, E. Murphy, S. Nicol and W. de la Mare. We dedicate this chapter to the memory of Dr Rodion Makarov who passed away so tragically on 12 August 1994.

References

Agnew, D.J. (1992) Attempts at a basic accounting for Subareas 48.1 and 48.2. *SC-CAMLR-XI, Annex 4, Appendix F.* CCAMLR, Hobart. pp. 174–76.

Agnew, D.J. (1993) Distribution of krill (*Euphausia superba*) catches in the South Shetlands and South Orkneys. In *Selected Papers Presented to the Scientific Committee of CCAMLR, 1992.* CCAMLR, Hobart. pp. 287–303.

Agnew, D.J. (1994) Further development of a krill fishery simulation model. Document WG-Joint-94/4. CCAMLR, Hobart. 10 pp.

Agnew, D.J. & Marin, V.H. (1994) Preliminary model of krill fishery behaviour in Subarea 48.1. *CCAMLR Sci.* **1**, 71–79.

Agnew, D.J. & Nicol, S. (1997) Marine disturbances – Commercial fishing. In *Foundations for*

Ecological Research West of the Antarctic Peninsula (R.M. Ross, E.E. Hofmann & L.B. Quetin, eds), pp. 417–35. Antarctic Research Series No. 70. American Geophysical Union, Washington.

Agnew, D.J. & Phegan, G. (1995) Development of a fine-scale model of land-based predator foraging demands in the Antarctic. *CCAMLR Sci.* **2**, 99–110.

Anon. (1978) The Antarctic krill resource: Prospects for commercial exploitation. *Tetra Technical Report.* TC-903. 149 pp.

Anon. (1986) Post-FIBEX acoustic workshop. Frankfurt, Federal Republic of Germany, September 1984. *BIOMASS Report Series* **40**, 106 pp.

Anon. (1995) *Commission for the Conservation of Antarctic Marine Living Resources: Basic Documents*, 6th edn. CCAMLR, Hobart. 122 pp.

Anon. (1997) *CEMP Standard Methods for Monitoring Studies 1991.* CCAMLR, Hobart. 222 pp.

Beddington, J.R. & Cooke, J.G. (1983) The potential yield of fish stocks. *FAO Fisheries Technical Paper* **242**, 47 pp.

Bengtson, J.L. (1984) Review of Antarctic marine fauna. In *Selected Papers Presented to the Scientific Committee of CCAMLR, 1982–1984.* Part I. CCAMLR, Hobart. pp. 1–126.

Bengtson, J.L., Croll, D.A. & Goebel, M.E. (1993) Diving behaviour of chinstrap penguins at Seal Island. *Antarctic Sci.* **5**, 9–15.

Boczek, B.A. (1983) The protection of the Antarctic ecosystem: A study in international environment law. *Ocean Development and International Law Journal* **13**, 347–425.

Budzinski, E., Bykowski, P. & Dutkiewcz, D. (1985) Possibilities of processing and marketing of products made from Antarctic krill. *FAO Fisheries Technical Paper* **268**, 46 pp.

Burukovskiy, R.N. & Yaragov, B.A. (1967) Studying the Antarctic krill for the purpose of organizing krill fisheries. In *Soviet Fishery Research on Antarctic Krill* (R.N. Burukovskiy, ed.), pp. 5–17. U.S. Department of Commerce (TT 67-32683), Washington D.C.

Butterworth, D.S. (1986) Antarctic ecosystem management. *Polar Rec.* **23**, 37–47.

Butterworth, D.S. (1989a) A simulation study of krill fishing by an individual Japanese trawler. In *Selected Papers Presented to the Scientific Committee of CCAMLR, 1988*, Part I, pp. 1–108. CCAMLR, Hobart.

Butterworth, D.S. (1989b) Some aspects of the relation between Antarctic krill abundance and CPUE measures in the Japanese krill fishery. In *Selected Papers Presented to the Scientific Committee of CCAMLR, 1988*, Part I, pp. 109–125. CCAMLR, Hobart.

Butterworth, D.S. (1990) Towards an initial operational management procedure for the krill fishery in Subareas 48.1, 48.2 and 48.3. In *Selected Papers Presented to the Scientific Committee of CCAMLR, 1989*, pp. 189–219. CCAMLR, Hobart.

Butterworth, D.S. & Thomson, R.B. (1994) Further calculations of the effects of krill fishing on krill predators. WG-Krill-94/24. CCAMLR, Hobart. 25 pp.

Butterworth, D.S. & Thomson, R.B. (1995) Possible effects of different levels of krill fishing on predators – some initial modelling attempts. *CCAMLR Sci.* **2**, 79–97.

Butterworth, D.S., Punt, A.E. & Basson, M. (1992) A simple approach for calculating the potential yield from biomass survey results. In *Selected Papers Presented to the Scientific Committee of CCAMLR, 1991*, pp. 207–215. CCAMLR, Hobart.

Butterworth, D.S., Gluckman, G.R., Thomson. R.B., Chalis, S., Hiramatsu, K. & Agnew, D.J. (1994) Further computations of the consequences of setting the annual krill catch limit to a fixed fraction of the estimate of krill biomass from a survey. *CCAMLR Sci.* **1**, 81–106.

CCAMLR (1990) *Report of the Ninth Meeting of the Commission.* CCAMLR, Hobart. 123 pp.

CCAMLR (1991) *Report of the Tenth Meeting of the Commission.* CCAMLR, Hobart. 101 pp.

CCAMLR (1992) *Report of the Eleventh Meeting of the Commission.* CCAMLR, Hobart. 99 pp.

CCAMLR (1994) *Report of the Thirteenth Meeting of the Commission.* CCAMLR, Hobart. 123 pp.

CCAMLR (1998) *Report of the Seventeenth Meeting of the Commission.* CCAMLR, Hobart. 166 pp.

Croxall, J.P. (1990) Uses of indices of predator status and performance in CCAMLR fishery management. In *Selected Papers Presented to the Scientific Committee of CCAMLR, 1989*, pp. 353–63. CCAMLR, Hobart.

Croxall, J.P., Ricketts, C. & Prince, P.A. (1984) The impact of seabirds on marine resources, especially krill, at South Georgia. In *Seabird Energetics* (C.G. Whittow & H. Rahn, eds), pp. 285–318. Plenum, New York.

Croxall, J.P., Ricketts, C. & Prince, P.A. (1985) Relationship between prey life-cycles and the extent, nature and timing of seal and seabird breeding in the Scotia Sea. In *Antarctic Nutrient Cycles and Food Webs* (P.R. Condy, W.R. Siegfried & R.M. Laws, eds), pp. 516–33. Springer-Verlag, Berlin, Heidelberg.

Croxall, J.P., McCann, T.S., Prince, P.A. & Rothery, P. (1988) Variation in reproductive performance of seabirds and seals at South Georgia, 1976–1986 and its implications for Southern Ocean monitoring studies. In *Antarctic Ocean and Resources Variability* (D. Sahrhage, ed.), pp. 261–85. Springer-Verlag, Berlin.

Croxall, J.P., Everson, I. & Miller, D.G.M. (1992) Management of the Antarctic krill fishery. *Polar Rec.* **28**, 64–66.

Daly, K.L. (1990) Overwintering development, growth, and feeding of larval *Euphausia superba* in the Antarctic marginal ice zone. *Limn. Oceanogr.* **35**, 1564–76.

de la Mare, W.K. (1986). Some principles for fisheries regulation from an ecosystem perspective. In *Selected Papers Presented to the Scientific Committee of CCAMLR, 1986*, pp. 323–39. CCAMLR, Hobart.

de la Mare, W.K. (1994a) Modelling krill recruitment. *CCAMLR Sci.* **1**, 49–54.

de la Mare, W.K. (1994b) Estimating krill variability and its recruitment. *CCAMLR Sci.* **1**, 55–69.

Dolzhenkov, V.N, Lubimova, T.G., Makarov, R.R., Parfenovich, S.S. & Spiridonov, V.A. (1989) Some specific features of the USSR krill fishery and possibilities of applying fishery statistics to studies of krill biology and stocks. in *Selected Papers Presented to the Scientific Committee of CCAMLR, 1988*, Part I, pp. 237–50. CCAMLR, Hobart.

Eddie, G.C. (1977) The harvesting of krill. Southern Ocean Fisheries Survey Programme. GLO/SO/77/2. FAO, Rome. 76 pp.

Edwards, D.M. & Heap, J.A. (1981) Convention on the conservation of Antarctic marine living resources: A commentary. *Polar Rec.* **20**, 353–62.

El-Sayed, S.Z. (1988) Antarctic marine living resources: The BIOMASS Program. *Oceanus.* **31**, 75–79.

El-Sayed, S.Z. & McWhinnie, M.A. (1979) Protein of the last frontier. *Oceanus* **22**, 13–20.

Everson, I. (1977) The living resources of the Southern Ocean. *Southern Ocean Fisheries Survey Programme.* GLO/SO/77/1. FAO, Rome. 156 pp.

Everson, I. (1978) Antarctic fisheries. *Polar Rec.* **19**, 233–51.

Everson, I. (1981) Antarctic krill. *BIOMASS Scientific Series* **2**, 31–45.

Everson, I. (1984) Marine interactions. In *Antarctic Ecology*, Vol. 2. (R.M. Laws, ed.), pp. 783–819. Academic Press, London.

Everson, I. (1988a) Can we satisfactorily estimate krill abundance? In *Antarctic Ocean and Resources Variability* (D. Sahrhage, ed.), pp. 199–208. Springer-Verlag, Berlin.

Everson, I. (1988b) Prey monitoring surveys. In *Selected Papers Presented to the Scientific Committee of CCAMLR, 1987*, pp. 323–38. CCAMLR, Hobart.

Everson, I. & de la Mare, W. (1996) Some thoughts on precautionary measures for the krill fishery. *CCAMLR Sci.* **3**, 1–12.

Everson, I. & Goss, C. (1991) Krill fishing activity in the southwest Atlantic. *Antarctic Sci.* **3**, 351–58.

Everson, I., Watkins, J.L., Bone, D.G. & Foote, K.G. (1990) Implications of new target strength measurements for abundance estimates of Antarctic krill. *Nature* **345**, 338–40.

FAO (1976) *FAO Yearbook of Fishery Statistics*, FAO, Rome. 336 pp.

Fedulov, P.P., Murphy, E.J. & Shulgovsky, K.E. (1996) Environment-krill relations in the South Georgia marine ecosystem. *CCAMLR Sci.* **3**, 13–30.

Foote, K.G., Everson, I., Watkins, J.L. & Bone, D.G. (1990) Target strength of Antarctic krill (*Euphausia superba*) at 38 and 120 kHz. *J. Acoust. Soc. America* **87**, 16–24.

Freestone, D. (1995) The conservation of marine ecosystems under international law. In *International Law and Conservation of Biodiversity* (C. Redgwell & M. Bowman, eds), pp. 91–107. Kluwer Law International, Great Britain.

Grantham, G.J. (1977) The utilization of krill. *Southern Ocean Fisheries Survey Programme*. GLO/SO/77/3. FAO, Rome. 61 pp.

Greene, C.H., Stanton, T.K., Wiebe, P.H. & McClatchie, S. (1990) Acoustic estimates of Antarctic krill. *Nature* **349**, 110.

Gulland, J.A. (1977) Antarctic marine living resources. In *Adaptations within Antarctic Ecosystems* (G.A. Llano, ed.), pp. 1135–44. Gulf Publishing, Houston.

Gulland, J.A. (1983) The development of fisheries and stock assessment of resources in the Southern Ocean. *Mem. Natl Inst. Polar Res.*, Special Issue **27**, 129–52.

Hamner, W.M., Hamner, P.P. & Obst, B.S. (1989) Field observations on the ontogeny of schooling of *Euphausia superba* furciliae and its relationship to ice in Antarctic waters. *Limn. Oceanogr.* **34**, 451–56.

Ichii, T. (1988) Observations of fishing operations on a krill trawler and distributional behaviour of krill off Wilkes Land during the 1985/86 season. In *Selected Papers Presented to the Scientific Committee of CCAMLR, 1987*, pp. 323–38. CCAMLR, Hobart.

Ichii, T. (1990) Distribution of Antarctic krill concentrations exploited by Japanese krill trawlers and minke whales. In *Proceedings of the National Institute of Polar Research Symposium on Polar Biology* **3**, pp. 36–56.

Ichii, T. & Naganobu, M. (1996) Surface water circulation in krill fishing areas near the South Shetland Islands. *CCAMLR Sci.* **3**, 125–36.

Ichii, T., Ishii, H. & Naganobu, M. (1993) Abundance, size and maturity of krill (*Euphausia superba*) in the krill fishing ground of Subarea 48.1 during the 1990/91 austral summer. In *Selected Papers Presented to the Scientific Committee of CCAMLR, 1992*, pp. 183–99. CCAMLR, Hobart.

Ichii, T., Naganobu, M. & Ogishima, T. (1994a) An assessment of the impact of the krill fishery on penguins in the South Shetland Islands. *CCAMLR Sci.* **1**, 107–13.

Ichii, T., Naganobu, M. & Ogishima, T. (1994b) A revised assessment of the impact of krill fishery on penguins in the south Shetland Islands. Document WG-Joint-94/17. CCAMLR, Hobart. 20 pp.

ICCAT (1985) *Basic Texts.* International Commission for the Conservation of Atlantic Tunas, Madrid. 99 pp.

ICSEAF (1984) *Handbook of Regulatory Measures.* International Commission for the Southeast Atlantic Fisheries, Madrid. pp. 1–49.

Ivanov, B.G. (1970) On the biology of the Antarctic krill (*Euphausia superba* Dana). *Mar. Biol.* **7**, 340–51.

Kawaguchi, S. & Satake, M. (1994) Relationship between recruitment of the Antarctic krill and the degree of ice cover near the South Shetland Islands. *Fish. Sci.* **60**, 123–24.

Kerry, K., Clarke, J.R. & Else, G.D. (1993) The foraging range of Adélie penguins at Béchervaise Island, MacRobertson Land, Antarctica, and its overlap with the krill fishery. In *Selected Papers Presented to the Scientific Committee of CCAMLR, 1992*, pp. 337–44. CCAMLR, Hobart.

Knox, G.A. (1984) The key rôle of krill in the ecosystem of the Southern Ocean with special reference to the Convention on the Conservation of Antarctic Marine Living Resources. *Ocean Management* **9**, 113–56.

Kock, K.-H. (1991) The state of exploited fish stocks in the Southern Ocean: A review. *Archiv für Fischereiwissenschaft* **41**, 1–66.

Kock, K.-H. (1992) *Antarctic Fish and Fisheries*, 359 pp. Cambridge University Press, Cambridge.

Laws, R.M. (1977) Seals and whales of the Southern Ocean. *Phil. Trans. Roy. Soc., London, Series B.* **279**, 81–96.

Laws, R.M. (1985) The ecology of the Southern Ocean. *Am. Scientist* **43**, 26–40.

Laws, R.M. (1989) *Antarctica: The last frontier*, 208 pp. Boxtree Publishers, Cambridge.

Lishman, G.S. (1985) The food and feeding of Adélie and Chinstrap penguins at Signy Island, South Orkney Islands. *J. Zool. Soc. London* **205**, 245–63.

Loeb, V., Siegel, V., Holm-Hansen, O., Hewitt, R., Fraser, W., Trivelpiece, W. & Trivelpiece, S. (1997) Effects of sea-ice extent and krill or salp dominance on the Antarctic food web. *Nature* **387**, 897–900.

Mackintosh, N.A. (1972) Life cycle of Antarctic krill in relation to ice and water conditions. *Discovery Rep.* **36**, 1–94.

Mackintosh, N.A. (1973) Distribution of post-larval krill in the Antarctic. *Discovery Rep.* **36**, 95–156.

Mangel, M. (1989) Analysis and modelling of the Soviet Southern Ocean krill fleet. In *Selected Papers Presented to the Scientific Committee of CCAMLR, 1988*, Part I, pp. 127–235. CCAMLR, Hobart.

Mangel, M. (1994) Spatial patterning in resource exploitation and conservation. *Phil. Trans. Roy. Soc., London, Series B* **343**, 93–98.

Mangel, M. & Switzer, P.V. (1998) A model at the level of the foraging trip for the indirect effects of krill (*Euphausia superba*) fisheries on krill predators. *Ecol. Model* **105**, 235–56.

Marin, V.H., Mujica, A. & Eberhard, P. (1992) Chilean krill fishery: Analysis of the 1991 season. In *Selected Papers Presented to the Scientific Committee of CCAMLR, 1991*, pp. 273–87. CCAMLR, Hobart.

Marr, J.W.S. (1962) The natural history and geography of the Antarctic krill (*Euphausia superba* Dana). *Discovery Rep.* **32**, 33–464.

May, R.M., Beddington, J.R., Clark, C.W., Holt, S.J. & Laws, R.M. (1979) Management of multispecies fisheries. *Science* **205**, 267–77.

McElroy, J.K. (1980) The economics of harvesting krill. In *The Management of the Southern Ocean* (B. Mitchell and R. Sandbrook, eds), pp. 60–80. International Institute for Environment and Development, London.

Miller, D.G.M. (1990) Commercial krill fisheries in the Antarctic, 1973–1988. In *Selected Papers Presented to the Scientific Committee of CCAMLR, 1989*, pp. 229–81. CCAMLR, Hobart.

Miller, D.G.M. (1991) Exploitation of Antarctic marine living resources: A brief history and a possible approach to managing the krill fishery. *S. Afr. J. Mar. Sci.* **10**, 321–39.

Miller, D.G.M. & Hampton, I. (1989) Biology and ecology of the Antarctic krill (*Euphausia superba* Dana). *BIOMASS Sci. Ser.* **9**, 166 pp.

Mitchell, B. & Sandbrook, R. (eds)(1980) The management of the Southern Ocean. In *A Report to the International Institute for Environmental Development*, 162 pp. International Institute for Environment and Development, London.

Mujica, A.R., Acuña, S. & Rivera, A.O. (1993) Krill population biology during the 1991 Chilean Antarctic krill fishery. In *Selected Papers Presented to the Scientific Committee of CCAMLR, 1992*, pp. 223–35. CCAMLR, Hobart.

Murphy, E.J. (1995) Spatial structure of the Southern Ocean ecosystem: Predator-prey linkages in Southern Ocean food webs. *J. Anim. Ecol.* **64**, 33–347.

Murphy, E.J., Morris, D.J., Watkins, J.L. & Priddle, J. (1988) Scales of interaction between Antarctic krill and the environment. In *Antarctic Ocean and Resources Variability* (D. Sahrhage, ed.), pp. 120–30. Springer-Verlag, Berlin.

Murphy, E.J., Clarke, A., Simon, C. & Priddle, J. (1995) Temporal variation in Antarctic sea-ice: Analysis of long-term fast-ice record from the South Orkney Islands. *Deep-Sea Res.* **42**, 1045–62.

Murphy, E.J., Constable, A. & Agnew, D.J. (1999) Modelling the dynamics of krill populations in the Antarctic Peninsula region. CCAMLR, Mimeo, WG-EMM-99/56. 18 pp.

Nemoto, T. & Nasu, K. (1975) Present status of exploitation and biology of krill in the Antarctic. In *Oceanology International 75*. BPS Exhibitions, London. 353–60.

Nicol, S. (1989) Who's counting on krill? *New Scientist* **11**, 38–41.

Nicol, S. (1991) CCAMLR and its approaches to management of the krill fishery. *Polar Rec.* **27**, 229–36.

Nicol, S. & de la Mare, W.K. (1993) Ecosystem management and the Antarctic krill. *Am. Scientist* **81**, 36–47.

Nicol, S. & Endo, Y. (1997) Krill fisheries of the world. *FAO Fish. Tech. Paper* **367**, 100 pp.

Pauly, T., Higginbottom, I., Nicol, S. & de la Mare, W. (1996) Results of a hydroacoustic survey of Antarctic krill populations in CCAMLR Division 58.4.1 carried out in January to April 1996. CCAMLR, Mimeo, WG-EMM-96/28. 22 pp.

Reid, W.V. (1988) Managing the Southern Ocean krill fishery. *Resources.* (Spring) 11–12.

Robinson, M.A. (1982) Prospects for world fisheries to 2000. FAO Fisheries Circular 722. 16.

SC-CAMLR (1984) *Report of the Third Meeting of the Scientific Committee.* CCAMLR, Hobart. 238 pp.

SC-CAMLR (1985) *Report of the Fourth Meeting of the Scientific Committee.* CCAMLR, Hobart. 275 pp.

SC-CAMLR (1986) *Report of the Fifth Meeting of the Scientific Committee.* CCAMLR, Hobart. 271 pp.

SC-CAMLR (1987) *Report of the Sixth Meeting of the Scientific Committee.* CCAMLR, Hobart. 263 pp.

SC-CAMLR (1988) *Report of the Seventh Meeting of the Scientific Committee.* CCAMLR, Hobart. 211 pp.

SC-CAMLR (1990) *Report of the Ninth Meeting of the Scientific Committee.* CCAMLR, Hobart. 345 pp.

SC-CAMLR (1991) *Report of the Tenth Meeting of the Scientific Committee.* CCAMLR, Hobart. 427 pp.

SC-CAMLR (1992) *Report of the Eleventh Meeting of the Scientific Committee.* CCAMLR, Hobart. 487 pp.

SC-CAMLR (1993) *Report of the Twelfth Meeting of the Scientific Committee.* CCAMLR, Hobart. 431 pp.

SC-CAMLR (1994) *Report of the Thirteenth Meeting of the Scientific Committee.* CCAMLR, Hobart. 453 pp.

SC-CAMLR (1995) *Report of the Fourteenth Meeting of the Scientific Committee.* CCAMLR, Hobart. 460 pp.

SC-CAMLR (1996) *Report of the Fifteenth Meeting of the Scientific Committee.* CCAMLR, Hobart. 456 pp.

SC-CAMLR (1997) *Report of the Sixteenth Meeting of the Scientific Committee.* CCAMLR, Hobart. 437 pp.

SC-CAMLR (1998) *Report of the Seventeenth Meeting of the Scientific Committee.* CCAMLR, Hobart. 456 pp.

SC-CAMLR (1999) *Report of the Eighteenth Meeting of the Scientific Committee.* CCAMLR, Hobart. 461 pp.

Shimadzu, Y. (1985) A brief summary of Japanese fishing activity relating to Antarctic krill, 1972/73 to 1982/83. In *Selected Papers Presented to the Scientific Committee of CCAMLR, 1982–1984.* Part I, pp. 439–71. CCAMLR, Hobart.

Siegel, V. (1992) Estimation of krill (*Euphausia superba*) mortality and production rate in the Antarctic Peninsula region. In *Selected Papers Presented to the Scientific Committee of CCAMLR, 1991*, pp. 159–76. CCAMLR, Hobart.

Siegel, V. & Loeb, V. (1995) Recruitment of Antarctic krill (*Euphausia superba*) and possible causes for its variability. *Mar. Ecol. Prog. Ser.* **123**, 45–56.

Siegel, V., Loeb, V. & Gröger, J. (1998) Krill (*Euphausia superba*) density, proportional and absolute recruitment and biomass in the Elephant Island region (Antarctic Peninsula) during the period from 1977 to 1997. *Polar Biol.* **19**, 393–98.

Stammerjohn, S.E. & Smith, R.C. (1997) Spatial and temporal variability in west Antarctic sea-ice coverage. In *Biology of the Antarctic Peninsula Area* (R.M. Ross, E.E. Hofmann & L.B. Quetin, eds). Antarctic Research Series No. 70, pp. 81–104. American Geophysical Union, Washington.

Statistical Bulletin (1990a) *Volume 1 (1970–1979).* CCAMLR, Hobart. 61 pp.

Statistical Bulletin (1990b) *Volume 2 (1980–1989).* CCAMLR, Hobart. 109 pp.

Statistical Bulletin (1991) *Volume 3 (1981–1990).* CCAMLR, Hobart. 119 pp.

Statistical Bulletin (1992) *Volume 4 (1982–1991).* CCAMLR, Hobart. 133 pp.
Statistical Bulletin (1993) *Volume 5 (1983–1992).* CCAMLR, Hobart. 123 pp.
Statistical Bulletin (1994) *Volume 6 (1984–1993).* CCAMLR, Hobart. 143 pp.
Statistical Bulletin (1995) *Volume 7 (1985–1994).* CCAMLR, Hobart. 147 pp.
Statistical Bulletin (1996) *Volume 8 (1986–1995).* CCAMLR, Hobart. 150 pp.
Statistical Bulletin (1997) *Volume 9 (1987–1996).* CCAMLR, Hobart. 157 pp.
Statistical Bulletin (1998) *Volume 10 (1988–1997).* CCAMLR, Hobart. 134 pp.

Stasenko, V.D. (1967) Determining the rational krill fishing methods and the commercial effectiveness of the chosen fishing gear. In *Soviet Fishery Research on Antarctic Krill* (R.N. Burukovskiy, ed.), Document (TT 67-32683), pp. 61–78. US Dept. of Commerce, Washington, D.C.

Sushin, V.A. & Myskov, A.S. (1993) Location and intensity of the Soviet krill fishery in the Elephant Island area (South Shetland Islands), 1988/89. In *Selected Papers Presented to the Scientific Committee of CCAMLR, 1992*, pp. 305–335. CCAMLR, Hobart.

Trathan, P.N., Agnew, D.J., Miller, D.G.M., Watkins, J.L., Everson, I., Thorley, M.R., Murphy, E.J., Murray, A.W.A. & Goss, C. (1993) Krill biomass in Area 48 and Area 58: Recalculations of FIBEX data. In *Selected Papers Presented to the Scientific Committee of CCAMLR, 1992*, pp. 157–81. CCAMLR, Hobart.

Trathan, P.N., Everson, I., Miller, D.G.M., Watkins, J.L. & Murphy, E.J. (1995) Krill biomass in the Antarctic. *Nature* **373**, 201–202.

Trathan, P.N., Everson, I., Murphy, E.J. & Parkes, G.B. (1998) Analysis of haul data from the South Georgia krill fishery. *CCAMLR Sci.* **5**, 9–30.

Trivelpiece, W., Trivelpiece, S.G. & Volkman, N.J. (1987) Ecological segregation of Adélie, gentoo and chinstrap penguins at King George Island, Antarctica. *Ecology* **68**, 351–61.

Vagin, A.V., Makarov, R.R. & Menshenina, L.L. (1993) Diurnal variations in biological characteristics of krill, *Euphausia superba* Dana, to the west of the South Orkney Islands, 24 March to 18 June 1990 – Based on data reported by a biologist observer. In *Selected Papers Presented to the Scientific Committee of CCAMLR, 1992*, pp. 201–222. CCAMLR, Hobart.

Watters, G. & Hewitt R.P. (1993) Alternative methods for determining subarea and local area catch limits for krill in Statistical Area 48. In *Selected Papers Presented to the Scientific Committee of CCAMLR, 1992*, pp. 237–49. CCAMLR, Hobart.

Wöehler, E.J. (compiler) (1993) *The Distribution and Abundance of Antarctic and Subantarctic Penguins.* Scientific Committee for Antarctic Research, Cambridge. 76 pp.

Chapter 13
Management of Krill Fisheries in Canadian Waters

Inigo Everson

13.1 British Columbia krill fishery

An experimental krill fishery began around 1970 off the east coast of Vancouver Island, British Columbia the annual catch from which, up to 1987, remained below 200 t. Since that time the fishery increased to a peak in 1990 of over 500 t with a value of Can$ 415 000. Annual catches are shown in Fig. 13.1. In 1992 some poor quality krill were sold with the result that the market for British Columbian krill became depressed and demand went down. A direct result was that landings in 1993 were very low (Nicol & Endo, 1997). Since that time landings have increased and in

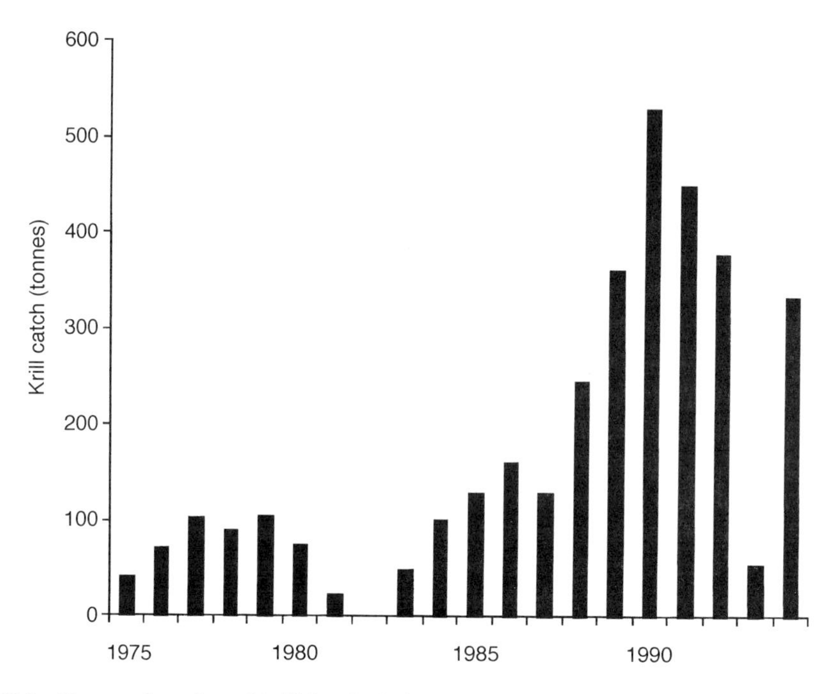

Fig. 13.1 Reported catches of krill for the krill fishery in British Columbia.

1994 were over 300 t with a value of Can$ 259 000 (Nicol & Endo, 1997, Pitcher & Chuenpagdee, 1996, Ichii, Chapter 9).

Fishing has been carried out by two types of vessel, small trawlers that have only a limited freezing capacity of around 5 t of krill a day and seine netters which land large quantities of krill for processing and freezing ashore (Nicol & Endo, 1997, Ichii, Chapter 9). This form of fishing practice effectively restricts the fishery to the coastal region where small vessels can operate and larger vessels can return to port within a few hours of making catches.

Currently the fishery is closely controlled and restricted to the Jervis Inlet, a small and narrow inlet which opens into the Malaspina Channel situated on the eastern side of the Strait of Georgia and shown in Fig. 13.2. The area is further subdivided into a series of five management units which are designated by narrowing of the channel or other topographical features.

In 1976 a total allowable catch (TAC) was set at 500 t for a fishing season extending from November through to March. The criterion for setting the TAC was that it should be no more than 3% of the estimated consumption of krill by all predators in the Strait of Georgia (Nicol & Endo, 1997). The status of the fishery and the justification for reviewing the TAC were discussed in 1995 at a workshop meeting in Vancouver (Pitcher & Chuenpagdee, 1996) and the main questions raised at that meeting concerned the ecosystem approach to management, how that might be applied to the region and issues associated with funding the research needed to develop this management advice.

13.2 Nova Scotian krill fisheries

Research studies investigating the feasibility of harvesting krill from the Nova Scotian region began in 1972 arising from which a proposal was made to harvest krill (Sameoto, 1975). Initially estimates of biomass were made using nets although recent studies using acoustics have indicated a standing stock of krill in the Gulf of St Lawrence of 8 to 96 kt (Simard & Lavoie, 1999, Everson, Chapter 3).

The Maritimes Region of the Department of Fisheries and Oceans Canada (DFO) in 1991 issued its first scientific permit to harvest zooplankton. That licence was for both krill and copepods, in the Gulf of St Lawrence. Exploratory fishing commenced in November 1993 in the St Lawrence Estuary off Ste-Anne-des-Monts. In 1994 so as to exercise control over any possible expansion in the fishery a TAC was set at 100 t for krill and 50 t for copepods, specifically *Calanus*. During that season the reported catches were only of 6.3 t of krill and 0.4 t of *Calanus*. In 1995 the TACs were increased to 300 t of krill and 2000 t of *Calanus* but catches were again very low, 2 t of krill and 1 t of *Calanus* (Harding, 1996).

In late 1995, DFO received a proposal to develop a 1000 t experimental fishery for krill, a forage species, on the Scotian Shelf and in the Bay of Fundy (Anon., 1996). This proposal was reviewed through the regional advisory process (RAP) whose report indicated that such a fishery was likely to have only a negligible effect on the

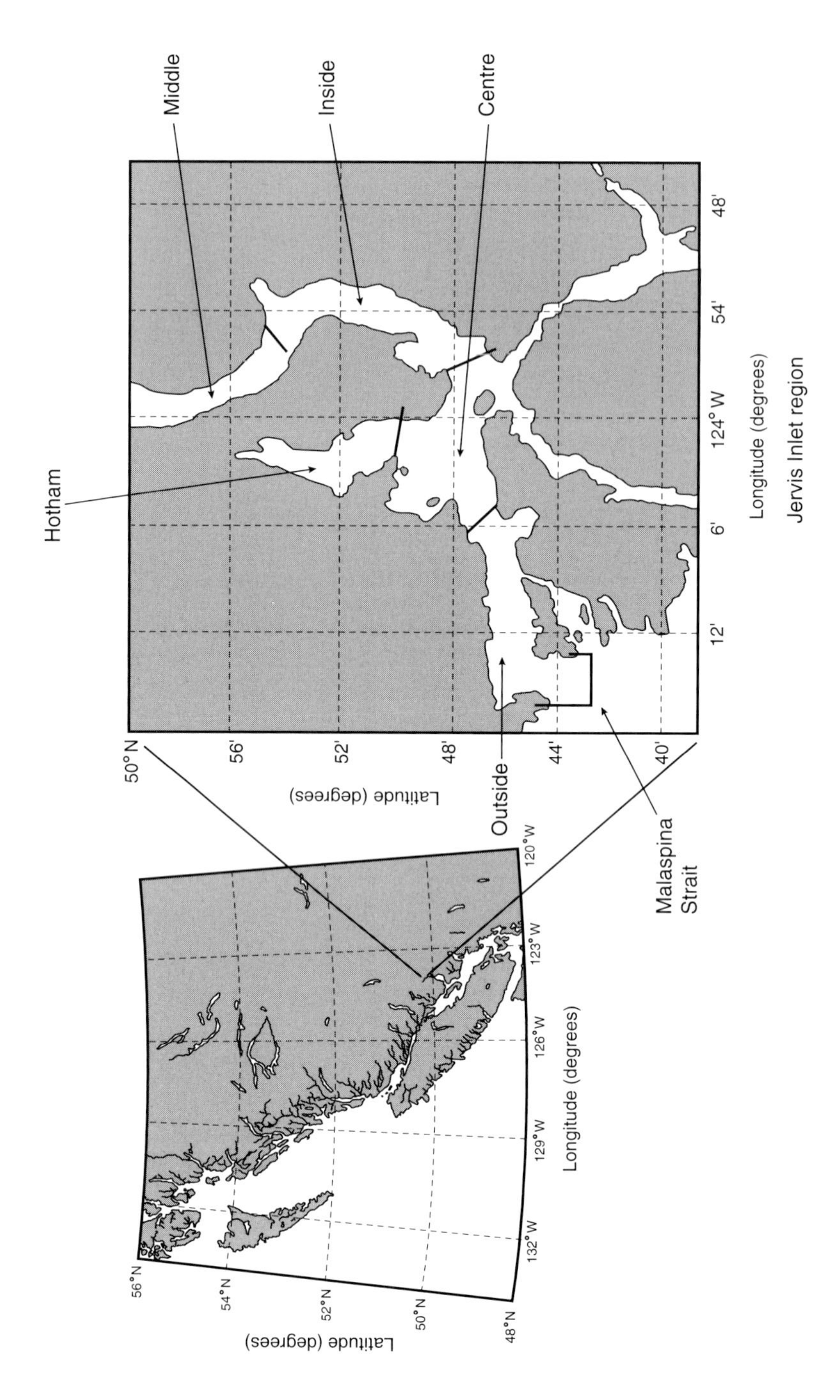

Fig. 13.2 The Jervis Inlet region showing the relationship of the fishery area to the coast of British Columbia. The five management areas within the inlet are marked.

ecosystem. At that time there was concern that although the 1000 t TAC might have no adverse effect on the ecosystem, should the fishery prove successful, there might be pressure to expand. The consequences of such expansion needed careful consideration before approval for such action might be given. At this point the debate was broadened to include what are now termed 'forage species'; these are species that form the diet of higher trophic levels. Arising from these considerations the RAP Steering Committee organised a workshop to consider the broader implications of fisheries on forage species. The broad objective of the workshop was to provide guidance for the future development of forage species fisheries in the DFO Maritimes Region. That workshop took place in April 1997 in Halifax (Head, 1997) and was attended by both national and international experts as well as local interested parties.

13.3 Krill fisheries management

The experiences from both the Atlantic and Pacific Canadian krill fisheries have been running in the same direction and it is convenient to consider the two together.

During the discussion phase of the 1995 Vancouver Workshop (Pitcher & Chuenpagdee 1996) a series of questions was identified by J. Morrison of DFO that needed to be addressed in order to develop revisions to the management plan. The first was aimed at the industry and asked specifically whether it wished to support future surveys in order to provide information that might be used to revise the TAC. The second question was aimed at scientists and managers seeking to determine the minimum information that would be required to revise the current TAC. The third question was directed at ecological modellers encouraging them to come up with models that take account of ecosystem interactions and also the risk associated with different management strategies.

In the course of the discussion two points emerged, firstly that an ecosystem approach, whereby dependent species were factored into the equation, was essential. The second point related to the resources available to provide assessment advice, a question that, put simply, reduces to comparing the value of the krill fishery to that of higher priced finfish such as salmon and hake. These points and other matters arising from the discussion were taken forward when the proposal to increase the krill TAC on the Atlantic seaboard was considered.

The 1997 Halifax Workshop (Head, 1997) considered key questions related to the overall objectives for forage fisheries in Canadian waters the first of which was:

What should be the conservation objectives related to fisheries on forage species?

This question can be looked at in the following ways:

- maintenance of ecological relationships/ecosystem integrity;
- minimisation of risk of irreversible decline;

- maintenance of ecosystems and constituent species within the bounds of natural fluctuation;
- maintenance of full recruitment potential (including genetic diversity); and
- allowance of fisheries which meet conservation objectives (and maximize knowledge returns).

Many of these topics are essentially a restatement of the requirements of Article II of the CCAMLR Convention that is operational in the Southern Ocean and is discussed in detail by Miller and Agnew (Chapter 12). The key difference is associated with the emphasis of the different components. In the Southern Ocean few of the dependent species are the subjects of commercial harvesting so that the fishery there is being targetted almost exclusively at one trophic level, that of the krill. By contrast the Canadian situation is one where many of the dependent species are themselves the subject of significant commercial fishing. Thus the following fish species, discussed by Tanasichuk (Chapter 7), are dependent to a greater or lesser extent on krill in Canadian waters: salmon (*Onchorynchus tsawyscha* and *O. kisutch*), hake (*Merluccius bilinearis*, *M. productus* and *Urophycis chuss*), cod (*Gadus morhua*), haddock (*Melanogrammus aeglefinus*), herring (*Clupea harengus* and *C. pallasi*), capelin (*Mallotus villosus*) and redfish (*Sebastes* spp). The management regime in this forum needs to take account of the status of the dependent species and furthermore come to a view as to the likely effect of different levels of krill harvest. Such answers have an economic as well as ecological connotation.

The second set of questions raised at the Halifax Workshop related to the requirements that would need to be met by management of the fishery in order for the conservation objectives to be satisfied. The conclusions that arose from this were summarised in the following key points:

- The knowledge base must be adequate to allow risk evaluation.
- There should be confidence that monitoring and controls are adequate to allow management decisions (i.e. responses to change) to be made and implemented in timely fashion.
- Monitoring of direct impact and indirect impact on dependent species and of by-catch is required.
- Ongoing assessment of natural levels of fluctuations in species and ecosystem is required.
- Proponents should bear the incremental costs of acquiring knowledge and managing the fishery.

Arising from this it was clear that the CCAMLR approach (Miller & Agnew, Chapter 12) provided a good framework against which to consider krill fisheries on the Scotian shelf. However in taking the matter forward there was a clear indication that that approach would need adjustment in order to address the specific requirements of Canadian fisheries.

These adjustments were not required for the krill-centred part of the system where the CCAMLR approach was seen as providing a good model to use but for the dependent species that were subject to a commercial harvest. The developments that were required were envisioned in the report of the RAP Steering Committee. That Committee recommended, amongst other things, that:

- Decisions made concerning these (forage) fisheries and their scientific justification must be credible to a broad range of stakeholders.

and

- The CCAMLR approach requires ongoing monitoring of the biological health of dependent species. In the case of krill, one such species would be hake. Steps would need to be taken to identify dependent species and initiate monitoring programmes for any emerging or existing forage fisheries.

The term 'stakeholder' is an important term in this context because it draws together those with interests in harvesting krill, harvesting fish which feed on krill, with those interested in maintaining a healthy ecosystem in perpetuity. The aims and aspirations of these separate and sometimes opposing interests need to be considered by the fisheries' managers as they develop policy. Where these recommendations differ from the CCAMLR approach is governed by two factors. Firstly the Canadian fisheries fall under national jurisdiction so that policy can be made, implemented and monitored; essentially this is funded through the licence revenue from the fisheries. The Canadian Government can make the rules and implement them. CCAMLR by contrast operates by consensus and is an international management organ; its income derives from payments by member states and these member states are responsible for ensuring that their own nationals comply with conservation measures. The second point is that many of the dependent species in Canadian waters are themselves subject to commercial harvest whereas those in the Southern Ocean, with the exception of the mackerel icefish, are not fished. It is this second point which is worth considering in a little more detail.

The two criteria that CCAMLR have agreed for the Antarctic krill fishery are that over a 20 year period the minimum spawning stock biomass should not fall below 20% of its unexploited level in any season, and at the end of the period, that the mean spawning stock biomass should not fall below 75% of its pre-exploitation level. These criteria were established in order to provide a precautionary catch limit that satisfied the requirements of Article II of the CCAMLR Convention. In the Canadian context, satisfying the requirements of Article II is only part of the story because reducing the mean spawning stock biomass will most likely have an effect on commercial fish species.

The scientific advice that would be sought in order to address these points focuses on the dependent species because information is then required on the effect of a 25% reduction in krill availability to, for example, hake. A likely outcome of such an exercise would be that hake production, and consequently its TAC, would go down. From the point of view of the stakeholders who are interested in fishing this reduces to the simple economic question of whether it is more profitable to fish for krill or hake.

Energetically, based on a simple 'rule of thumb', we might assume that ten tonnes of krill lead to the production of one tonne of hake. If one assumes a conversion ratio of perhaps 10% for the production of krill meal or 50% for krill products for direct human consumption on the one hand and perhaps 60% for headed and gutted hake then we have a basis for comparing tonnages. These can then be considered alongside the market value of the different products in assessing the value of the different fisheries. Further development of these ideas is clearly necessary but falls outside the bounds of the current chapter.

In summary the Canadian approach has been one to impose a high level of caution from the outset specifically to address ecosystem considerations. In developing the approach the experiences of CCAMLR have been accepted and are being further developed to consider the implications of harvesting at different trophic levels in order to develop a sustainable management regime that provides a satisfactory return to stakeholders.

References

Anon. (1996) *Krill on the Scotian Shelf.* DFO Atlantic Fisheries Stock Status Report 96/106E.

Harding, G.C.H. (1996) *Ecological factors to be considered in establishing a new krill fishery in the Maritimes Region.* DFO Atlantic Fisheries Research Document 96/99, 11 pp.

Head, E. (1997) Proceedings of the workshop on *Ecosystem considerations for krill and other forage fisheries.* Canadian Stock Assessment Proceedings Series 97/5, 61 pp.

Nicol, S. & Endo, Y. (1997) Krill fisheries of the world. *FAO Fish. Tech. Paper* **367**, 1–100.

Pitcher, A.J. & Chuenpagdee, R. (eds) (1996) Harvesting krill: ecological impact, assessment, products and markets. Fisheries Centre Research Reports 3 (3). Fisheries Centre, University of British Columbia, Canada. 82 pp.

Sameoto, D.D. (1975) A krill fishery in the Gulf of St Lawrence. *Can. Fisherman Ocean Sci.* **61**, 16–17.

Simard, Y. & Lavoie, D. (1999) The rich krill aggregation of the Saguenay–St Lawrence Marine Park: hydroacoustic and geostatistical biomass estimates, structure, variability and significance for whales. *Can. J. Fish. Aquat. Sci.* **56**, 1182–97.

Chapter 14
Krill Fisheries and the Future

Inigo Everson, David Agnew and Denzil Miller

The foregoing chapters have brought together much information on the biology, ecology and fisheries for krill in different parts of the world. Against that background we can now reflect where that information is likely to lead in the future. We consider this from the several different perspectives.

We have seen in Chapter 10 that there are several main types of product derived from krill which can be categorised by their market value and volume. On the one hand we have the lower value high volume products such as fish-meal and aquaculture feed and for which the technology is quite well established. At the other extreme are products for direct human consumption and for these there is a requirement for krill of a specified quality to be processed in an efficient manner. This leads to low volume high quality products. In addition there are the products that are produced as by-products from one of the other processing systems. By-products themselves are unlikely to have much influence in guiding the fisheries except by offering a marginal increase in revenue so we need not consider these further in a consideration of future developments. High value low volume products tend to be produced as a result of some key technological development; this means that any developments in this line are more likely to appear in Patent files than texts such as this one. The high volume products and in particular those supporting the aquaculture industry can provide further food for thought.

Over the past thirty or so years there has been a rapid increase in the aquaculture industry in many parts of the world which has resulted in a requirement for large amounts of feed. Krill products have been used increasingly for this purpose largely because of their nutritional content and also because the carotene pigments enhance the colour of the farmed fish. It is interesting to note that the main centres of krill fishing provide a poor match for the regions where aquaculture is at its greatest. The aquaculture industry is particularly strong in northern Europe, north America and south-east Asia in the northern hemisphere and yet the only major krill fishery is in Japan. In the southern hemisphere the largest krill fishery is in the Southern Ocean, a region not amenable to the needs of aquaculture, whereas the main aquaculture interests appear to be centred in Tasmania and Chile.

Krill concentrations are known to occur close to many of the marine locations where the aquaculture industry has developed and it is worth considering why more use is not made of the local krill. The answer lies probably in the seasonal cycle of the krill. Aquaculture requires a consistent supply of feed year-round and this is unlikely to come from a local source that has large-scale seasonal cycles of abundance. Improvements in the efficiency of bulk transport mean that sufficient feed can be transported over great distances and, with careful planning, this can smooth out variation in krill supply to ensure a steady supply of feed. Thus the main advantage to be gained from using local krill for aquaculture feed is only when there is a need for the krill to be fresh. In spite of this there are concentrations of krill close to areas where aquaculture is a key industry such as Scotland and Norway where *Meganyctiphanes norvegica* is seasonally present and likewise for *Nyctiphanes australis* off Tasmania (Nicol & Endo, 1999). That fisheries have not developed in these areas may in part be due to the ready supply of krill products from the established fisheries but also may be a result of considerations for fisheries management such as are currently under discussion for the Canadian fisheries.

The debate on the Canadian krill fisheries has taken the ecosystem approach of CCAMLR a step further by extending the philosophy to encompass dependent species that are themselves the subject of commercial fishing. Although in the background, due to the concerns over the impact of Antarctic whaling, when the CCAMLR Convention was developed the current scenario of dependent species being harvested is of marginal concern in that forum. The only example currently is the mackerel icefish *Champsocephalus gunnari* (Miller & Agnew, Chapter 12). The Canadian approach has introduced the more complex consideration of harvesting at two trophic levels within the same system. In order to achieve this there are the considerations of maintaining populations of the harvested and dependent species sustainably and in perpetuity, the current CCAMLR approach, allied to assessments of the relative values of the products from the different trophic levels (Everson, Chapter 13). That part of the equation which relates to the CCAMLR approach can be developed effectively by a scientific analysis of the system; this has been largely achieved and the approach is being refined (Miller & Agnew, Chapter 12).

Harvesting at more than one trophic level can also introduce socio-economic factors into the equation. At the economic level, this might be viewed as a simple trade-off between the values of individual products adjusted for volume and production costs. The social component comes in when a desire is indicated to harvest krill and to do so would entail a reduction in total allowable catch (TAC) of a 'traditional' finfish species. The resolution of that situation requires scientific advice on the impact of different levels of harvesting at the different trophic levels leading to a political decision to decide how the division should be made. This is in effect a restatement of the fact that it is not possible to harvest two species from different trophic levels at their individual levels of maximum sustainable yield (MSY) (Beddington & May, 1980). The political decision determining which fishery will be permitted the greater proportion of its individual MSY will very likely be further affected to take account of lobbying groups whose primary concerns might not be associated directly with fishing. Such considerations might include, for example, the tourist potential from whale watching or the scientific issues associated with maintaining biodiversity.

Two points arise from these considerations. Firstly, there is the need to understand the interactions between the species that are being harvested at the two trophic levels. The second point indicates that there is likely to be more scope for krill fishing when there is no harvesting at higher trophic levels.

Understanding the interactions between different trophic levels, particularly when that information is likely to be taken forward in determining management policy, is a daunting task. In the case of CCAMLR and its development of an Ecosystem Monitoring Programme (CEMP) indicator species and sampling regimes had to be developed *de novo*. This has meant that it has taken over a decade to develop the database sufficiently to be able to begin to understand the intricacies of the situation. Ironically when the dependent species is a commercial fish there is likely to be a database already established. The fishery can also provide information on total catch, catch per unit of effort (CPUE) and biological information on the species in question. Fish stock assessments traditionally require estimates of standing stock, recruitment, growth and mortality. If the species is of significant commercial importance then other ecological components such as diet, spawning cycles, condition, distribution and migrations will also have been studied albeit with less intensity. Much information may already be available therefore with which to study trophic interactions. Examples taken from the chapters of this volume where this approach might be considered are hake in Canadian waters and walleye pollock and blue eye in Japanese waters (see Tanasichuk, Chapter 7 and Endo, Chapter 7).

Returning to the potential for expansion in krill fisheries, it is clear that, where the intention is to harvest at two trophic levels, the forage fishery scenario, the process of developing a management framework is more complicated than that for the CCAMLR type system. The scope for expansion envisioned by the difference between dependent species' consumption of krill and the current fishery in the Southern Ocean means that, should there be a desire to expand the world krill fishery, there is capacity within the CCAMLR region.

Integrating krill fisheries into the general category of forage fisheries, as has been done in Canada and Alaska (there is currently a prohibition on exploiting forage fish in Alaskan waters), provides for some interesting avenues for future consideration of fisheries on mid-trophic level animals. In some ways krill are similar to squid and small pelagics, both of which are subject to successful fisheries in many parts of the world. These animals are characterised by having a short life-span and a seasonal migration pattern that is largely driven by oceanographic circulation (e.g. Agnew *et al.*, in prep). Consideration of these animals as food for dependent top-level predators (mammals and birds) has thus been extended to other exploited species (i.e. fish) which also depend upon them. There is still a considerable way to go in generating suitable management regimes for these mid-trophic level species, since the systems under consideration usually respond to change in very complex ways.

Even the simplest apparent causal relationship, such as that of the effects of the Alaskan pollock fishery on Steller sealion populations (Lowell *et al.*, 1995), or the North Sea sand eel fishery on seabirds (Wright, 1996) is often extremely difficult to confirm. For instance, in the case of sand eels, it appears that recruitment variability, which appears to be independent of the adult fishery, has caused the observed fluctuations in seabird breeding success. Nevertheless, the underlying issues are often similar for different fisheries, and the work done in establishing management regimes for krill can arguably be seen as providing useful information for consideration in other fora. Likewise there are lessons to be learnt for krill fishery managers by careful scrutiny of the practices in other fisheries. This is true particularly where harvesting is occurring simultaneously at two or more trophic levels and the problems are compounded by the addition of environmental and social concerns regarding the effects of such harvesting on all trophic levels within an ecosystem. A need to move towards 'holistic' fisheries management, considering the effects of harvesting on the ecosystem as a whole, is likely to assume much more prominence in the 21st century, and is likely to have a significant effect on the way krill and other fisheries are managed.

References

Agnew, D.J., Hill, S. & Beddington, J.R. (in prep) Predicting the recruitment strength of an annual squid stock: *Loligo gahi* around the Falkland Islands.

Beddington, J.R. & May, R. (1980) Maximum sustainable yields in systems subject to harvesting at more than one trophic level. *Math. Biosci.* **51**, 261–81.

Lowell, F.W., Ferrero, R.C. & Berg, R.J. (1995) The threatened status of Steller sea lions, *Eumetopias jubatus*, under the Endangered Species Act: Effects on Alaska groundfish fisheries management. *Mar. Fish. Rev.* **57**, 14–27.

Nicol, S. & Endo, Y. (1999) Krill fisheries: development, management and ecosystem implications. *Aquat. Liv. Res.* **12**, 105–120.

Wright, P.J. (1996) Is there a conflict between sandeel fisheries and seabirds? A case study at Shetland. In *Aquatic Predators and their Prey* (S.P.R. Greenstreet & M.L. Tasker, eds), pp. 154–65. Fishing News Books, Oxford.

Glossary

abdomen posterior part of the body

abiotic pertaining to non-living things

aggregation a large concentration of krill which may represent a group of swarms

Alcidae family of birds belonging to the order Charadriiformes which includes auks, puffins, razorbill and guillemots

allometry the relative growth rate of part of an organism in relation to another part or to the whole

Amphipoda laterally compressed Peracarid Crustacea in which the cephalothorax lacks a carapace and the whole head is fused to one, rarely two, thoracomeres

Antarctic Polar Front the frontal zone at the northern extremity of the Southern Ocean. Recognised by an increase in surface water temperature of a few degrees as it is crossed in a northerly direction

antioxidant a compound added to food products to reduce the effects of oxidation and the accompanying degradation of the product

austral at high latitude in the Southern Hemisphere (cf. boreal)

autodigestion post mortem degradation due to the action of enzymes present within the body

autoproteolysis use of natural proteolytic enzymes to produce a high yield protein concentrate

baleen whale the group of whales characterised by having horny plates derived from the mucous membrane of the palate which act as a strainer to filter plankton from the water

beam trawl a frame net having a beam to hold the net open during fishing operations

benthic living on the seabed

benthopelagic mainly associated with the seabed but making excursions up into the water column

biomass the total mass of biological material of either a species or a group present in an area. Synonymous with standing stock

biotic relating to life or living creatures

boreal of the north. The boreal zone is the region of short summers and long cold winters

brood pouch exoskeletal shield to protect eggs or larvae

carapace an exoskeleton shield covering part of the dorsal surface

cephalothorax the region of the body formed by the fusion of the head and thorax

chitin a nitrogenous polysaccharide with the formula $(C_8H_{13}N_5)_n$ occurring as skeletal material in crustacea

chumbait ground-bait thrown into the water by fishers working with hook and line fishing devices

C:N ratio carbon to nitrogen ratio. Used as an indicator of the amount of protein present in a tissue

cod end the narrow end of a trawl net in which the catch is retained

cohort group of animals arising from a single spawning event. If the spawning events are annual these correspond to year classes

condition index the ratio of measured mass to estimated mass; the estimated mass coming from an overall long-term dataset

Copepoda a Subclass of the order Crustacea

combined standardised index an indication of the availability of krill to dependent species and obtained by combining several CEMP indices from one season (CCAMLR)

critical period distance a CCAMLR index to indicate the degree of overlap between fishing activity and dependent species foraging activity

crystalline cone the outer refractive body of a crustacean eye which acts as a light guide

cytoplasm that part of the cell which exists outside of the nucleus but inside the cell wall

Decapoda an order of Malacostracan Crustacea with 3 pairs of thoracic limbs modified as maxillipeds and 5 as walking legs (shrimps, prawns, crabs, lobsters etc.)

deep scattering layer an oceanic layer characterised by a high volume backscatter and showing up as an echo-trace

demersal living on or near the seabed

diel synonymous with diurnal in the sense that it pertains to a period of 24 hours, not in the sense of *by day* rather than *by night*

Discovery Investigations research programme undertaken by the UK during the inter-war period mainly in the Southern Ocean to study whales and their environment

diurnal activity over a 24-hour time-scale. Comparison of day and night activity

dual beam transducer echo-sounder transducer having two co-axial beam patterns, a narrow and a wide beam. Targets seen in the narrow beam are assumed to be on the axis of the wide beam

echo-chart paper output from an echo-sounder

echogram as for echo-chart

echo-sounder	device for sending pulses of high frequency sound into the water and listening for return echoes. The time delay to receipt of the echo provides an indication of target range
echo-integration	accumulating echo signals over a period of time. The integrated signal is assumed proportional to the density of the targets
echo-trace	marks on an echogram to indicate the distribution by depth and time of targets
ejaculatory ducts	a narrow muscular tube forming the lower part of the vas deferens and leading into the copulatory organ (petasma)
El Niño	a complex set of changes in water temperature in the Eastern Pacific Equatorial region producing a warm current; it occurs annually to some degree between October and February
endogenous	a general term used to suggest that the causes of some physical condition are due to internal factors
epipelagic	in the upper part of the pelagic zone, i.e. near to the sea surface
exogenous	resulting from causes external to an organism
exopodites	the outer ramus of a crustacean appendage
exoskeleton	hard supporting or protective structures that are external to and secreted by the ectoderm
eye stalk	a paired stalk arising close to the median line on the dorsal surface of the head of many Crustacea, bearing an eye
facultative schooler	having the ability to live and adapt to schooling while not being restricted to that mode of life
fecundity	the number of young or eggs produced by a species or individual
forage fishery	fishery based on species that form the food of other species higher up the food chain
gadoid	member of the cod family
geometric scattering	the situation where an acoustic target is large relative to the wavelength of the sound such that there is little change in target strength with frequency
geostatistics	branch of statistics dealing with the analysis of data of highly contagious distributions
germinal zone	a layer of epithelium from which oöcytes develop
gravid	carrying ripe eggs, ready to spawn
growth	an irreversible change in an organism accompanied by the utilisation of material and resulting in a change of volume
gyre	circular pattern of water movement generally arising due to topography or else the counter movement of two water masses
hepatopancreas	a glandular diverticulum of the mesenteron, believed to carry out the functions of the liver and pancreas of higher vertebrates

hydrolysate	a krill product produced by controlled autoproteolysis and thermal precipitation of the water soluble protein and thus extract it from the exoskeleton
ice-algae	algae which grow on and in pack-ice
instantaneous growth rate	growth rate determined over a short period of time such as a day or week
isolume	the same light level
iteroparous	reproducing more than once during a lifetime
krill flux	rate of movement of krill through a region
length-at-age	a table to show the range of length of krill of a particular age
length frequency distribution	frequency distribution of lengths of krill obtained from a sample or group of samples
length density distribution	length frequency distribution of a population and calculated as a weighted length frequency distribution from all relevant samples
lipid	any of a group of fat or fat-like substances
lipofuscin	a fatty hydrocarbon substance that accumulates in certain animal tissues upon ageing
longevity	life span
luminous organs	light producing organs
mode	the value represented by the greatest number of individuals in a frequency distribution
mortality	generally applied to the rate at which individuals in a population die. Frequently described by the formula: $N_t = N_0\,e^{-Zt}$ where N is the number at times t and zero and Z is the coefficient of total mortality
moult	the act of casting off the outer layer of the integument
Myctophidae	Lantern fishes
Mysidacea	Opossum shrimps, an order Pericarida, carnivorous shrimp-like crustaceans having biramous thoracic appendages
nearest neighbour distance	the distance between adjacent krill in a swarm
neritic	the shallow-water coastal region
net selectivity	the degree to which animals of different sizes are retained by a fishing net
net-sounder	an echo-sounder operating from a fishing net to determine the depth of the net from the surface or seabed and also the amount of fish entering it
oöcyte	female gametocyte
oögonia	an egg-mother-cell
ova	eggs or egg-cells
ovary	female gonad
pack-ice	ice formed by freezing of the sea-surface
pelagic	in the water column, i.e. above the seabed
perivitelline space	the space between the vitelline membrane and the capsule of an egg

petasma	male external sexual character, situated on the first pair of abdominal pleopods
photoperiod	light to dark regime in which an animal lives
photophore	light producing organ
phototaxis	movement in response to light
phytoplankton	microscopic drifting plant life
plankton	drifting life. Essentially living organisms that are unable to make significant movements against oceanic currents
pleopod	an abdominal appendage adapted for swimming
polymodal	a frequency distribution having more than one peak
polyna	area of open water within the pack-ice zone
Procellariiformes	an order of birds including albatrosses, petrels, shear-waters and fulmar, characterised by tube-like nostrils on their beaks
proteolysis	the degradation of proteins into peptides and amino acids by cleavage of their peptide bonds
pycnocline	density discontinuity
Rayleigh scattering	the situation where the strength of the echo from an acoustic target is strongly proportional to the size of the target or else is in a resonant region where echo strength varies strongly with target size
recruitment	first arrival of juvenile or young individuals into the population. In fishery terms arrival into the exploited phase of the population
rostrum	a median anterior projection of the carapace
salp	a transparent pelagic member of the order Tunicata (sea squirts)
schooling	the process by which krill aggregations and swarms are formed
sea-ice	generally synonymous with pack-ice
selectivity	characteristics of a net describing the proportion of individuals of different sizes that are retained by a net
setae	long thin bristle-like structures on the end of euphausiid limbs
shelf	relatively shallow area adjacent to a land mass
shelf-break	the transition zone where the shelf ends and the water becomes very much deeper in the open ocean
spermatophore	a packet of spermatozoa enclosed within a capsule
split-beam transducer	an echo-sounder transducer which is divided into four quadrants. Differences in phase between return echoes from the same target are used to determine the position of the target on the main axis of the transducer
Southern Ocean	the area of ocean surrounding the Antarctic Continent. The northern limit is generally taken to be the Antarctic Polar Front
standing stock	synonymous with biomass

stock group of animals of a single species which can be treated as a single unit for management purposes

submersible instrument package for *in situ* sampling of the underwater environment

super-swarm extremely large swarm of krill

swarm dense concentration of krill which appear to be organised and/or moving in a coherent manner

sympatric two species or populations having common or overlapping geographical distributions

target strength an acoustic ratio describing the proportion of incident sound that is reflected back by a target

telson the post-segmental region of the abdomen

testis male gonad

thelycum female external sexual character

thorax the region of the body lying between the head and abdomen. Often used erroneously as a synonym for cephalothorax

topographic detailed description of the seabed

total allowable catch fishery term to limit the total catch of a species within a management area

transect line sampled during a survey

triacylglycerol a triglyceride. The product of the esterification of a glycerol molecule with three fatty acid molecules

unimodal a frequency distribution having a single peak

upwelling water circulation pattern which brings deep water to the surface

uropod an appendage of the abdominal somite preceding the telson

vasa deferentia ducts leading from the testes to the petasma

vitellogenesis the process of yolk deposition in the ova

volume backscattering strength acoustic term to describe the proportion of the incident energy that is reflected by a layer or volume of seawater

warm-core rings areas of warm water that are entrapped within cooler water masses by local circulation patterns

zooplankton animal plankton

Index

Taxonomic Index

List of Abbreviations

ACC	Antarctic Circum-Polar Current
ADCP	Acoustic Doppler Current Profiler
ANARE	Australian National Antarctic Research Expedition
APFZ	Antarctic Polar Frontal Zone
AUV	Autonomous Underwater Vehicle
AVHRR	Advanced Very High Resolution Radiometer
B_0	Pre-exploitation biomass
BAS	British Antarctic Survey
BIOMASS	Biological Investigations of Marine Antarctic Systems and Stocks
BIONESS	Bedford Institute of Oceanography Net and Environmental Sampling System
CDW	Circum-Polar Deep Water
CEMP	CCAMLR Ecosystem Monitoring Programme
CCAMLR	Commission for the Conservation of Antarctic Marine Living Resources (also sometimes used for Convention for the Conservation of Antarctic Marine Living Resources)
CM	Conservation Measure (CCAMLR)
CPD	Critical Period Distance (CCAMLR)
CPH	Catch Per Hour fishing
CPUE	Catch Per Unit of Effort
CSI	Combined Standardised Index
DFO	Department of Fisheries and Oceans (Canada)
DHA	Dodecahexaenoic acid
EEZ	Exclusive Economic Zone
EPA	Eicosapentaenoic acid
ENSO	El Niño Southern Oscillation
FAO	Food and Agriculture Organisation of the United Nations
FFT	Fast Fourier Transform
FIBEX	First International BIOMASS Experiment
FRAM	Fine Resolution Antarctic Model
GIS	Geographical Information System
GLOBEC	Global Ocean Ecosystems Dynamics (an international programme having individual national programmes such as US-GLOBEC)
GRT	Gross Registered Tonnage (shipping)
GZ	Germinal Zone
ICCAT	International Commission for the Conservation of Atlantic Tunas

ICSEAF	International Commission for the Southeast Atlantic Fisheries
IGR	Instantaneous Growth Rate
k	Coefficient of catabolism in VGBF
KYM	Krill Yield Model (CCAMLR)
LTER	Long Term Ecological Research Programme (USA)
LLHPR	Large Longhurst-Hardy Plankton Recorder
M	Coefficient of natural mortality
MIX	Interactive computer programme to analyse a length frequency distribution and determine modal values
MLMIX	An extension of MIX to provide a length density distribution
MOCNESS	Multiple Opening and Closing Net and Environmental Sensing System
MSY	Maximum Sustainable Yield
NOAA	National Oceanographic and Atmospheric Administration (USA)
NND	Nearest Neighbour Distance
OPC	Optical Plankton Counter
PVS	Perivitelline space
R	Recruitment
RAP	Regional Advisory Process (Canada)
RMT	Rectangular Midwater Trawl
ROV	Remotely Operated Vehicles
SCAR	Scientific Committee for Antarctic Research
SC-CAMLR	Scientific Committee for the Conservation of Antarctic Marine Living Resources
SCOR	Scientific Committee for Oceanic Research
TAC	Total Allowable Catch
TS	Target Strength
UNCLOS	United Nations Commission on the Law Of the Sea
UV-B	Ultra-Violet light in the frequency range 280–350 nannometres
VBGF	von Bertalanffy growth function
WG-EMM	Working Group on Ecosystem Monitoring and Management (CCAMLR)

Fish and Aquatic Resources Series

Series Editor: Professor Tony. J. Pitcher

Director, Fisheries Centre, University of British Columbia, Canada

The *Blackwell Science Fish and Aquatic Resources Series* is an initiative aimed at providing key books in this fast-moving field, published to a high international standard.

The Series includes books that review major themes and issues in the science of fishes and the interdisciplinary study of their exploitation in human fisheries. Volumes in the Series combine a broad geographical scope with in-depth focus on concepts, research frontiers and analytical frameworks. These books will be of interest to research workers in the biology, zoology, ichthyology, ecology, physiology of fish and the economics, anthropology, sociology and all aspects of fisheries. They will also appeal to non-specialists such as those with a commercial or industrial stake in fisheries.

It is the aim of the editorial team that books in the *Blackwell Science Fish and Aquatic Resources Series* should adhere to the highest academic standards through being fully peer reviewed and edited by specialists in the field. The Series books are produced by Blackwell Science in a prestigious and distinctive format. The Series Editor, Professor Tony J. Pitcher is an experienced international author, and founding editor of the leading journal in the field of fish and fisheries.

The Series Editor and Publisher at Blackwell Science, Nigel Balmforth, will be pleased to discuss suggestions, advise on scope, and provide evaluations of proposals for books intended for the Series. Please see contact details listed below.

Titles currently included in the Series

1. *Effects of Fishing on Marine Ecosystems and Communities* (S. Hall) 1999
2. *Salmonid Fishes* (Edited by Y. Altukhov *et al.*) 2000
3. *Percid Fishes* (J. Craig) 2000
4. *Fisheries Oceanography* (Edited by P. Harrison & T. Parsons) 2000
5. *Sustainable Fishery Systems* (A. Charles) 2000
6. *Krill* (Edited by I. Everson) 2000
7. *Tropical Estuarine Fishes* (S.J.M. Blaber) 2000

For further information concerning books in the series, please contact:

Nigel Balmforth, Professional Division, Blackwell Science, Osney Mead, Oxford OX2 0EL, UK
Tel: +44 (0) 1865 206206; Fax +44 (0) 1865 721205
e-mail: nigel.balmforth@blacksci.co.uk